Mechanical Properties
of
Solid Polymers

Second Edition

Mechanical Properties
of
Solid Polymers

Second Edition

I. M. Ward

Department of Physics,
University of Leeds

A Wiley–Interscience Publication

JOHN WILEY & SONS
Chichester · New York · Brisbane · Toronto · Singapore

Copyright © 1971, 1983 by John Wiley & Sons, Ltd.

Reprinted May 1974
Reprinted August 1979.

Library of Congress Cataloging in Publication Data:
Ward, I. M. (Ian Macmillan), 1928–
  Mechanical properties of solid polymers.

  Includes index.
  1. Polymers and polymerization.   I. Title.
TA455.P58W37   1983   620.1'92   82–11086

ISBN   0 471 90011 7

British Library Cataloguing in Publication Data:

Ward, I. M.
  Mechanical properties of solid polymers.—
  2nd ed.
  1. Polymers and polymerization
  I. Title
  620.1'92   TA455.P58

ISBN   0 471 90011 7

Printed in England by
J. W. Arrowsmith Ltd., Bristol BS3 2NT

ERRATA

p. 28   *l9 should read*

$$\frac{\overline{ds}^2}{ds^2} = \frac{(dx + du)^2 + (dy + dv)^2 + (dz + dw)^2}{ds^2}$$

p. 240   *l11 equation of the form (9.20) should read*

equation of the form (9.23)

p. 304   Caption to figure. First two lines missing. These are

Figure 10.37.   Model of the morphology of oriented
and annealed sheets of low density poly-

p. 407   *Caption to Table 12.1 should read*

Fracture surface energies (in joules per square meter $\times 10^2$)

p. 412   *ordinate scale in Figure 12.10 should read*

Surface energy $(Jm^{-2} \times 10^2)$

p. 421   *Caption should read*

Variation of $\mathcal{T}_0^*$

p. 423   *Figure 12.16 Caption permission* from

Preface to the First Edition

This text-book is a somewhat extended version of a lecture course given to MSc students studying the Physics of Materials in the Physics Department of the University of Bristol during the period 1965–1970. It is therefore written for those whose background is a first degree in physics, chemistry, engineering, metallurgy or some related discipline. Although it is intended to be self-contained, there is much contingent reading particularly in mechanics and in polymer science if the reader is to derive full benefit.

The book follows the approach of developing the mechanics of the behaviour first, and then discussing molecular and structural interpretations. The relative detail with which this is done reflects the current state of the art. In this respect, then, the treatment of the different aspects of mechanical properties is often very different in style. For example, the expositions of rubber-like behaviour and linear viscoelastic behaviour are mostly didactic, as befits subjects of long standing. On the other hand, the discussions of yield and fracture take the form of extensive reviews of progress.

Each chapter has been written as a separate unit and I hope that this will appeal to those who would not wish to read the book as a whole in the order of presentation. In this respect it is hoped that the book may prove helpful to research workers in Polymer Science who are seeking an introduction to a part of the subject which is other than their own speciality.

Many of my colleagues have given considerable assistance by reading and commenting on the draft manuscript. The whole text was considered by Professors G. Allen, D. W. Saunders and R. S. Stein who made many helpful suggestions. In addition I wish to thank the following for reading various chapters: G. R. Davies, R. A. Duckett, J. S. Foot, J. S. Harris, N. H. Ladizesky, S. Rabinowitz, E. B. Ranby, T. G. Rogers and J. Scanlan.

I owe particular debt of gratitude to Mr Ralph Davis who undertook the production of the figures.

Finally, I am especially grateful to my wife, without whose constant encouragement this book would not have been written.

I. M. WARD

Preface to the Second Edition

In one major respect, the preparation of this second edition proved a more difficult task than that of its predecessor. The subject has expanded, so that a much greater degree of selectivity had to be adopted, if the book were not to expand to unmanageable proportions. In the event, this edition retains a form similar to the previous one. The first six chapters are essentially didactic, and intended to introduce the reader to the subject by discussion of the better defined and more established topics of rubber elasticity and linear viscoelastic behaviour. The chapters on rubber elasticity have been extended and modified, to take into account recent developments. The chapters on linear viscoelastic behaviour have been altered very little, only adding a section on the latest theories of the dynamics of chain molecules.

As in the first edition, the last four chapters are essentially self-contained reviews of four major subject areas. The changes to these chapters are more substantial and reflect not only the very great expansion in the subject, but also the rather different emphasis which is gradually emerging. In 1971, when the first edition was published, the mechanical behaviour of polymers was only beginning to fit into the scheme of materials science. It had just been recognized that a formal description of several aspects of the mechanical behaviour of polymers could be made using the techniques of solid mechanics. Whilst it is still important to retain a sound theoretical framework, there is increasing interest in bridging the gap between the formal mathematical description in terms of solid mechanics, and a molecular or morphological understanding. The links between structure and properties are perhaps most developed in the case of anisotropic mechanical behaviour, but connections are also beginning to emerge in non-linear viscoelasticity, yield and fracture. The recognition of this movement within the subject has caused some alteration to the style and emphasis in the last four chapters.

Finally, as on the previous occasion, I am greatly indebted to several of my colleagues for reading and commenting on the draft manuscript. I would especially like to thank Professor N. Brown, Dr R. A. Duckett, Dr N. H. Ladizesky and Professor L. R. G. Treloar for their very constructive criticism of those chapters concerned with their own particular interests.

I also owe a considerable debt of gratitude to my wife and family, who have been most forebearing and supportive, to Mrs J. M. Crowther for typing the manuscript and to Mr C. Morath for assistance with the new diagrams.

I. M. WARD

Contents

1

The Structure of Polymers

Although this book is primarily concerned with the mechanical properties of polymers, it seems desirable at the outset to introduce a few elementary ideas concerning their structure.

1.1 CHEMICAL COMPOSITION

1.1.1 Polymerization

The first essential point is that polymers consist of long molecular chains of covalently bonded atoms. One of the simplest polymers is polyethylene, which consists of long chains of the $-CH_2$ repeat unit (Figure 1.1(a) and (b)). This is an *addition* polymer and is made by polymerizing the monomer ethylene $CH_2=CH_2$ to form the polymer

$$[-CH_2-CH_2-]_n.$$

A well known class of polymers is made from the compounds

$$\overset{\displaystyle X}{\underset{\displaystyle CH_2=CH}{|}}$$

where X represents a chemical group. These are the *vinyl* polymers and familiar examples are polypropylene

$$\overset{\displaystyle CH_3}{\underset{\displaystyle [-CH_2-CH-]_n}{|}}$$

polystyrene

$$\overset{\displaystyle C_6H_5}{\underset{\displaystyle [-CH_2-CH-]_n}{|}}$$

and polyvinyl chloride

$$\overset{\displaystyle Cl}{\underset{\displaystyle [-CH_2-CH-]_n}{|}}$$

Condensation polymers are made by reacting difunctional molecules, with the elimination of water. One example is the formation of polyethylene

2

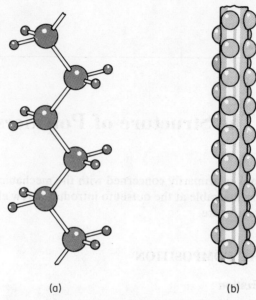

(a)   (b)

Figure 1.1. (a) The polyethylene chain $(CH_2)_n$ in schematic form (larger spheres, carbon; smaller spheres, hydrogen); (b) sketch of a molecular model of a polyethylene chain.

terephthalate (Terylene or Dacron) from ethylene glycol and terephthalic acid:

$$n\text{HO}-\text{CH}_2-\text{CH}_2-\text{OH} + n\text{HOOC}\langle\bigcirc\rangle\text{COOH}$$

$$\rightarrow \text{H}[-\text{O}-\text{CH}_2-\text{CH}_2-\text{O}-\underset{\text{O}}{\overset{\text{O}}{\underset{\|}{\text{C}}}}\langle\bigcirc\rangle\overset{\text{O}}{\underset{\|}{\text{C}}}-]_n\text{OH} + n\text{H}_2\text{O}.$$

Nylon 66,

$$-\text{NH}-(\text{CH}_2)_6-\text{NH}-\overset{\text{O}}{\underset{\|}{\text{C}}}-(\text{CH}_2)_4-\overset{\text{O}}{\underset{\|}{\text{C}}}-$$

is another condensation polymer.

1.1.2 Cross-linking and Chain-branching

We have so far considered linear chains where the monomer units are joined into a single continuous ribbon, and each molecular chain is a separate unit. In some cases, however, the chains are joined by other chains at points along their length to make a cross-linked structure (Figure 1.2), which is the situation for rubbers and for a thermoset such as bakelite.

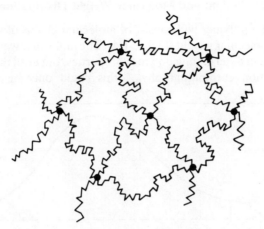

Figure 1.2. Schematic diagram of a cross-
linked polymer.

A similar but less extreme complication is chain-branching, where a secondary chain initiates from a point on the main chain. This is illustrated in Figure 1.3 for polyethylene. Low density polyethylene, as distinct from the high density or linear polyethylene shown in Figure 1.1, possesses on average one long branch per molecule and a larger number of small branches, mainly ethyl $[-CH_2-CH_3]$ or butyl $[-(CH_2)_3-CH_3]$ side groups. We will see that the presence of these branch points leads to considerable differences between the mechanical behaviour of low and high density polyethylene.

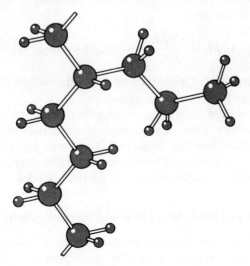

Figure 1.3. A chain branch in polyethylene.

4

1.1.3 Molecular Weight and Molecular Weight Distribution

Each sample of a polymer will consist of molecular chains of varying lengths, i.e. of varying molecular weight (Figure 1.4). The molecular weight distribution is of importance in determining polymer properties, but until the recent advent of gel permeation chromatography[1,2], this could only be determined by

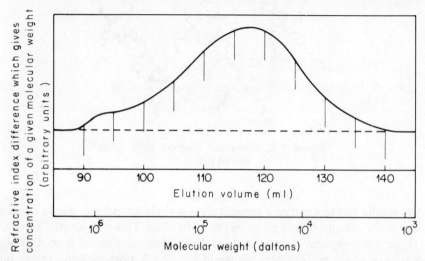

Figure 1.4. The gel permeation chromatograph trace gives a direct indication of the molecular distribution. (Result obtained in Marlex 6009 by Dr T. Williams.)

extremely tedious fractionation procedures. Most investigations were therefore concerned with different types of average molecular weight, the number average $\bar{M}_n$ and the weight average $\bar{M}_w$, which are determined by osmotic pressure and light-scattering measurements respectively, on dilute solutions of the polymers. These are

$$\text{number average, } \bar{M}_n = \frac{\Sigma N_i M_i}{\Sigma N_i}; \quad \text{weight average, } \bar{M}_w = \frac{\Sigma (N_i M_i) M_i}{\Sigma N_i M_i},$$

where N_i is the number of molecules of molecular weight M_i and Σ denotes summation over all i molecular weights.

The molecular weight distribution is important in determining flow properties. It may therefore affect the mechanical properties of a solid polymer indirectly by influencing the final physical state. Direct correlations of molecular weight to viscoelastic behaviour and brittle strength have also been obtained, and this is a developing area of polymer science.

1.1.4 Chemical and Steric Isomerism and Stereoregularity

A further complication in the chemical structure of polymers is the possibility of different chemical isomeric forms for the repeat unit, or for a series of repeat units. A particularly simple example is presented by the possibility of either head-to-head or head-to-tail addition of vinyl monomer units

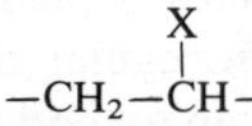

$$-CH_2-\overset{\overset{X}{|}}{CH}-.$$

We can have

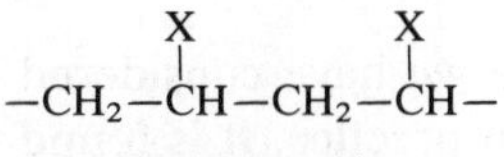

$$-CH_2-\overset{\overset{X}{|}}{CH}-CH_2-\overset{\overset{X}{|}}{CH}-$$
(head-to-tail)

or

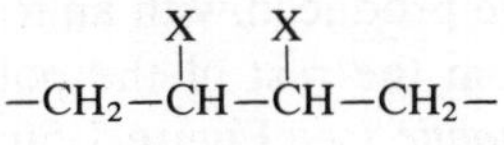

$$-CH_2-\overset{\overset{X}{|}}{CH}-\overset{\overset{X}{|}}{CH}-CH_2-$$
(head-to-head)

Most vinyl polymers show predominantly head-to-tail substitution, but the presence of the alternative situation leads to a loss of regularity which can reflect itself in a reduced degree of crystallization and hence affect mechanical properties.

A rather more complex case is that of steric isomerism and the question of stereoregularity. Consider the simplest type of vinyl polymer in which a substituent group X is attached to every alternate carbon atom. For illustrative purposes let us suppose that the polymer chain is a planar zig-zag. Then, as shown in Figure 1.5(a) and (b), there are two very simple regular polymers which can be constructed. In the first of these the substituent groups are all added in an identical manner as we proceed along the chain (Figure 1.5(a)). This is the *isotactic* polymer. In the second regular polymer, there is an inversion of the manner of substitution for each monomer unit. In this planar

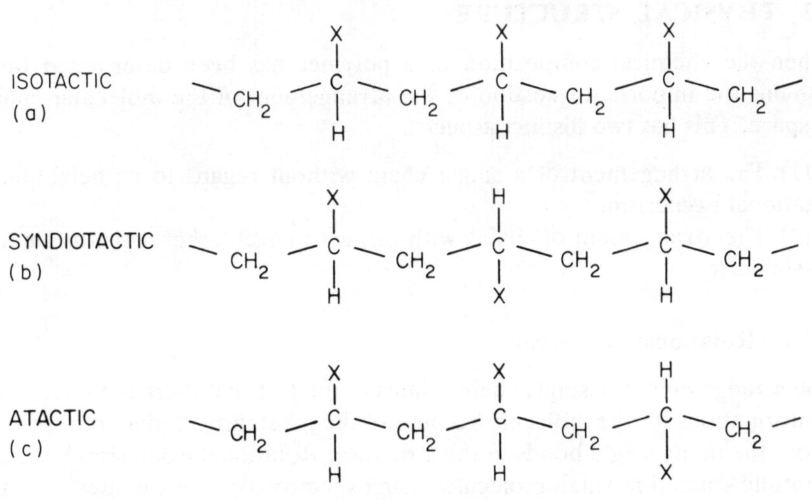

Figure 1.5. A substituted α olefin can take three stereosubstituted forms.

model, the substituent groups alternate regularly on opposite sides of the chain. This polymer is called *syndiotactic* (Figure 1.5(b)). The importance of this regular addition is that it makes it possible for crystallization to occur. This is stereoregularity, and stereoregular polymers are crystalline and can possess high melting points. This can extend their working range appreciably, as compared with amorphous polymers, whose range is limited by the lower softening point.

The two structures which we have considered are those for ideal isotactic or syndiotactic polymers. In practice, it is found that although stereospecific catalysts can produce polymer chains which are predominantly isotactic or syndiotactic, there are always a considerable number of faulty substitutions. Thus a number of chains are produced with an irregular substitution pattern. These can be separated from the rest of the polymer by solvent extraction and this fraction is called *atactic* (see Figure 1.5(c)).

1.1.5 Blends, Grafts and Copolymers

Blending, grafting and copolymerization are commonly used to increase the ductility and toughness of brittle polymers or to increase the stiffness of rubbery polymers.

A *blend* is a mixture of two or more polymers.

A *graft* is where long side chains of a second polymer are chemically attached to the base polymer.

A *copolymer* is where chemical combination exists in the main chain between two polymers $[A]_n$ and $[B]_n$. A copolymer can be a *block* copolymer $[AAA \ldots]$ $[BB \ldots]$ or a *random* copolymer ABAABAB, the latter having no long sequences of A or B units.

1.2 PHYSICAL STRUCTURE

When the chemical composition of a polymer has been determined there remains the important question of the arrangement of the molecular chains in space. This has two distinct aspects:

(1) The arrangement of a single chain without regard to its neighbours: rotational isomerism.

(2) The arrangement of chains with respect to each other: orientation and crystallinity.

1.2.1 Rotational Isomerism

The arrangement of a single chain relates to the fact that there are alternative conformations for the molecule because of the possibility of hindered rotation about the many single bonds in the structure. Rotational isomerism has been carefully studied in small molecules using spectroscopic techniques[3] and the *trans* and *gauche* isomerism of disubstituted ethanes. Figure 1.6 shows a

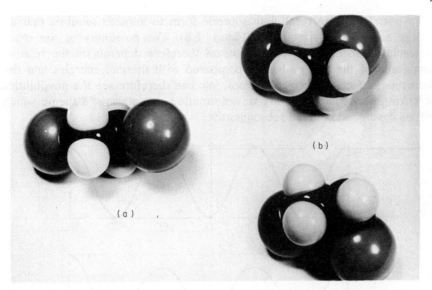

Figure 1.6. (a) The *trans* and (b) the *gauche* conformations of dibromoethane.

familiar example. Similar considerations apply to polymers, and the situation for polyethylene terephthalate[4], where there are two possible conformations of the glycol residue, is illustrated in Figure 1.7.

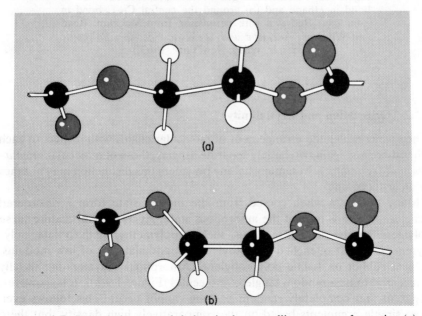

Figure 1.7. Polyethylene terephthalate in the crystalline *trans* conformation (a) and the *gauche* conformation (b) which is present in 'amorphous' regions (after Grime and Ward[4]).

8

To pass from one rotational isomeric form to another requires that an energy barrier be surmounted (Figure 1.8). The possibility of the chain molecules changing their conformations therefore depends on the relative magnitude of the energy barrier compared with thermal energies and the perturbing effects of applied stresses. We can therefore see the possibilities of linking molecular flexibility to deformation mechanisms, a theme which will be developed in detail subsequently.

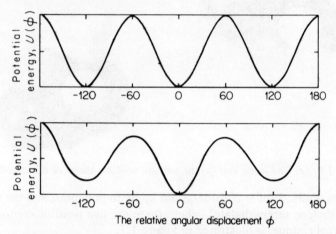

Figure 1.8. Potential energy for rotation (a) around the C—C bond in ethane and (b) around the central C—C bond in *n*-butane. [Redrawn with permission from McCrum, Read and Williams, *Anelastic and Dielectric Effects in Polymeric Solids*, Wiley, New York, 1963.]

1.2.2 Orientation and Crystallinity

When we consider the arrangement of molecular chains with respect to each other there are again two largely separate aspects, those of molecular *orientation* and *crystallinity*. In semicrystalline polymers this distinction may at times be an artificial one.

Many polymers when cooled from the molten state form a disordered structure, which is termed the amorphous state. At room temperature these polymers may be of high modulus, such as polymethyl methacrylate, polystyrene and melt-quenched polyethylene terephthalate, or of low modulus, such as rubber or atactic polypropylene. Amorphous polymers are usually considered to be a random tangle of molecules (Figure 1.9(a)). It is, however, apparent that completely random packing cannot occur. This follows even from simple arguments based on the comparatively high density[5] but there is no distinct structure as, for example, revealed by X-ray diffraction techniques.

Polymethyl methacrylate, polystyrene and melt-quenched polyethylene terephthalate are examples of amorphous polymers. If such a polymer is stretched the molecules may be preferentially aligned along the stretch direction. In polymethyl methacrylate and polystyrene such molecular orientation may be detected by optical measurements; but X-ray diffraction measurements still reveal no sign of three-dimensional order. The structure is therefore regarded as a somewhat elongated tangled skein (Figure 1.9(b)). We would say that such a structure is oriented amorphous, but not crystalline.

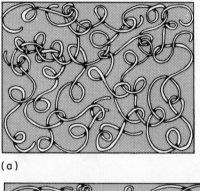

(a)

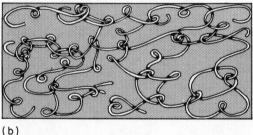

(b)

Figure 1.9. Schematic diagrams of (a) unoriented amorphous polymer and (b) oriented amorphous polymer.

In polyethylene terephthalate, however, stretching produces both molecular orientation and small regions of three-dimensional order, namely crystallites. The simplest explanation of such behaviour is that the orientation processes have brought the molecules into adequate juxtaposition for them to take up positions of three-dimensional order, and hence crystallize.

Many polymers, including polyethylene terephthalate, also crystallize if they are cooled slowly from the melt. In this case we may say that they are crystalline but oriented. Although such specimens are unoriented in the macroscopic sense, i.e. they possess isotropic bulk mechanical properties, they are not homogeneous in the microscopic sense and often show a spherulitic structure under a polarizing microscope.

10

The crystal structures of the crystalline regions can be determined from the wide angle X-ray diffraction patterns of polymers in the stretched crystalline form. Such structures have been obtained for all the well known crystalline polymers, e.g. polyethylene, nylon, polyethylene terephthalate, polypropylene (Figure 1.10).

Such information is extremely valuable in gaining an understanding of the structure of a polymer. It was early recognized, however, that in addition to the discrete reflections from the crystallites the diffraction pattern of a semi-crystalline polymer also shows much diffuse scattering and this was attributed to the 'amorphous' regions.

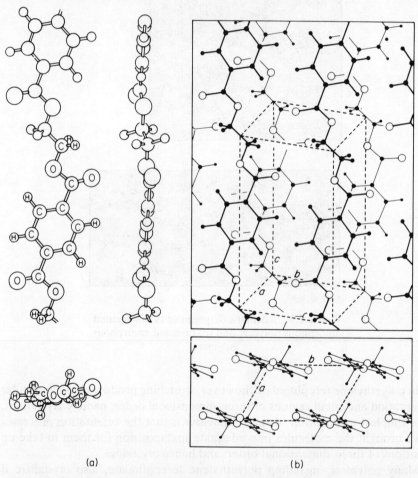

(a)   (b)

Figure 1.10. (a) Configuration of atoms in the molecule and (b) arrangement of the molecules in crystalline polyethylene terephthalate. Upper diagrams show projections normal to the 010 plane; lower diagrams show projections along the *c* axis (larger dots, carbon; smaller dots, hydrogen; open circles, oxygen. [Redrawn with permission from Daubery, Bunn and Brown, *Proc. R. Soc.*, *A*, **226**, 531 (1954).]

This led to the so-called fringed micelle model (Figure 1.11) for the structure of a semicrystalline polymer, which is a natural development of the imagined situation in an amorphous polymer. The molecular chains alternate between regions of order (the crystallites) and disorder (the amorphous regions).

Figure 1.11. The fringed micelle representation of crystalline polymers. [Redrawn with permission from Bunn, in *Fibres from Synthetic Polymers* (R. Hill, ed.), Elsevier, Amsterdam, 1953, p. 243.]

This fringed micelle model was called into question by the discovery of polymer single crystals grown from solution[6-8]. Linear polyethylene, when crystallized from dilute solution, forms single crystal lamellae, with lateral dimensions of the order of 10–20 μm and of the order of 10 nm thick. Electron diffraction shows that the molecular chains are approximately normal to the lamellar surface and since the molecules are usually of the order of 1 μm in length, it can be deduced that they must be folded back and forth within the crystals. The initial proposal suggested that the folds were sharp and regular (adjacent re-entry), as shown schematically in Figure 1.12. Recent neutron

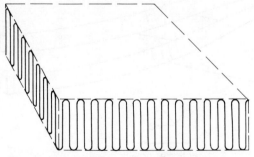

Figure 1.12. Diagrammatic representation of chain folding in polymer crystals with the folds drawn sharp and regular. [Redrawn with permission from Keller, *Progr. Rep. Phys.*, **31**, 623 (1969).]

scattering experiments[9] suggest that the chain-folded ribbon itself doubles up beyond a certain length, which has been termed *superfolding* and that there is a definite departure from strict adjacent re-entry. This is an area of some controversy and one view is that the folds are predominantly non-adjacent[10]. Single crystals have been isolated for most crystalline polymers, including nylon and polypropylene.

The influence of chain folding on the structure of bulk crystalline polymers is a matter of further controversy, although there is much evidence to support the existence of a lamellar morphology, and this is sometimes vital to the understanding of mechanical properties. From a molecular point of view neutron diffraction experiments have again provided key information in showing that only small changes occur in the radius of gyration of the chain molecules in several polymers in crystallization from the molten state[11]. We

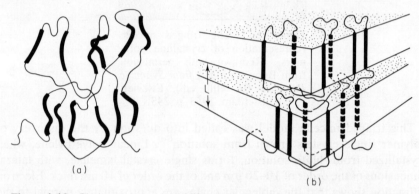

Figure 1.13. Chain conformation (a) in the melt and (b) in the crystal according to the solidification model. [Redrawn with permission from Stamm, Fischer, Dettenmaier and Convert, *Faraday Disc.*, **68**, 263 (1979).]

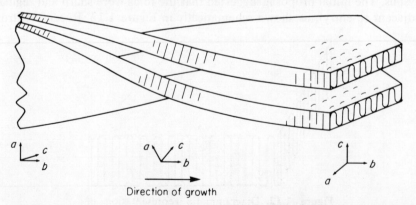

Direction of growth

Figure 1.14. A model of the lamellar arrangement in a polyethylene spherulite. The small diagrams of the *a*, *b*, *c* axes show the orientation of the unit cell at various points. [Redrawn with permission from Takayanagi, *Mem. Fac. Eng. Kyushu Univ.*, **23**, 1 (1963).]

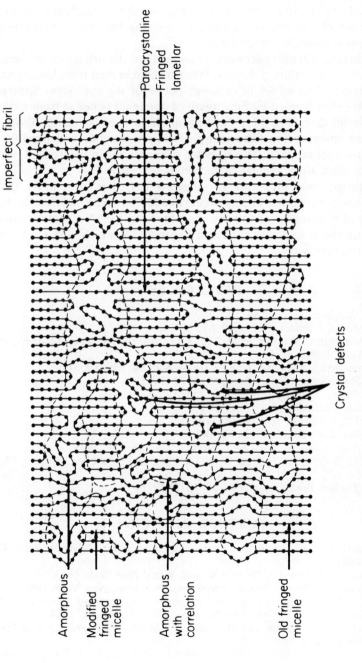

Figure 1.15. Schematic composite diagram of different types of order and disorder in oriented polymers. [Redrawn with permission from Hosemann, *Polymer* **3**, 349 (1962).]

are bound to conclude that the highly entangled topology of the chains which exists in the melt must be substantially retained in the semicrystalline state. One model (the solidification model[12]) proposes that crystallization occurs by straightening of coil sequences without involving long range diffusion processes. This is shown in Figure 1.13.

Nevertheless, it is still necessary to understand the origins of the lamellar morphology of crystalline polymers. When a polymer melt is cooled, crystallization is initiated at nuclei in different points of the specimen. Spherulitic textures are then formed by the growth of dominant lamellae from a central nucleus in all directions by a twisting of these lamellae along fibrils, the intervening spaces being filled in by subsidiary lamellae and possibly low molecular weight material. This is shown schematically in Figure 1.14, where, for ease of illustration, regular chain folding is sketched.

The present state of knowledge suggests that it is most likely that in crystalline polymers chain folding occurs in addition to the more conventional threading of molecules through the crystalline regions. A schematic attempt to illustrate the situation is shown in Figure 1.15, which also shows other types of irregularity.

REFERENCES

1. M. F. Vaughan, *Nature*, **188**, 55 (1960).
2. J. C. Moore, *J. Polymer Sci.*, **A**, **2**, 835 (1964).
3. S.-I. Mizushima, *Structure of Molecules and Internal Rotation*, Academic Press, New York, 1954.
4. D. Grime and I. M. Ward, *Trans. Faraday Soc.*, **54**, 959 (1958).
5. R. E. Robertson, *J. Phys. Chem.*, **69**, 1575 (1965).
6. E. W. Fischer, *Naturforschung*, **12a**, 753 (1957).
7. A. Keller, *Phil. Mag.*, **2**, 1171 (1957).
8. P. H. Till, *J. Polymer Sci.*, **24**, 301 (1957).
9. A. Keller, *Disc. Faraday Soc.*, **68**, 145 (1979).
10. D. Y. Yoon and P. J. Flory, *Disc. Faraday Soc.*, **68**, 288 (1979).
11. D. G. H. Ballard, P. Cheshire, G. W. Longman and J. Schelten, *Polymer*, **19**, 379 (1978).
12. E. W. Fischer, *Pure Appl. Chem.*, **50**, 1319 (1978).

FURTHER READING

D. C. Bassett, *Principles of Polymer Morphology*, Cambridge University Press, Cambridge, 1981.
F. W. Billmeyer, *Textbook of Polymer Science*, Wiley, New York, 1963.
P. H. Geil, *Polymer Single Crystals*, Interscience Publishers, New York, 1963.
P. Meares, *Polymers: Structure and Bulk Properties*, Van Nostrand, London, 1968.
H. Tadokoro, *Structure of Crystalline Polymers*, Wiley, New York, 1979.
I. M. Ward (ed.), *Structure and Properties of Oriented Polymers*, Applied Science Publishers, London, 1979.
B. Wunderlich, *Macromolecular Physics*, Vols 1 and 2, Academic Press, New York, 1973 and 1976.

2

The Mechanical Properties of Polymers: General Considerations

2.1 OBJECTIVES

Discussions of the mechanical properties of solid polymers often contain two interrelated objectives. The first of these is to obtain an adequate macroscopic description of the particular facet of polymer behaviour under consideration. The second objective is to seek an explanation of this behaviour in molecular terms, which may include details of the chemical composition and physical structure. In this book we will endeavour, where possible, to separate these two objectives and, in particular, to establish a satisfactory macroscopic or phenomenological description before discussing molecular interpretations.

This should make it clear that many of the established relationships are purely descriptive, and do not necessarily have any implications with regard to an interpretation in structural terms. For engineering applications of polymers this is sufficient, because a description of the mechanical behaviour under conditions which simulate their end use is often all that is required, together with empirical information concerning their method of manufacture.

2.2 THE DIFFERENT TYPES OF MECHANICAL BEHAVIOUR

It is difficult to classify polymers as particular types of materials such as a glassy solid or a viscous liquid, since their mechanical properties are so dependent on the conditions of testing, e.g. the rate of application of load, temperature, amount of strain.

A polymer can show all the features of a glassy, brittle solid or an elastic rubber or a viscous liquid depending on the temperature and time scale of measurement. Polymers are usually described as *viscoelastic* materials, a generic term which emphasizes their intermediate position between viscous liquids and elastic solids. At low temperatures, or high frequencies of measurement, a polymer may be glass-like with a Young's modulus of 10^9–10^{10} N m^{-2} and will break or flow at strains greater than 5%. At high temperatures or low frequencies, the same polymer may be rubber-like with a modulus of 10^6–10^7 N m^{-2}, withstanding large extensions ($\sim 100\%$) without permanent deformation. At still higher temperatures, permanent deformation occurs under load, and the polymer behaves like a highly viscous liquid.

16

In an intermediate temperature or frequency range, commonly called the glass transition range, the polymer is neither glassy nor rubber-like. It shows an intermediate modulus, is viscoelastic and may dissipate a considerable amount of energy on being strained. The glass transition manifests itself in several ways, for example by a change in the volume coefficient of expansion, which can be used to define a glass transition temperature T_g. The glass transition is central to a great deal of the mechanical behaviour of polymers for two reasons. First, there are the attempts to link the time-temperature equivalence of viscoelastic behaviour with the glass transition temperature T_g. Secondly, glass transitions can be studied at a molecular level by such techniques as nuclear magnetic resonance and dielectric relaxation. In this way it is possible to gain an understanding of the molecular origins of the viscoelasticity.

The different features of polymer behaviour such as creep and recovery, brittle fracture, necking and cold-drawing are usually considered separately, by comparative studies of different polymers. It is customary, for example, to compare the brittle fracture of polymethyl methacrylate, polystyrene, and other polymers which show similar behaviour at room temperature. Similarly comparative studies have been made of the creep and recovery of polyethylene, polypropylene and other polyolefins. Such comparisons often obscure the very important point that the whole range of phenomena can be displayed by a single polymer as the temperature is changed. Figure 2.1 shows load-elongation curves for a polymer at four different temperatures. At temperatures well below the glass transition (curve A), where brittle fracture occurs, the load rises to the breaking point linearly with increasing elongation, and rupture occurs at low strains (~10%). At high temperatures (curve D), the polymer is rubber-like and the load rises to the breaking point with a

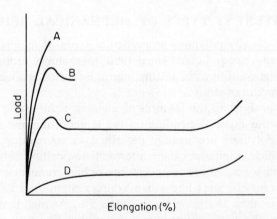

Figure 2.1. Load–elongation curves for a polymer at different temperatures. Curve A, brittle fracture; curve B, ductile failure; curve C, cold-drawing; curve D, rubber-like behaviour.

sigmoidal relationship to the elongation, and rupture occurs at very high strains (~30–1000%).

In an intermediate temperature range below the glass transition (curve B), the load–deformation relationship resembles that of a ductile metal, showing a load maximum, i.e. a yield point before rupture occurs. At slightly higher temperatures (curve C), still below the glass transition, the remarkable phenomenon of necking and cold-drawing is observed. Here the conventional load–elongation curve again shows a yield point and a subsequent decrease in conventional stress. However, with a further increase in the applied strain, the load falls to a constant level at which deformations of the order of 300–1000% are accomplished. At this stage a neck has formed and the strain in the specimen is not uniform. (This is discussed in detail in Chapter 11.) Eventually the load begins to rise again and finally fracture occurs.

It is usual to discuss the mechanical properties in the different temperature ranges separately, because different approaches and mathematical formalisms are adopted for the different features of mechanical behaviour. This conventional treatment will be followed here, although it is recognized that it somewhat arbitrarily isolates particular facets of the mechanical properties of polymers.

2.3 THE ELASTIC SOLID AND THE BEHAVIOUR OF POLYMERS

Mechanical behaviour is in most general terms concerned with the deformations which occur under loading. In any specific case the deformations depend on details such as the geometrical shape of the specimen or the way in which the load is applied. Such considerations are the province of the plastics engineer, who is concerned with predicting the performance of a polymer in a specified end use. In our discussion of the mechanical properties of polymers we will ignore such questions as these, which relate to solving particular problems of behaviour in practice. We will concern ourselves only with the generalized equations termed *constitutive relations*, which relate stress and strain for a particular type of material. First it will be necessary to find constitutive relations which give an adequate description of the mechanical behaviour. Secondly, where possible, we will obtain a molecular understanding of this behaviour by a molecular model which predicts the constitutive relations.

One of the simplest constitutive relations is Hooke's law, which relates the stress σ to the strain e for the uniaxial deformation of an ideal elastic isotropic solid. Thus

$$\sigma = Ee,$$

where E is the Young's modulus.

There are five important ways in which the mechanical behaviour of a polymer may deviate from that of an ideal elastic solid obeying Hooke's law. First, in an elastic solid the deformations induced by loading are independent of the history or rate of application of the loads, whereas in a polymer the deformations can be drastically affected by such considerations. This means that the simplest constitutive relation for a polymer should in general contain time or frequency as a variable in addition to stress and strain. Secondly, in an elastic solid all the situations pertaining to stress and strain can be reversed. Thus, if a stress is applied, a certain deformation will occur. On removal of the stress, this deformation will disappear exactly. This is not always true for polymers. Thirdly, in an elastic solid obeying Hooke's law, which in its more general implications is the basis of small-strain elasticity theory, the effects observed are *linearly* related to the influences applied. This is the essence of Hooke's law; stress is exactly proportional to strain. This is not generally true for polymers, but applies in many cases only as a good approximation for very small strains; in general the constitutive relations are non-linear. It is important to note that *non-linearity* is not related to *recoverability*. In contrast to metals, polymers may recover from strains beyond the proportional limit without any permanent deformation.

Fourthly, the definitions of stress and strain in Hooke's law are only valid for small deformations. When we wish to consider larger deformations a new theory must be developed in which both stress and strain are defined more generally.

Finally, in many practical applications (such as films and synthetic fibres) polymers are used in an oriented or anisotropic form, which requires a considerable generalization of Hooke's law.

It will be convenient to discuss these various aspects separately as follows: (1) behaviour at large strains in Chapters 3 and 4 (finite elasticity and rubber-like behaviour, respectively); (2) time-dependent behaviour (viscoelastic behaviour) in Chapters 5, 6, 7 and 8; (3) non-linearity in Chapter 9 (non-linear viscoelastic behaviour); (4) the behaviour of oriented polymers in Chapter 10 (mechanical anisotropy); (5) the non-recoverable behaviour in Chapter 11 (plasticity and yield). It should be recognized, however, that we cannot hold to an exact separation and that there are many places where these aspects overlap and can be brought together by the physical mechanisms which underlie the phenomenological description.

2.4 STRESS AND STRAIN

It is desirable at this juncture to outline very briefly the concepts of stress and strain. For a more comprehensive discussion the reader is referred to standard text-books on the theory of elasticity[1,2].

2.4.1 The State of Stress

The components of stress in a body can be defined by considering the forces acting on an infinitesimal cubical volume element (Figure 2.2) whose edges

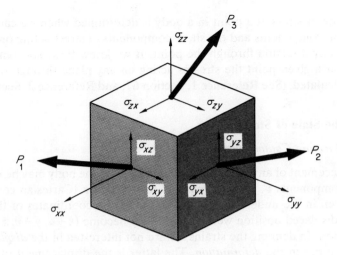

Figure 2.2. The stress components.

are parallel to the coordinate axes x, y, z. In equilibrium, the forces per unit area acting on the cube faces are

$$P_1 \text{ on the } yz \text{ plane,}$$

$$P_2 \text{ on the } zx \text{ plane,}$$

$$P_3 \text{ on the } xy \text{ plane.}$$

These three forces are then resolved into their nine components in the x, y and z directions as follows:

$$P_1: \quad \sigma_{xx}, \quad \sigma_{xy}, \quad \sigma_{xz},$$

$$P_2: \quad \sigma_{yx}, \quad \sigma_{yy}, \quad \sigma_{yz},$$

$$P_3: \quad \sigma_{zx}, \quad \sigma_{zy}, \quad \sigma_{zz}.$$

The first subscript refers to the direction of the normal to the plane on which the stress acts, and the second subscript to the direction of the stress. In the absence of body torques the total torque acting on the cube must also be zero, and this implies three equalities:

$$\sigma_{xy} = \sigma_{yx}, \qquad \sigma_{xz} = \sigma_{zx}, \qquad \sigma_{yz} = \sigma_{zy}.$$

The components of stress are therefore defined by six independent quantities, σ_{xx}, σ_{yy} and σ_{zz}, the normal stresses, and σ_{xy}, σ_{yz} and σ_{zx}, the shear stresses. These form the six independent components of the stress tensor σ_{ij}:

$$\sigma_{ij} = \begin{bmatrix} \sigma_{xx} & \sigma_{xy} & \sigma_{xz} \\ \sigma_{yx} & \sigma_{yy} & \sigma_{yz} \\ \sigma_{zx} & \sigma_{zy} & \sigma_{zz} \end{bmatrix}.$$

The state of stress at a point in a body is determined when we can specify the normal components and the shear components of stress acting on a plane drawn in any direction through the point. If we know these six components of stress at a given point the stresses acting on any plane through this point can be calculated. (See Reference 1, Section 67, and Reference 2, Section 47.)

2.4.2 The State of Strain

The Engineering Components of Strain

The displacement of any point P_1 (see Figure 2.3) in the body may be resolved into its components u, v and w parallel to x, y and z (Cartesian coordinate axes chosen in the undeformed state) so that if the coordinates of the point in the undisplaced position were (x, y, z) they become $(x+u, y+v, z+w)$ on deformation. In defining the strains we are not interested in the *displacement* or rotation but in the *deformation*. The latter is the displacement of a point relative to adjacent points. Consider a point P_2, very close to P_1, which in the undisplaced position had coordinates $(x+dx, y+dy, z+dz)$ and let the displacement which it has undergone have components $(u+du, v+dv, w+dw)$. The quantities required are then du, dv and dw, the *relative* displacements.

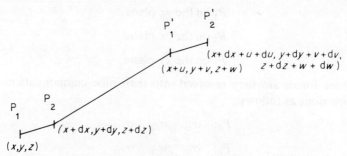

Figure 2.3. The displacements produced by deformation.

If dx, dy and dz are sufficiently small i.e. infinitesimal,

$$du = \frac{\partial u}{\partial x}dx + \frac{\partial u}{\partial y}dy + \frac{\partial u}{\partial z}dz,$$

$$dv = \frac{\partial v}{\partial x}dx + \frac{\partial v}{\partial y}dy + \frac{\partial v}{\partial z}dz,$$

$$dw = \frac{\partial w}{\partial x}dx + \frac{\partial w}{\partial y}dy + \frac{\partial w}{\partial z}dz.$$

Thus we require to define the nine quantities

$$\frac{\partial u}{\partial x}, \quad \frac{\partial u}{\partial y}, \quad \ldots, \quad \text{etc.}$$

For convenience these nine quantities are regrouped and denoted as follows:

$$e_{xx} = \frac{\partial u}{\partial x}, \qquad e_{yy} = \frac{\partial v}{\partial y}, \qquad e_{zz} = \frac{\partial w}{\partial z},$$

$$e_{yz} = \frac{\partial w}{\partial y} + \frac{\partial v}{\partial z}, \qquad e_{zx} = \frac{\partial u}{\partial z} + \frac{\partial w}{\partial x}, \qquad e_{xy} = \frac{\partial v}{\partial x} + \frac{\partial u}{\partial y},$$

$$2\bar{\omega}_x = \frac{\partial w}{\partial y} - \frac{\partial v}{\partial z}, \qquad 2\bar{\omega}_y = \frac{\partial u}{\partial z} - \frac{\partial w}{\partial x}, \qquad 2\bar{\omega}_z = \frac{\partial v}{\partial x} - \frac{\partial u}{\partial y}.$$

The first three quantities e_{xx}, e_{yy} and e_{zz} correspond to the fractional expansions or contractions along the x, y and z axes of an infinitesimal element at P_1. The second three quantities e_{yz}, e_{zx}, and e_{xy} correspond to the components of shear strain in the yz, zx and xy planes respectively. The last three quantities $\bar{\omega}_x$, $\bar{\omega}_y$, and $\bar{\omega}_z$ do not correspond to a deformation of the element at P_1, but are the components of its rotation as a rigid body.

The concept of shear strain can be conveniently illustrated by a diagram showing the two-dimensional situation of shear in the yz plane (Figure 2.4).

ABCD is an infinitesimal square which has been displaced and deformed into the rhombus A'B'C'D', θ_1 and θ_2 being the angles which A'D' and A'B' make with the y and z axes respectively.

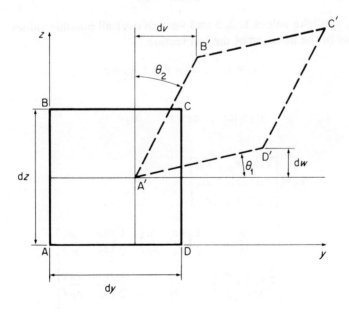

Figure 2.4. Shear strains. [Redrawn with permission from Kolsky, *Stress Waves in Solids*, Dover, New York, 1963.]

Now

$$\tan \theta_1 \doteqdot \theta_1 = \frac{dw}{dy} \rightarrow \frac{\partial w}{\partial y},$$

$$\tan \theta_2 \doteqdot \theta_2 = \frac{dv}{dz} \rightarrow \frac{\partial v}{\partial z}.$$

The shear strain in the yz plane is given by

$$e_{yz} = \frac{\partial w}{\partial y} + \frac{\partial v}{\partial z} = \theta_1 + \theta_2.$$

$2\bar{\omega}_x = \theta_1 - \theta_2$ does not correspond to a deformation of ABCD but to twice the angle through which AC has been rotated.

The deformation is therefore defined by the first six quantities e_{xx}, e_{yy}, e_{zz}, e_{yz}, e_{zx}, e_{xy}, which are called the components of strain. It is important to note that *engineering strains* have been defined (see below).

The Tensor Components of Strain

In the tensor notation the components of strain are defined as the components of the strain tensor

$$\varepsilon_{ij} = \frac{1}{2}\left(\frac{\partial u_i}{\partial x_j} + \frac{\partial u_j}{\partial x_i}\right),$$

where i, j can take values 1, 2, 3 and we sum over all possible values.

In terms of the notation of the last section,

$$x_1 = x, \qquad x_2 = y, \qquad x_3 = z$$

and

$$u_1 = u, \qquad u_2 = v, \qquad u_3 = w.$$

Then

$$\varepsilon_{ij} = \begin{bmatrix} \varepsilon_{xx} & \varepsilon_{xy} & \varepsilon_{xz} \\ \varepsilon_{xy} & \varepsilon_{yy} & \varepsilon_{yz} \\ \varepsilon_{xz} & \varepsilon_{yz} & \varepsilon_{zz} \end{bmatrix}$$

$$= \begin{bmatrix} \dfrac{\partial u}{\partial x} & \dfrac{1}{2}\left(\dfrac{\partial v}{\partial x} + \dfrac{\partial u}{\partial y}\right) & \dfrac{1}{2}\left(\dfrac{\partial w}{\partial x} + \dfrac{\partial u}{\partial z}\right) \\ \dfrac{1}{2}\left(\dfrac{\partial v}{\partial x} + \dfrac{\partial u}{\partial y}\right) & \dfrac{\partial v}{\partial y} & \dfrac{1}{2}\left(\dfrac{\partial v}{\partial z} + \dfrac{\partial w}{\partial y}\right) \\ \dfrac{1}{2}\left(\dfrac{\partial w}{\partial x} + \dfrac{\partial u}{\partial z}\right) & \dfrac{1}{2}\left(\dfrac{\partial v}{\partial z} + \dfrac{\partial w}{\partial y}\right) & \dfrac{\partial w}{\partial z} \end{bmatrix}$$

in our previous notation and

$$\varepsilon_{ij} = \begin{bmatrix} e_{xx} & \frac{1}{2}e_{xy} & \frac{1}{2}e_{xz} \\ \frac{1}{2}e_{xy} & e_{yy} & \frac{1}{2}e_{yz} \\ \frac{1}{2}e_{xz} & \frac{1}{2}e_{yz} & e_{zz} \end{bmatrix}$$

in terms of the engineering components of strain.

2.5 THE GENERALIZED HOOKE'S LAW

The most general *linear* relationship between stress and strain is obtained by assuming that each of the tensor components of stress is linearly related to all the tensor components of strain and vice versa. Thus

$$\sigma_{xx} = a\varepsilon_{xx} + b\varepsilon_{yy} + c\varepsilon_{zz} + d\varepsilon_{xz} + \dots \text{etc.}$$

and

$$\varepsilon_{xx} = a'\sigma_{xx} + b'\sigma_{yy} + c'\sigma_{zz} + d'\sigma_{xz} + \dots \text{etc.,}$$

where $a, b, \dots, a', b', \dots$ are constants. This is the generalized Hooke's law.

In tensor notation we relate the second-rank tensor σ_{ij} to the second-rank strain tensor ε_{ij} by fourth-rank tensors c_{ijkl} and s_{ijkl}. Thus

$$\sigma_{ij} = c_{ijkl}\varepsilon_{kl}$$

or equivalently

$$\varepsilon_{ij} = s_{ijkl}\sigma_{kl},$$

where

$$\sigma_{ij} = \sigma_{xx}, \sigma_{yy}, \dots, \text{etc.}$$

and

$$\varepsilon_{ij} = \varepsilon_{xx}, \varepsilon_{yy}, \dots, \text{etc.}$$

The fourth-rank tensors s_{ijkl} and c_{ijkl} contain the compliance and stiffness constants respectively, with i, j, k, l taking values 1, 2, 3. These equations use 1, 2, 3 as being synonymous with x, y, z respectively. The compliances and stiffness are written in terms of 1, 2, 3 and the stresses and strains in terms of x, y, z.

It is customary to adopt an abbreviated nomenclature in which the generalized Hooke's law relates the six independent components of stress to the six independent components of the engineering strains.

We have

$$\sigma_p = c_{pq}e_q \quad \text{and} \quad e_p = s_{pq}\sigma_q,$$

where σ_p represents $\sigma_{xx}, \sigma_{yy}, \sigma_{zz}, \sigma_{xz}, \sigma_{yz}$ or σ_{xy}, and e_q represents $e_{xx}, e_{yy}, e_{zz}, e_{xz}, e_{yz}$ or e_{xy}. We form matrices for c_{pq} and s_{pq} in which p and q take the values 1, 2, ..., 6. In the case of the stiffness constants the values of p and q are obtained in terms of i, j, k, l by substituting 1 for 11, 2 for 22, 3 for 33, 4 for 23, 5 for 13 and 6 for 12. For the compliance constants rather more complicated rules apply owing to the occurrence of the factor-2 difference between the definition of the tensor shear strain components and the definition of engineering shear strains. Thus

$$s_{ijkl} = s_{pq} \text{ when } p \text{ and } q \text{ are 1, 2 or 3,}$$

$$2s_{ijkl} = s_{pq} \text{ when either } p \text{ or } q \text{ are 4, 5 or 6,}$$

$$4s_{ijkl} = s_{pq} \text{ when both } p \text{ and } q \text{ are 4, 5 or 6.}$$

A typical relationship between stress and strain is now written as

$$\sigma_{xx} = c_{11}e_{xx} + c_{12}e_{yy} + c_{13}e_{zz} + c_{14}e_{xz} + c_{15}e_{yz} + c_{16}e_{xy}.$$

The existence of a strain-energy function (see Reference 2, p. 149) provides the relationships

$$c_{pq} = c_{qp}, \qquad s_{pq} = s_{qp}$$

and reduces the number of independent constants from 36 to 21. We then have

$$c_{pq} = \begin{pmatrix} c_{11} & c_{12} & c_{13} & c_{14} & c_{15} & c_{16} \\ c_{12} & c_{22} & c_{23} & c_{24} & c_{25} & c_{26} \\ c_{13} & c_{23} & c_{33} & c_{34} & c_{35} & c_{36} \\ c_{14} & c_{24} & c_{34} & c_{44} & c_{45} & c_{46} \\ c_{15} & c_{25} & c_{35} & c_{45} & c_{55} & c_{56} \\ c_{16} & c_{26} & c_{36} & c_{46} & c_{56} & c_{66} \end{pmatrix}$$

and similarly

$$s_{pq} = \begin{pmatrix} s_{11} & s_{12} & s_{13} & s_{14} & s_{15} & s_{16} \\ s_{12} & s_{22} & s_{23} & s_{24} & s_{25} & s_{26} \\ s_{13} & s_{23} & s_{33} & s_{34} & s_{35} & s_{36} \\ s_{14} & s_{24} & s_{34} & s_{44} & s_{45} & s_{46} \\ s_{15} & s_{25} & s_{35} & s_{45} & s_{55} & s_{56} \\ s_{16} & s_{26} & s_{36} & s_{46} & s_{56} & s_{66} \end{pmatrix}.$$

These matrices define the relationships between stress and strain in a general elastic solid, whose properties vary with direction, i.e. an anisotropic elastic solid. In most of this book we will be concerned with isotropic polymers; all discussion of anisotropic mechanical properties will be reserved for Chapter 10.

It is then most straightforward to use the compliance constants matrix, and note that measured quantities, such as the Young's modulus E, Poisson's ratio ν and the torsional or shear modulus G, relate directly to the compliance constants.

For an isotropic solid, the matrix s_{pq} reduces to

$$
s_{pq} = \begin{pmatrix}
s_{11} & s_{12} & s_{12} & 0 & 0 & 0 \\
s_{12} & s_{11} & s_{12} & 0 & 0 & 0 \\
s_{12} & s_{12} & s_{11} & 0 & 0 & 0 \\
0 & 0 & 0 & 2(s_{11}-s_{12}) & 0 & 0 \\
0 & 0 & 0 & 0 & 2(s_{11}-s_{12}) & 0 \\
0 & 0 & 0 & 0 & 0 & 2(s_{11}-s_{12})
\end{pmatrix}.
$$

It can be shown that the Young's modulus is given by

$$ E = 1/s_{11}, $$

the Poisson's ratio by

$$ \nu = -s_{12}/s_{11} $$

and the torsional modulus by

$$ G = 1/2(s_{11}-s_{12}). $$

Thus we obtain the stress–strain relationships which are the starting point in many elementary text-books of elasticity (Reference 1, pp. 7–9):

$$ e_{xx} = \frac{1}{E}\sigma_{xx} - \frac{\nu}{E}(\sigma_{yy}+\sigma_{zz}), $$

$$ e_{yy} = \frac{1}{E}\sigma_{yy} - \frac{\nu}{E}(\sigma_{xx}+\sigma_{zz}), $$

$$ e_{zz} = \frac{1}{E}\sigma_{zz} - \frac{\nu}{E}(\sigma_{xx}+\sigma_{yy}), $$

$$ e_{xz} = \frac{1}{G}\sigma_{xz}, $$

$$ e_{yz} = \frac{1}{G}\sigma_{yz}, $$

$$ e_{xy} = \frac{1}{G}\sigma_{xy}, $$

where

$$ G = \frac{E}{2(1+\nu)}. $$

Another basic quantity is the bulk modulus K, which determines the dilatation $\Delta = e_{xx}+e_{yy}+e_{zz}$ produced by a uniform hydrostatic pressure. Using the stress–strain relationships above, it can be shown that the strains produced by a uniform hydrostatic pressure p are given by

$$e_{xx} = (s_{11} + 2s_{12})p,$$
$$e_{yy} = (s_{11} + 2s_{12})p,$$
$$e_{zz} = (s_{11} + 2s_{12})p.$$

Then

$$K = \frac{p}{\Delta} = \frac{1}{3(s_{11} + 2s_{12})} = \frac{E}{3(1 - 2\nu)}.$$

This completes our introduction to linear elastic behaviour at small strains. The extension to large strains will be considered in the next chapter on finite elasticity.

REFERENCES

1. S. Timoshenko and J. N. Goodier, *Theory of Elasticity*, McGraw–Hill, New York, 1951.
2. A. E. H. Love, *A Treatise on the Mathematical Theory of Elasticity*, 4th edn, Macmillan, New York, 1944.

3

The Behaviour of Polymers in the Rubber-like State: Finite Strain Elasticity

In the rubber-like state, a polymer may be subjected to large deformations and still show complete recovery. The behaviour of a rubber band stretching to two or three times its original length and, when released, recovering essentially instantaneously to its original shape, is a matter of common experience. To a good approximation this is elastic behaviour at large strains. The first stage in developing an understanding of this behaviour is to seek a generalized definition of strain which will not suffer the restriction of that derived in Chapter 2, that the strains are small. This is followed by a rigorous definition of stress for the situation where the deformations are not small. These considerations are the basis of finite elasticity theory. This subject has been considered in several notable texts[1,2] and it is not intended to duplicate the elegant treatments presented elsewhere. For the most part the development of finite strain elasticity theory has been made using tensor calculus. In this book the treatment will be at a more elementary level, and it is hoped that in this way the exposition will be clear to those who only require an outline of the subject in order to understand the relevant mechanical properties of polymers.

3.1 THE GENERALIZED DEFINITION OF STRAIN

The generalized theory of strain considers the ratio of the length of the line joining two points in the undeformed solid to the length of the line joining the same two points in the deformed solid. It will be shown that this comparison, suitably defined, can lead to a strain tensor with six independent components which reduce to the expressions of Section 2.4.2 for small strains when the deformation is small.

Consider a system of rectangular coordinate axes x, y and z (Figure 3.1). The point P has coordinates (x, y, z) and the neighbouring point Q has coordinates $(x + dx, y + dy, z + dz)$. When the body is deformed, P and Q become $\bar{P}$ and $\bar{Q}$ with coordinates $(x + u, y + v, z + w)$ and $(x + dx + u + du,$

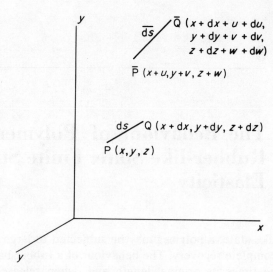

Figure 3.1. Finite deformation transforms the line PQ to the line $\bar{P}\bar{Q}$.

$y + dy + v + dv$, $z + dz + w + dw$) respectively, where

$$
\left.\begin{aligned}
du &= \frac{\partial u}{\partial x}dx + \frac{\partial u}{\partial y}dy + \frac{\partial u}{\partial z}dz, \\
dv &= \frac{\partial v}{\partial x}dx + \frac{\partial v}{\partial y}dy + \frac{\partial v}{\partial z}dz, \\
dw &= \frac{\partial w}{\partial x}dx + \frac{\partial w}{\partial y}dy + \frac{\partial w}{\partial z}dz.
\end{aligned}\right\} \tag{3.1}
$$

The line PQ has length ds, where $ds^2 = dx^2 + dy^2 + dz^2$. The line $\bar{P}\bar{Q}$ has length $\overline{ds}$, where $(\overline{ds})^2 = (dx + du)^2 + (dy + dv)^2 + (dz + dw)^2$.

We wish to relate $\overline{ds}$, the length of the line $\bar{P}\bar{Q}$ in the deformed state, to ds, the length of the line in the undeformed state PQ. Then

$$
\frac{\overline{ds}^2}{ds^2} = \frac{(dx + du)^2 + (dy + dv)^2 + (dz + dw)^2}{ds^2}
$$

$$
= \left(\frac{dx}{ds} + \frac{du}{ds}\right)^2 + \left(\frac{dy}{ds} + \frac{dv}{ds}\right)^2 + \left(\frac{dz}{ds} + \frac{dw}{ds}\right)^2.
$$

Substituting for du/ds, dv/ds and dw/ds from equations (3.1),

$$
\left(\frac{\overline{ds}}{ds}\right)^2 = \left(\frac{dx}{ds} + \frac{\partial u}{\partial x}\frac{dx}{ds} + \frac{\partial u}{\partial y}\frac{dy}{ds} + \frac{\partial u}{\partial z}\frac{dz}{ds}\right)^2
$$

$$
+ \left(\frac{dy}{ds} + \frac{\partial v}{\partial x}\frac{dx}{ds} + \frac{\partial v}{\partial y}\frac{dy}{ds} + \frac{\partial v}{\partial z}\frac{dz}{ds}\right)^2
$$

$$
+ \left(\frac{dz}{ds} + \frac{\partial w}{\partial x}\frac{dx}{ds} + \frac{\partial w}{\partial y}\frac{dy}{ds} + \frac{\partial w}{\partial z}\frac{dz}{ds}\right)^2. \tag{3.2}
$$

The deformation is being described with respect to a system of coordinates in the *undeformed* state. The line ds has direction cosines

$$l = \frac{dx}{ds}, \qquad m = \frac{dy}{ds}, \qquad n = \frac{dz}{ds}.$$

Equation (3.2) can therefore be written as

$$\left(\frac{\overline{ds}}{ds}\right)^2 = \left\{ l\left(1 + \frac{\partial u}{\partial x}\right) + m\frac{\partial u}{\partial y} + n\frac{\partial u}{\partial z}\right\}^2$$

$$+ \left\{ l\frac{\partial v}{\partial x} + m\left(1 + \frac{\partial v}{\partial y}\right) + n\frac{\partial v}{\partial z}\right\}^2$$

$$+ \left\{ l\frac{\partial w}{\partial x} + m\frac{\partial w}{\partial y} + n\left(1 + \frac{\partial w}{\partial z}\right)\right\}^2.$$

Finally we write

$$\left(\frac{\overline{ds}}{ds}\right)^2 = (1 + 2\mathbf{e}_{xx})l^2 + (1 + 2\mathbf{e}_{yy})m^2 + (1 + 2\mathbf{e}_{zz})n^2$$

$$+ 2\mathbf{e}_{yz}mn + 2\mathbf{e}_{zx}nl + 2\mathbf{e}_{xy}lm, \tag{3.3}$$

where $\mathbf{e}_{xx}$, $\mathbf{e}_{yy}$, etc., are the six components of finite strain and are given by

$$\mathbf{e}_{xx} = \frac{\partial u}{\partial x} + \frac{1}{2}\left\{\left(\frac{\partial u}{\partial x}\right)^2 + \left(\frac{\partial v}{\partial x}\right)^2 + \left(\frac{\partial w}{\partial x}\right)^2\right\},$$

$$\mathbf{e}_{yy} = \frac{\partial v}{\partial y} + \frac{1}{2}\left\{\left(\frac{\partial u}{\partial y}\right)^2 + \left(\frac{\partial v}{\partial y}\right)^2 + \left(\frac{\partial w}{\partial y}\right)^2\right\},$$

$$\mathbf{e}_{zz} = \frac{\partial w}{\partial z} + \frac{1}{2}\left\{\left(\frac{\partial u}{\partial z}\right)^2 + \left(\frac{\partial v}{\partial z}\right)^2 + \left(\frac{\partial w}{\partial z}\right)^2\right\},$$

$$\mathbf{e}_{yz} = \frac{\partial w}{\partial y} + \frac{\partial v}{\partial z} + \frac{\partial u}{\partial y}\frac{\partial u}{\partial z} + \frac{\partial v}{\partial y}\frac{\partial v}{\partial z} + \frac{\partial w}{\partial y}\frac{\partial w}{\partial z},$$

$$\mathbf{e}_{zx} = \frac{\partial u}{\partial z} + \frac{\partial w}{\partial x} + \frac{\partial u}{\partial z}\frac{\partial u}{\partial x} + \frac{\partial v}{\partial z}\frac{\partial v}{\partial x} + \frac{\partial w}{\partial z}\frac{\partial w}{\partial x},$$

$$\mathbf{e}_{xy} = \frac{\partial v}{\partial x} + \frac{\partial u}{\partial y} + \frac{\partial u}{\partial x}\frac{\partial u}{\partial y} + \frac{\partial v}{\partial x}\frac{\partial v}{\partial y} + \frac{\partial w}{\partial x}\frac{\partial w}{\partial y}.$$

Equation (3.3) for $(\overline{ds}/ds)^2$, defines the ratio of the deformed to the undeformed length of an elemental line originally at a point (x, y, z) having any specified direction defined by direction cosines l, m, n in the undeformed state.

This is known as the Lagrangian measure of strain, as distinct from the Eulerian measure where the coordinates are convected with the deformation. $(1 + 2\mathbf{e}_{xx})$, $(1 + 2\mathbf{e}_{yy})$ and $(1 + 2\mathbf{e}_{zz})$ give the values of $(\overline{ds}/ds)^2$ for elements which have directions parallel to the x, y and z axes respectively in the undeformed state. This observation forms the link between the generalized definition of finite strain and the simple ideas of extension ratios which are used in the molecular theories of rubber elasticity.

A deformation is called pure strain if the three orthogonal lines which are chosen as the system of coordinates in the undeformed state are not rotated by the deformation (Reference 3, p. 68). This means that if we choose the system of Cartesian coordinate axes to coincide with these three orthogonal lines the shear strain components are zero.

Thus for an homogeneous pure strain where λ_1, λ_2 and λ_3 are the lengths in the deformed state of linear elements parallel to the x, y and z axes respectively which have unit length in the undeformed state, we have

$$\lambda_1^2 = 1 + 2\mathbf{e}_{xx}, \quad \lambda_2^2 = 1 + 2\mathbf{e}_{yy} \quad \text{and} \quad \lambda_3^2 = 1 + 2\mathbf{e}_{zz}$$

and

$$\mathbf{e}_{yz} = \mathbf{e}_{zx} = \mathbf{e}_{xy} = 0.$$

In the general case the shear components of strain $\mathbf{e}_{yz}$, $\mathbf{e}_{zx}$, $\mathbf{e}_{xy}$, are not zero in the chosen system of coordinates. A series of suitable rotations is then required to find a system of coordinates in which the shear components of strain are zero, thus defining the principal axes of strain. λ_1, λ_2 and λ_3 are the principal extension ratios. This is the well known eigenvalue/eigenvector problem in which λ_1, λ_2 and λ_3 are found by taking the roots of the determinant

$$\begin{vmatrix} 1 + 2\mathbf{e}_{xx} - \lambda^2 & \mathbf{e}_{yx} & \mathbf{e}_{zx} \\ \mathbf{e}_{xy} & 1 + 2\mathbf{e}_{yy} - \lambda^2 & \mathbf{e}_{zy} \\ \mathbf{e}_{xz} & \mathbf{e}_{yz} & 1 + 2\mathbf{e}_{zz} - \lambda^2 \end{vmatrix} = 0$$

(see Reference 4).

A rubber has a very high bulk modulus compared with the other moduli. To a good approximation the changes in volume on deformation may therefore be neglected. This gives the following important relationship between the principal extension ratios:

$$\lambda_1 \lambda_2 \lambda_3 = 1.$$

The finite strain components include terms such as $(\partial u/\partial y)^2$, $(\partial v/\partial y)^2$, $(\partial u/\partial y)(\partial u/\partial z)$, $(\partial v/\partial y\ \partial v/\partial z)$, etc., which are of second order and can be neglected when the strains are small. The strain components then reduce to the expressions derived previously for small strains (Section 2.4.2 above), e.g.

$$\mathbf{e}_{xx} \to e_{xx} = \frac{\partial u}{\partial x}, \qquad \mathbf{e}_{yz} \to e_{yz} = \frac{\partial w}{\partial y} + \frac{\partial v}{\partial z}.$$

This can also be seen directly from equation (3.3). Consider the case of simple elongation in the x direction. Then

$$(\bar{d}s)^2 = ds^2(1 + 2\mathbf{e}_{xx}) \quad \text{i.e.} \quad (\bar{d}s/ds)^2 - 1 = 2\mathbf{e}_{xx}.$$

For small strains

$$\frac{\overline{ds}}{ds} = 1 + e_{xx} \quad \text{and} \quad (1 + e_{xx})^2 - 1 = 2\mathbf{e}_{xx},$$

i.e. $e_{xx} \simeq \mathbf{e}_{xx}$ when $e_{xx} \ll 1$ as required.

It is important to note that we have defined *engineering strains* as for the small strain case in Chapter 2. The components of finite strain, defined as the components of the strain tensor, are given by $\mathbf{e}_{xx}$, $\mathbf{e}_{yy}$, $\mathbf{e}_{zz}$, $\frac{1}{2}\mathbf{e}_{xz}$, $\frac{1}{2}\mathbf{e}_{yz}$, $\frac{1}{2}\mathbf{e}_{xy}$, i.e. as for small strains, the tensor shear strains are one-half the engineering shear strains.

In much of the literature the index notation is used. This notation is extremely useful for the further development of the theory. We will therefore summarize briefly the definition of finite strain in this notation.

The point P (x, y, z) has coordinates x_1, x_2, x_3 which are written as x_i.

The neighbouring point Q $(x + dx, y + dy, z + dz)$ is similarly written as $x_i + dx_i$.

When the body is deformed these points become

$$x_i + u_i \quad \text{and} \quad x_i + u_i + dx_i + du_i$$

which we have written for $x_1 + u_1$, $x_2 + u_2$, $x_3 + u_3$, etc. The line PQ has length ds where

$$ds^2 = (dx_1)^2 + (dx_2)^2 + (dx_3)^2$$

$$= dx_i \, dx_i = \delta_{ij} \, dx_i \, dx_j,$$

where δ_{ij} is the Kronecker δ defined as

$$\delta_{ij} = 1 \quad \text{for } i = j \quad \text{and} \quad \delta_{ij} = 0 \quad \text{for } i \neq j.$$

The summation convention is that a product involving several variables such as $\delta_{ij} \, dx_i \, dx_j$ represents the sum of each of the products involving all possible combinations of the repeated indices. In this case both i and j can take the values 1, 2 and 3. The line $\overline{PQ}$ has length $\overline{ds}$ where

$$(\overline{ds})^2 = (dx_1 + du_1)^2 + (dx_2 + du_2)^2 + (dx_3 + du_3)^2$$

and

$$du_1 = \frac{\partial u_1}{\partial x_1} dx_1 + \frac{\partial u_1}{\partial x_2} dx_2 + \frac{\partial u_1}{\partial x_3} dx_3,$$

and similarly for du_2 and du_3. Because u_i is quite generally a function of x_j, i.e. each displacement u_1, etc., depends on x_1, x_2, x_3, we must write

$$(\overline{ds})^2 = \delta_{ij}(dx_i + du_i)(dx_j + du_j)$$

$$= \delta_{ij}\left(dx_i + \frac{\partial u_i}{\partial x_h} dx_h\right)\left(dx_j + \frac{\partial u_j}{\partial x_k} dx_k\right).$$

Note that h and k can take the values 1, 2 and 3 as in the use of i and j.

In the index notation

$$\frac{\partial u_i}{\partial x_h} = u_{i,h}$$

Then

$$(\bar{d}s)^2 = \delta_{ij}(dx_i + u_{i,h} \, dx_h)(dx_j + u_{j,k} \, dx_k)$$
$$= \delta_{ij} \, dx_i \, dx_j + \delta_{ij} \, dx_i u_{j,k} \, dx_k$$
$$+ \delta_{ij} u_{i,h} \, dx_h \, dx_j + \delta_{ij} u_{i,h} \, dx_h u_{j,k} \, dx_k.$$

Changing some of the dummy indices this gives

$$(\bar{d}s)^2 = \delta_{hk} \, dx_h \, dx_k + \delta_{hj} \, dx_h u_{j,k} \, dx_k$$
$$+ \delta_{ik} u_{i,h} \, dx_h \, dx_k + \delta_{ij} u_{i,h} u_{j,k} \, dx_h \, dx_k$$
$$= [\delta_{hk} + u_{h,k} + u_{k,h} + u_{j,h} u_{j,k}] \, dx_h \, dx_k,$$

remembering that $\delta_{hj} = 0$ unless $h = j$, etc. This gives

$$(\bar{d}s)^2 = [\delta_{hk} + 2e_{hk}] \, dx_h \, dx_k, \qquad (3.4)$$

where

$$2e_{hk} = u_{h,k} + u_{k,h} + u_{j,h} u_{j,k}.$$

If $e_{hk} = 0$ then $(\bar{d}s)^2 = ds^2$, i.e. there is no deformation. Otherwise e_{hk} describes the deformation and the quantities e_{hk} are the strain components. There are nine strain components $e_{11}, e_{22}, e_{33}, e_{12}, e_{13}, e_{23}, e_{31}, e_{21}, e_{32}$. It is immediately apparent from the definition of e_{hk} above that $e_{12} = e_{21}$, etc., leaving only six independent components:

$$e_{11} = e_{xx}, \qquad e_{22} = e_{yy}, \qquad e_{33} = e_{zz},$$
$$e_{13} = \tfrac{1}{2} e_{xz}, \qquad e_{23} = \tfrac{1}{2} e_{yz}, \qquad e_{12} = \tfrac{1}{2} e_{xy}.$$

3.2 THE DEFINITION OF COMPONENTS OF STRESS

In small strain elasticity theory, the components of stress in the deformed body are defined by considering the equilibrium of an elemental cube within the body. When the strains are small the dimensions of the body are to a first approximation unaffected by the strains. It is thus of no consequence whether the components of stress are referred to an elemental cube in the deformed body or to an elemental cube in the undeformed body. For finite strains, however, this is no longer true. We will *choose* to define the components of stress with reference to the equilibrium of a cube in the *deformed* body. We will also refer the stress components to a point of the material which is at (x, y, z) in the undeformed state, i.e. to a point which is at $x' = x + u$, $y' = y + v$, $z' = z + w$, in the deformed state. To distinguish these stress components from those in the small strain case we will write $\boldsymbol{\sigma}_{xx}$, $\boldsymbol{\sigma}_{yy}$, etc., instead of σ_{xx}, σ_{yy}, etc.

Thus the stress component $\boldsymbol{\sigma}_{xx}$ denotes the force parallel to the x axis, per unit area of the deformed material, this area being normal to the x axis in the deformed state. The terms $\boldsymbol{\sigma}_{yy}$ and $\boldsymbol{\sigma}_{zz}$ are defined similarly.

The stress component $\boldsymbol{\sigma}_{yz}$ denotes the force parallel to the z axis, per unit area of the deformed material, which in the deformed state is normal to the y axis. The stress components $\boldsymbol{\sigma}_{zy}$, $\boldsymbol{\sigma}_{zx}$, $\boldsymbol{\sigma}_{xz}$, $\boldsymbol{\sigma}_{xy}$ and $\boldsymbol{\sigma}_{yx}$ are similarly defined.

This defines nine stress components. Considering the equilibrium of the cube gives the condition that these forces have zero moment. Thus by taking moments about the edges of the elemental cube, it can be shown that $\boldsymbol{\sigma}_{xy} = \boldsymbol{\sigma}_{yx}$,

$\sigma_{xz} = \sigma_{zx}$, $\sigma_{yz} = \sigma_{zy}$ as for the small strain case in Section 2.4.1 above. There are therefore six independent stress components: three normal components of stress and three shear components of stress.

3.3 THE STRESS–STRAIN RELATIONSHIPS

Using these definitions of finite strain and of stress it is clearly possible to construct constitutive relations for finite strain elastic behaviour which are analogous to the generalized Hooke's law for small strain elastic behaviour. In principle, each component of stress can be a general function of each component of strain and vice versa. The restriction similar to Hooke's law would be that each component of stress is a *linear* function of each component of strain and vice versa, e.g.

$$\sigma_{xx} = a\mathbf{e}_{xx} + b\mathbf{e}_{yy} + c\mathbf{e}_{zz} + d\mathbf{e}_{xz} + e\mathbf{e}_{yz} + f\mathbf{e}_{xy}.$$

We could use this as a starting point for a theory of finite elasticity. It would be desirable to reduce the number of independent coefficients a, b, c, etc., by considerations such as that of material symmetry. In principle it is possible to develop a theory from this basis to solve problems in finite elasticity in a similar manner to those solved in small strain elasticity. It would be necessary, for example, to satisfy conditions of stress equilibrium and compatibility of strain. The latter are more complex for finite strain than for small strain, as can be imagined from the inclusion of second-order terms in displacement derivatives, and involve the Riemann–Christoffel tensors[1].

In this text we will not be concerned with the development of a general theory of finite elasticity. Instead, we will seek to describe the phenomenological behaviour of rubbers from the most elementary considerations. There are then two bases for simplification:

(1) A rubber is isotropic in the undeformed state.
(2) The changes in volume on deformation are very small and may be neglected, i.e. a rubber is incompressible.

First consider the simplifications which these assumptions make to small strain elasticity theory.

For an isotropic solid Hooke's law can be written in terms of Young's modulus E and Poisson's ratio ν:

$$e_{xx} = \frac{1}{E}\sigma_{xx} - \frac{\nu}{E}\sigma_{yy} - \frac{\nu}{E}\sigma_{zz},$$

$$e_{yy} = -\frac{\nu}{E}\sigma_{xx} + \frac{1}{E}\sigma_{yy} - \frac{\nu}{E}\sigma_{zz},$$

$$e_{zz} = -\frac{\nu}{E}\sigma_{xx} - \frac{\nu}{E}\sigma_{yy} + \frac{1}{E}\sigma_{zz},$$

$$e_{yz} = \frac{2}{E}(1+\nu)\sigma_{yz},$$

$$e_{xz} = \frac{2}{E}(1+\nu)\sigma_{xz},$$

$$e_{xy} = \frac{2}{E}(1+\nu)\sigma_{xy}.$$

Rewrite the first three expressions in a more symmetric form as

$$e_{xx} = \frac{1+\nu}{E}\left[\sigma_{xx} - \frac{\nu}{1+\nu}(\sigma_{xx} + \sigma_{yy} + \sigma_{zz})\right],$$

$$e_{yy} = \frac{1+\nu}{E}\left[\sigma_{yy} - \frac{\nu}{1+\nu}(\sigma_{xx} + \sigma_{yy} + \sigma_{zz})\right],$$

$$e_{zz} = \frac{1+\nu}{E}\left[\sigma_{zz} - \frac{\nu}{1+\nu}(\sigma_{xx} + \sigma_{yy} + \sigma_{zz})\right].$$

As a further simplification put

$$\frac{\nu}{1+\nu}(\sigma_{xx} + \sigma_{yy} + \sigma_{zz}) = p.$$

Because p is the same for each direction, it is equivalent to a hydrostatic pressure.

This gives

$$e_{xx} = \frac{1+\nu}{E}(\sigma_{xx} - p),$$

$$e_{yy} = \frac{1+\nu}{E}(\sigma_{yy} - p),$$

$$e_{zz} = \frac{1+\nu}{E}(\sigma_{zz} - p).$$

At this stage the second assumption of incompressibility is introduced:

$$e_{xx} + e_{yy} + e_{zz} = 0$$

gives

$$\frac{1+\nu}{E}\{\sigma_{xx} + \sigma_{yy} + \sigma_{zz} - 3p\} = 0.$$

Substituting the definition of p gives $\nu = \frac{1}{2}$. For an isotropic, incompressible elastic solid, Hooke's law therefore reduces to

$$e_{xx} = \frac{3}{2E} (\sigma_{xx} - p),$$

$$e_{yy} = \frac{3}{2E} (\sigma_{yy} - p),$$

$$e_{zz} = \frac{3}{2E} (\sigma_{zz} - p),$$

$$e_{yz} = \frac{3}{E} \sigma_{yz},$$

$$e_{xz} = \frac{3}{E} \sigma_{xz}$$

and

$$e_{xy} = \frac{3}{E} \sigma_{xy}.$$

It is to be noted that if the stresses are specified, p is determined by the relationship $p = \frac{1}{3}(\sigma_{xx} + \sigma_{yy} + \sigma_{zz})$. If, on the other hand, the strains are specified, we can only find the quantities $\sigma_{xx} - p, \sigma_{yy} - p, \sigma_{zz} - p$, i.e. the normal components of stress are indeterminate to the extent of an arbitrary hydrostatic pressure p. On reflection this is to be expected; incompressibility just means that an arbitrary hydrostatic pressure produces no change in volume.

We will now propose by analogy the constitutive relations for the case where the deformations are finite. The form of these relations must be such that they reduce to those derived above when the strains are small.

To simplify matters we will consider the case where the shear strain components are zero. As will be apparent from the previous discussion there is no loss of generality in so doing because any general strain can be reduced to three principal strains by suitable choice of the coordinate axes.

For small strains the constitutive relations are then

$$\left. \begin{aligned} 2e_{xx} &= \frac{3}{E} (\sigma_{xx} - p), \\[6pt] 2e_{yy} &= \frac{3}{E} (\sigma_{yy} - p), \\[6pt] 2e_{zz} &= \frac{3}{E} (\sigma_{zz} - p), \\[6pt] e_{yz} &= e_{zx} = e_{xy} = 0. \end{aligned} \right\} \qquad (3.5)$$

Following Rivlin[5] we propose that the analogous relations for the finite strain case are

$$
\left.
\begin{aligned}
1 + 2\mathbf{e}_{xx} &= \frac{3}{E}(\boldsymbol{\sigma}_{xx} - \mathbf{p}), \\[4pt]
1 + 2\mathbf{e}_{yy} &= \frac{3}{E}(\boldsymbol{\sigma}_{yy} - \mathbf{p}), \\[4pt]
1 + 2\mathbf{e}_{zz} &= \frac{3}{E}(\boldsymbol{\sigma}_{zz} - \mathbf{p}), \\[4pt]
\mathbf{e}_{yz} = \mathbf{e}_{zx} &= \mathbf{e}_{xy} = 0.
\end{aligned}
\right\}
\tag{3.6}
$$

The difference between the quantities such as the stress components σ_{xx} and $\boldsymbol{\sigma}_{xx}$, and the hydrostatic pressures p and $\mathbf{p}$ are indicated by the use of bold type for the finite deformation case. Again we have the situation that if the strains are specified we can only find the quantities $(\boldsymbol{\sigma}_{xx} - \mathbf{p})$, etc., i.e. the normal components of stress are indeterminate to the extent of a hydrostatic pressure $\mathbf{p}$. It should be pointed out that although in the small strain case, if the stresses are specified, $p = \frac{1}{3}(\sigma_{xx} + \sigma_{yy} + \sigma_{zz})$, the corresponding relationship does not hold for finite deformations, i.e. if $\boldsymbol{\sigma}_{xx}$, $\boldsymbol{\sigma}_{yy}$ and $\boldsymbol{\sigma}_{zz}$ are given, $\mathbf{p} \neq \frac{1}{3}(\boldsymbol{\sigma}_{xx} + \boldsymbol{\sigma}_{yy} + \boldsymbol{\sigma}_{zz})$. This is because the stress–strain relations are not the same for the two cases.

Consider the relationships (3.6) for the simple case of extension parallel to the x-axis under an applied stress $\boldsymbol{\sigma}^*$. This gives a pure homogeneous deformation with λ_1, λ_2, λ_3 as the extension ratios in the x, y and z directions respectively. Then

$$
\left.
\begin{aligned}
\lambda_1^2 = 1 + 2\mathbf{e}_{xx}, \qquad \lambda_2^2 &= 1 + 2\mathbf{e}_{yy}, \qquad \lambda_3^2 = 1 + 2\mathbf{e}_{zz}, \\[4pt]
\mathbf{e}_{xz} = \mathbf{e}_{yz} &= \mathbf{e}_{xy} = 0, \\[4pt]
\boldsymbol{\sigma}_{xx} = \boldsymbol{\sigma}^*, \boldsymbol{\sigma}_{yy} &= \boldsymbol{\sigma}_{zz} = 0, \\[4pt]
\boldsymbol{\sigma}_{xz} = \boldsymbol{\sigma}_{yz} &= \boldsymbol{\sigma}_{xy} = 0.
\end{aligned}
\right\}
\tag{3.7}
$$

Substituting in equations (3.6) gives

$$
\lambda_1^2 = \frac{3}{E}(\boldsymbol{\sigma}^* - \mathbf{p}), \qquad \lambda_2^2 = -\frac{3}{E}\mathbf{p}, \qquad \lambda_3^2 = -\frac{3}{E}\mathbf{p}.
$$

The condition that the solid is incompressible gives $\lambda_1 \lambda_2 \lambda_3 = 1$; thus

$$
\lambda_2^2 = \lambda_3^2 = -\frac{3\mathbf{p}}{E} = \frac{1}{\lambda_1}
$$

and

$$
\mathbf{p} = -\frac{E}{3}\frac{1}{\lambda_1}.
$$

Therefore

$$\sigma^* = \lambda_1^2 \frac{E}{3} + \mathbf{p} = \frac{E}{3}\left(\lambda_1^2 - \frac{1}{\lambda_1}\right).$$

It is sometimes convenient to consider the *nominal stress f*, the force per unit area of unstrained cross-section.

This gives

$$f = \frac{E}{3}\left(\lambda_1 - \frac{1}{\lambda_1^2}\right). \tag{3.8}$$

For small strains $e_{xx} = \lambda_1 - 1$ and the stress–strain relationship is

$$\sigma^* = \frac{E}{3}\left\{(1 + e_{xx})^2 - \frac{1}{1 + e_{xx}}\right\}$$

$$= Ee_{xx} \text{ (ignoring third-order terms)},$$

which is Hooke's law. The constitutive relationships which we have proposed therefore satisfy the three major requirements of (1) material isotropy; (2) incompressibility; (3) reduction to Hooke's law for small strains. For simplicity only the case where the shear strains are zero has been considered. The relationship may, however, readily be generalized.

We have already reached an important conclusion. This is that the familiar relationship

$$f = \text{const.}\left(\lambda_1 - \frac{1}{\lambda_1^2}\right),$$

which is more usually presented as a consequence of the molecular theories of rubber networks, follows from purely phenomenological considerations as a simple constitutive equation for the finite deformation of an isotropic, incompressible solid. Materials which obey this relationship are sometimes called neo-Hookean.

This discussion also hints at the possibility that the most general constitutive relationships for a rubber-like material might in fact be more complex than equations (3.6). Just as some materials deviate from Hooke's law at small strains and show non-linear behaviour, it must now be asked what is the equivalent situation in finite elasticity. A better starting point for the development of a more complex theory is the introduction of the stored energy or strain-energy function and this will now be considered.

3.4 THE USE OF A STRAIN-ENERGY FUNCTION

3.4.1 Thermodynamic Considerations

An alternative approach to the theory of elasticity can be made in terms of a strain-energy function. As in our previous treatment originating in the generalized Hooke's law, we will consider first the case of small strain elasticity. Because we will wish to follow the phenomenological treatment by one based on statistical mechanics, it is important at the outset to examine the different types of strain-energy function which can be defined, depending on the experimental conditions. This introduces thermodynamic considerations.

In the first place we are required to relate the work done in the deformation of an elastic solid to the components of stress and strain in the solid. For the one-dimensional situation this can be done in the following manner. Consider the uniaxial extension of an elastic solid of length l and cross-sectional area A, fixed at one end. The origin of a system of Cartesian coordinates is situated at this fixed end and the x-axis is chosen to coincide with the length direction of the solid.

In equilibrium under the action of a tensile force f the other end of the solid is displaced from x_0 to x_1. The solid is subjected to a homogeneous extensional strain. In terms of the components of strain we have

$$e_{xx} = \frac{x_1 - x_0}{l}.$$

The stress in the solid is also homogeneous and is defined as in Section 2.4.1 in terms of the equilibrium of a cubical element at a given point. There is only one non-vanishing component of stress, the normal component $\sigma_{xx} = f/A$.

Now imagine a change in strain which involves only an infinitesimal change in the length of the solid. This is achieved by movement of the tensile force f, through a distance $x_2 - x_1$ parallel to the x axis. The total work done is $f(x_2 - x_1)$, and the work done per unit volume is

$$\left(\frac{f}{A}\right) \frac{(x_2 - x_1)}{l}.$$

(We assume that $x_2 \sim x_1 \ll l$.)

The new extensional strain is

$$e'_{xx} = \frac{x_2 - x_0}{l}.$$

The infinitesimal change in strain is therefore

$$de_{xx} = e'_{xx} - e_{xx} = \frac{(x_2 - x_0)}{l} - \frac{(x_1 - x_0)}{l} = \frac{x_2 - x_1}{l}.$$

But

$$\sigma_{xx} = f/A.$$

The work done by external forces *per unit volume* may therefore be written as $\sigma_{xx}\, de_{xx}$. For a general deformation of an elastic solid this may be generalized to give the work done by external forces per unit volume as

$$\sigma_{xx}\, \mathrm{d}e_{xx} + \sigma_{yy}\, \mathrm{d}e_{yy} + \sigma_{zz}\, \mathrm{d}e_{zz} + \sigma_{yz}\, \mathrm{d}e_{yz} + \sigma_{xz}\, \mathrm{d}e_{xz} + \sigma_{xy}\, \mathrm{d}e_{xy}.$$

Now consider a small strain deformation of unit volume of an elastic solid occurring under adiabatic conditions at constant volume. The first law of thermodynamics gives

$$\mathrm{d}W = \mathrm{d}U - \mathrm{d}Q,$$

relating the work done on the solid $\mathrm{d}W$ to the increase in internal energy $\mathrm{d}U$ and the mechanical value of the heat supplied $\mathrm{d}Q$. For an adiabatic change of state $\mathrm{d}Q = 0$ and $\mathrm{d}W = \mathrm{d}U$. We can imagine that the deformation is produced by independent changes in each of the components of strain, i.e.

$$\mathrm{d}W = \mathrm{d}U = \frac{\partial U}{\partial e_{xx}}\,\mathrm{d}e_{xx} + \frac{\partial U}{\partial e_{yy}}\,\mathrm{d}e_{yy} + \frac{\partial U}{\partial e_{zz}}\,\mathrm{d}e_{zz} + \frac{\partial U}{\partial e_{yz}}\,\mathrm{d}e_{yz} + \frac{\partial U}{\partial e_{xz}}\,\mathrm{d}e_{xz} + \frac{\partial U}{\partial e_{xy}}\,\mathrm{d}e_{xy}.$$

But $\mathrm{d}W$ is the work done by the external forces, i.e.

$$\mathrm{d}W = \mathrm{d}U = \sigma_{xx}\, \mathrm{d}e_{xx} + \sigma_{yy}\, \mathrm{d}e_{yy} + \sigma_{zz}\, \mathrm{d}e_{zz} + \sigma_{yz}\, \mathrm{d}e_{yz} + \sigma_{xz}\, \mathrm{d}e_{xz} + \sigma_{xy}\, \mathrm{d}e_{xy}.$$

This means that

$$\sigma_{xx} = \frac{\partial U}{\partial e_{xx}}, \qquad \sigma_{yy} = \frac{\partial U}{\partial e_{yy}}, \qquad \sigma_{zz} = \frac{\partial U}{\partial e_{zz}},$$

$$\sigma_{yz} = \frac{\partial U}{\partial e_{yz}}, \qquad \sigma_{xz} = \frac{\partial U}{\partial e_{xz}}, \qquad \sigma_{xy} = \frac{\partial U}{\partial e_{xy}}.$$

We can therefore define a strain-energy function or stored energy function which defines the energy stored in the body as a result of the strain. For an adiabatic change of state at constant volume we will call this U_1 and note that it is identical with the thermodynamic internal energy function U.

Then for changes at constant volume under adiabatic conditions

$$\sigma_{xx} = \frac{\partial U_1}{\partial e_{xx}}, \qquad \sigma_{yy} = \frac{\partial U_1}{\partial e_{yy}}, \qquad \sigma_{zz} = \frac{\partial U_1}{\partial e_{zz}},$$

$$\sigma_{yz} = \frac{\partial U_1}{\partial e_{yz}}, \qquad \sigma_{xz} = \frac{\partial U_1}{\partial e_{xz}}, \qquad \sigma_{xy} = \frac{\partial U_1}{\partial e_{xy}}.$$

It is more usual experimentally to observe changes of state at constant pressure rather than at constant volume. It is then customary to introduce the thermodynamic enthalpy function $H = U + PV$.

For an adiabatic change of state at constant pressure

$$\mathrm{d}W = \mathrm{d}U + P\, \mathrm{d}V = \mathrm{d}(U + PV)_P = \mathrm{d}H$$

and we can define a second strain-energy function U_2 which is identical with the enthalpy H.

Thus for deformations occurring at constant pressure under adiabatic conditions we have

$$\sigma_{xx} = \frac{\partial U_2}{\partial e_{xx}}, \qquad \sigma_{yy} = \frac{\partial U_2}{\partial e_{yy}}, \qquad \sigma_{zz} = \frac{\partial U_2}{\partial e_{zz}},$$

$$\sigma_{yz} = \frac{\partial U_2}{\partial e_{yz}}, \qquad \sigma_{xz} = \frac{\partial U_2}{\partial e_{xz}}, \qquad \sigma_{xy} = \frac{\partial U_2}{\partial e_{xy}}.$$

Next, consider *isothermal* changes of state, first at constant volume.

Here $dW = dU - dQ$ but $dQ \neq 0$.

To deal with these changes of state the Helmholtz free energy $A = U - TS$ is introduced. For an isothermal change of state at constant volume

$$dW = dU - dQ = (dU - T\, dS) = d(U - TS)_{T,V} = dA,$$

where $dQ = T\, dS$, S being the entropy.

A further strain-energy function U_3 can then be defined which is identical with the Helmholtz free energy A and the components of stress are the derivatives of U_3 with respect to the corresponding components of strain. It will be shown that this strain-energy function can be calculated for a polymer in the rubbery state from statistical mechanical considerations in terms of the strains existing in the material.

Finally we note that the most usual experimental procedure will be to measure the stress–strain relations at constant temperature and pressure. To deal with this situation the Gibbs free energy function $G = U + PV - TS$ is introduced.

For isothermal changes of state at constant pressure

$$dW = dU + P\, dV - T\, dS = d(U + PV - TS)_{T,P} = dG.$$

This leads to the definition of a strain energy function U_4 which is identical with the Gibbs free energy.

We will see that it is important to distinguish clearly between these various strain-energy functions when we come to relate the mechanics to the statistical molecular theories of the rubber-like state. Experimentally the stress–strain relations usually give the components of stress as derivatives of the strain-energy function U_4 with respect to the corresponding strains whereas it is most straightforward to calculate the value of U_3. In our subsequent discussion of the mechanics we will refer to the strain-energy function as U with no subscript, bearing in mind that for a particular experimental procedure we obtain U_1, U_2, U_3 or U_4.

3.4.2 The Form of the Strain-energy Function

The generalized Hooke's law states that the stress components are linear functions of the strain components:

$$\sigma_{xx} = c_{11}e_{xx} + c_{12}e_{yy} + \ldots \qquad \text{etc.,}$$

where c_{11}, c_{12}, etc., are the stiffness constants. It therefore follows that the strain-energy function U must be a homogeneous quadratic function of the strain components.

For a general anisotropic solid there are 21 independent coefficients (Section 2.5 above). For an isotropic solid these are reduced to two. This result is forced upon us by the symmetry of the material, and does not depend on the existence of a strain-energy function[3].

It may also be obtained from considerations of the form of the strain-energy function. We make two premises:

(1) The strain-energy function must be a homogeneous quadratic function of the strain components.

(2) For an isotropic solid the strain-energy function must take a form which does not depend on the choice of the direction of the coordinate axes. This means that the strain-energy function must be a function of the strain invariants.

3.4.3 The Strain Invariants

For a second-rank tensor such as the strain tensor

$$e_{ij} = \begin{bmatrix} e_{xx} & \frac{1}{2}e_{xy} & \frac{1}{2}e_{xz} \\ \frac{1}{2}e_{yx} & e_{yy} & \frac{1}{2}e_{yz} \\ \frac{1}{2}e_{zx} & \frac{1}{2}e_{zy} & e_{zz} \end{bmatrix}$$

there are three quantities which are independent of our choice of the directions of the Cartesian coordinate axes. In the case of the strain tensor these quantities are called the strain invariants. They are given in Table 3.1 both

Table 3.1. The strain invariants.

Index notation	Cartesian notation	Number of terms
e_{ii}	$e_{xx} + e_{yy} + e_{zz}$	3
$e_{ij}e_{ji}$	$e_{xx}^2 + e_{yy}^2 + e_{zz}^2$ $+ \frac{1}{4}e_{xy}^2 + \frac{1}{4}e_{yz}^2 + \frac{1}{4}e_{zx}^2$ $+ \frac{1}{4}e_{yx}^2 + \frac{1}{4}e_{zy}^2 + \frac{1}{4}e_{xz}^2$	9
$e_{ij}e_{jk}e_{ki}$	$e_{xx}^3 + \frac{1}{4}e_{xx}e_{xy}^2 + \frac{1}{4}e_{xx}e_{xz}^2$ $+ \frac{1}{4}e_{xy}^2 e_{xx} + \frac{1}{4}e_{xy}^2 e_{yy} + \frac{1}{8}e_{xy}e_{yz}e_{zx}$ etc.	27

The strain invariants are sometimes given as

$$e_{xx} + e_{yy} + e_{zz}, \qquad e_{yy}e_{zz} + e_{zz}e_{xx} + e_{xx}e_{yy} - \frac{1}{4}(e_{yz}^2 + e_{zx}^2 + e_{xy}^2),$$

and

$$e_{xx}e_{yy}e_{zz} + \frac{1}{4}(e_{yz}e_{zx}e_{xy} - e_{xx}e_{yz}^2 - e_{yy}e_{zx}^2 - e_{zz}e_{xy}^2),$$

which follows from adding or subtracting simple functions of the three invariants of Table 3.1.

in the Cartesian coordinate notation and in the index notation, the latter being more convenient in some of the further development of the theory.

For an isotropic elastic solid the strain-energy function U must be a homogeneous quadratic function of the strains and a function of the strain invariants. It follows that U is a function of the first two strain invariants of Table 3.1 only, i.e.

$$2U = A'(e_{xx}+e_{yy}+e_{zz})^2 + B'[(e_{xx}^2+e_{yy}^2+e_{zz}^2)+\tfrac{1}{2}(e_{xy}^2+e_{yz}^2+e_{zx}^2)],$$

where A' and B' are constants possessing the dimensions of modulus, i.e. newtons per square metre.

It can be shown that this equation may be written as

$$2U = A(e_{xx}+e_{yy}+e_{zz})^2 + B(e_{yz}^2+e_{zx}^2+e_{xy}^2-4e_{yy}e_{zz}-4e_{zz}e_{xx}-4e_{xx}e_{yy}),$$

where A and B are new constants and we have constructed a new strain invariant

$$(e_{yz}^2+e_{zx}^2+e_{xy}^2-4e_{yy}e_{zz}-4e_{zz}e_{xx}-4e_{xx}e_{yy}) = 2[(e_{xx}^2+e_{yy}^2+e_{zz}^2)+\tfrac{1}{2}(e_{xy}^2+e_{yz}^2+e_{zx}^2)]$$
$$-2(e_{xx}+e_{yy}+e_{zz})^2.$$

This is the form of the strain-energy function given by Love (Reference 3, p. 102).

So far we have only considered small strain elasticity, but it can readily be appreciated that a formal treatment of finite strain elasticity can be made along similar lines. For an isotropic solid, the strain-energy function U is again a function of the strain invariants, and the algebra will be formally identical, since this part of the argument is based on the invariants of a second-rank tensor.

Thus for finite strains we have that the quantities e_{ii}, $e_{ij}e_{ji}$ and $e_{ij}e_{jk}e_{ki}$ are still invariant.

For reasons which will shortly be made apparent, it is convenient to define three alternative strain invariants in terms of these three basic invariants.

These are

$$I_1 = 3 + 2e_{rr},$$

$$I_2 = 3 + 4e_{rr} + 2(e_{rr}e_{ss} - e_{rs}e_{sr})$$

and

$$I_3 = |\delta_{rs} + 2e_{rs}|.$$

Our argument is then summarized by the equation

$$U = f(I_1, I_2, I_3).$$

We will now evaluate the invariants I_1, I_2, I_3 for the case of a homogeneous pure strain in which the extension ratios parallel to the three coordinate axes are $\lambda_1, \lambda_2, \lambda_3$ respectively.

From the initial discussion (equations (3.7))

$$\lambda_1^2 = 1+2e_{xx}, \qquad \lambda_2^2 = 1+2e_{yy}, \qquad \lambda_3^2 = 1+2e_{zz},$$

then

$$I_1 = 3 + 2e_{rr} = 3 + 2(e_{xx} + e_{yy} + e_{zz})$$
$$= 3 + (\lambda_1^2 - 1) + (\lambda_2^2 - 1) + (\lambda_3^2 - 1),$$

giving

$$I_1 = \lambda_1^2 + \lambda_2^2 + \lambda_3^2. \tag{3.9}$$

Secondly,

$$I_2 = 3 + 4e_{rr} + 2(e_{rr}e_{ss} - e_{rs}e_{sr}).$$

For homogeneous pure strain $e_{rs} = r_{sr} = 0$. (The shear-strain components are zero.) Thus

$$I_2 = 3 + 4e_{rr} + 2e_{rr}e_{ss}$$
$$= 3 + 4(e_{xx} + e_{yy} + e_{zz}) + 4(e_{xx}e_{yy} + e_{xx}e_{zz} + e_{yy}e_{zz})$$
$$= 3 + 4[\tfrac{1}{2}(\lambda_1^2 - 1) + \tfrac{1}{2}(\lambda_2^2 - 1) + \tfrac{1}{2}(\lambda_3^2 - 1)]$$
$$+ 4[\tfrac{1}{4}(\lambda_1^2 - 1)(\lambda_2^2 - 1) + \tfrac{1}{4}(\lambda_1^2 - 1)(\lambda_3^2 - 1) + \tfrac{1}{4}(\lambda_2^2 - 1)(\lambda_3^2 - 1)],$$

giving

$$I_2 = \lambda_1^2\lambda_2^2 + \lambda_1^2\lambda_3^2 + \lambda_2^2\lambda_3^2 \tag{3.10}$$

and

$$I_3 = |\delta_{rs} + 2e_{rs}| = \begin{vmatrix} 1+2e_{xx} & & \\ & 1+2e_{yy} & \\ & & 1+2e_{zz} \end{vmatrix}$$

$$= \begin{vmatrix} \lambda_1^2 & & \\ & \lambda_2^2 & \\ & & \lambda_3^2 \end{vmatrix} = \lambda_1^2\lambda_2^2\lambda_3^2 \tag{3.11}$$

If we make the further simplifying assumption that there is no change in volume on deformation

$$I_3 = \lambda_1^2\lambda_2^2\lambda_3^2 = 1.$$

Thus the strain-energy function U for an isotropic incompressible solid undergoing a pure homogeneous deformation is given by

$$U = \text{Function}\left(I_1 = \lambda_1^2 + \lambda_2^2 + \lambda_3^2, \, I_2 = \frac{1}{\lambda_1^2} + \frac{1}{\lambda_2^2} + \frac{1}{\lambda_3^2}\right)$$

with

$$\lambda_1^2\lambda_2^2\lambda_3^2 = 1.$$

Let us represent U as a polynomial function of the strain invariants. If U is to vanish at zero strain this implies that

$$U = \sum_{i=0, j=0}^{\infty} C_{ij} (I_1 - 3)^i (I_2 - 3)^j \tag{3.12}$$

with $C_{00} = 0$, i.e. the strain-energy function involves powers of $(I_1 - 3)$ and $(I_2 - 3)$.

A neo-Hookean material, that corresponding to the statistical theory for a Gaussian network (see p. 70), corresponds to the first term in this series, i.e.

$$U = C_1(I_1 - 3) = C_1(\lambda_1^2 + \lambda_2^2 + \lambda_3^2 - 3) \tag{3.13}$$

which corresponds exactly to the network theory if we put

$$C_1 = \tfrac{1}{2} NKT.$$

Another form of particular interest is

$$U = C_1(I_1 - 3) + C_2(I_2 - 3). \tag{3.14}$$

This is the most general first-order relationship for U in terms of I_1 and I_2. This form of the strain-energy function was first derived by Mooney[6] on the assumption of a linear stress–strain relationship in shear. (Shear deformation will be considered below.) In general the Mooney equation gives a closer approximation to the actual behaviour of rubbers than the simpler equation involving only I_1.

3.4.4 The Stress–Strain Relations

Assuming that a satisfactory strain-energy function has been defined, it is now necessary to consider how the stress–strain relations are obtained.

For the small-strain situation the stress components are the first derivatives of the appropriate strain-energy function with respect to the corresponding strain components, i.e.

$$\sigma_{xx} = \frac{\partial U}{\partial e_{xx}}, \qquad \sigma_{yy} = \frac{\partial U}{\partial e_{yy}}, \qquad \text{etc.,}$$

and using the strain-energy function for an elastic solid with any anisotropy

$$2U = c_{11}e_{xx}^2 + 2c_{12}e_{xx}e_{yy} + 2c_{13}e_{xx}e_{zz} + \ldots$$

the generalized Hooke's law is obtained. To solve a specific problem in small strain elasticity we find stress components which satisfy the equilibrium conditions, and the stress components and strain components which satisfy the boundary conditions and the compatibility conditions. This approach can be called the *direct* method of solution.

Solutions to problems in finite elasticity, on the other hand, are most easily obtained by *inverse* methods. The strain components are specified, and the stress components obtained using the strain-energy function. If we assume that the material is incompressible the stress components are then indeterminate with respect to an arbitrary hydrostatic pressure.

If the problem is treated in its most general form a considerable amount of algebra is involved in obtaining the stress components σ_{ij} in terms of derivatives of the strain-energy function with respect to the strain components e_{ij} defined with respect to the *undeformed* body. The reader is referred to standard texts for the development of these procedures[1,2].

For most purposes, however, it is only necessary to derive the result for homogeneous pure strain, with extension ratios λ_1, λ_2, λ_3, and this is obtained in the following elementary manner.

A cube of unit dimensions in the undeformed state deforms under the applied loads to the rectangular parallelepiped shown in Figure 3.2, which has edges λ_1, λ_2, λ_3, in the x, y, z directions respectively. In the deformed state the forces acting on the faces are f_1, f_2, f_3 with f = force per unit of undeformed cross-section, i.e. the forces are calculated in terms of the applied loads per unit cross-section in the undeformed state.

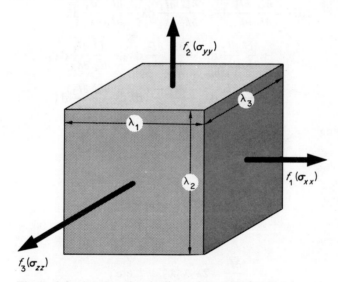

Figure 3.2. A cube of unit dimensions transforms to a rectangular parallelepiped of edges λ_1, λ_2 and λ_3 under the applied loads f_1, f_2 and f_3.

The corresponding stress components as defined above in the deformed state are σ_{xx}, σ_{yy}, σ_{zz}, where

$$\sigma_{xx} = \frac{f_1}{\lambda_2\lambda_3} = \lambda_1 f_1, \qquad \sigma_{yy} = \frac{f_2}{\lambda_1\lambda_3} = \lambda_2 f_2, \qquad \sigma_{zz} = \frac{f_3}{\lambda_1\lambda_2} = \lambda_3 f_3. \quad (3.15)$$

The work done (per unit of initial undeformed volume) in an infinitesimal displacement from the deformed state where λ_1, λ_2, λ_3 change to $\lambda_1 + d\lambda_1$,

$\lambda_2 + d\lambda_2$, $\lambda_3 + d\lambda_3$ is

$$dU = f_1\, d\lambda_1 + f_2\, d\lambda_2 + f_3\, d\lambda_3$$

$$= \frac{\sigma_{xx}}{\lambda_1}\, d\lambda_1 + \frac{\sigma_{yy}}{\lambda_2}\, d\lambda_2 + \frac{\sigma_{zz}}{\lambda_3}\, d\lambda_3, \tag{3.16}$$

$$U = f(I_1, I_2, I_3),$$

where

$$I_1 = \lambda_1^2 + \lambda_2^2 + \lambda_3^2,$$

$$I_2 = \frac{1}{\lambda_1^2} + \frac{1}{\lambda_2^2} + \frac{1}{\lambda_3^2},$$

$$I_3 = \lambda_1^2 \lambda_2^2 \lambda_3^2.$$

Therefore

$$dU = \frac{\partial U}{\partial I_1}\frac{\partial I_1}{\partial \lambda_1}\, d\lambda_1 + \frac{\partial U}{\partial I_1}\frac{\partial I_1}{\partial \lambda_2}\, d\lambda_2 + \frac{\partial U}{\partial I_1}\frac{\partial I_1}{\partial \lambda_3}\, d\lambda_3$$

$$+ \frac{\partial U}{\partial I_2}\frac{\partial I_2}{\partial \lambda_1}\, d\lambda_1 + \frac{\partial U}{\partial I_2}\frac{\partial I_2}{\partial \lambda_2}\, d\lambda_2 + \frac{\partial U}{\partial I_2}\frac{\partial I_2}{\partial \lambda_3}\, d\lambda_3$$

$$+ \frac{\partial U}{\partial I_3}\frac{\partial I_3}{\partial \lambda_1}\, d\lambda_1 + \frac{\partial U}{\partial I_3}\frac{\partial I_3}{\partial \lambda_2}\, d\lambda_2 + \frac{\partial U}{\partial I_3}\frac{\partial I_3}{\partial \lambda_3}\, d\lambda_3.$$

Substituting

$$\frac{\partial I_1}{\partial \lambda_1} = 2\lambda_1, \qquad \text{etc.,}$$

$$\frac{\partial I_2}{\partial \lambda_1} = -\frac{2}{\lambda_1^3}, \qquad \text{etc.,}$$

and

$$\frac{\partial I_3}{\partial \lambda_1} = 2\lambda_1 \lambda_2^2 \lambda_3^2, \qquad \text{etc.,}$$

we have

$$dU = 2\left\{ \lambda_1 \frac{\partial U}{\partial I_1}\, d\lambda_1 + \lambda_2 \frac{\partial U}{\partial I_1}\, d\lambda_2 + \lambda_3 \frac{\partial U}{\partial I_1}\, d\lambda_3 \right\}$$

$$- 2\left\{ \frac{1}{\lambda_1^3}\frac{\partial U}{\partial I_2}\, d\lambda_1 + \frac{1}{\lambda_2^3}\frac{\partial U}{\partial I_2}\, d\lambda_2 + \frac{1}{\lambda_3^3}\frac{\partial U}{\partial I_2}\, d\lambda_3 \right\}$$

$$+ 2I_3\left\{ \frac{1}{\lambda_1}\frac{\partial U}{\partial I_3}\, d\lambda_1 + \frac{1}{\lambda_2}\frac{\partial U}{\partial I_3}\, d\lambda_2 + \frac{1}{\lambda_3}\frac{\partial U}{\partial I_3}\, d\lambda_3 \right\}. \tag{3.17}$$

In equations (3.16) and (3.17), λ_1, λ_2 and λ_3 are independent variables.

We can therefore equate the coefficients of $d\lambda_1$, $d\lambda_2$, and $d\lambda_3$ in these equations to find the stress components.

This gives

$$\sigma_{xx} = 2\left\{\lambda_1^2 \frac{\partial U}{\partial I_1} - \frac{1}{\lambda_1^2} \frac{\partial U}{\partial I_2} + I_3 \frac{\partial U}{\partial I_3}\right\}, \qquad \text{etc.}$$

If the solid is incompressible $I_3 = 1$ and $U = f(I_1, I_2)$ only. In this case the stresses are now indeterminate with respect to an arbitrary hydrostatic pressure, $\mathbf{p}$, because this pressure does not produce any changes in the deformation variables λ_1, λ_2, λ_3. Then

$$\sigma_{xx} = 2\left\{\lambda_1^2 \frac{\partial U}{\partial I_1} - \frac{1}{\lambda_1^2} \frac{\partial U}{\partial I_2}\right\} + \mathbf{p}. \tag{3.18}$$

In index notation the stresses are given as

$$\sigma_{ii} = 2\left\{\lambda_i^2 \frac{\partial U}{\partial I_1} - \frac{1}{\lambda_i^2} \frac{\partial U}{\partial I_2}\right\} + \mathbf{p}, \qquad \sigma_{ij} = 0. \tag{3.19}$$

Because any homogeneous strain can be produced by a homogeneous pure strain followed by a suitable rotation, we do not lose generality by restricting the discussion to pure homogeneous strain.

3.5 EXPERIMENTAL STUDIES OF FINITE ELASTIC BEHAVIOUR IN RUBBERS

The experimental applications of finite elasticity are primarily confined to the behaviour of rubbers. One of the most definitive series of experiments was undertaken by Rivlin and Saunders[7] on sheets of vulcanized rubber. They examined homogeneous pure strain in a thin sheet with stresses maintained along the edges only (Figure 3.3), and assumed that the rubber was incompress-

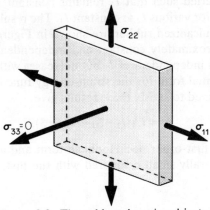

Figure 3.3. The rubber sheet is subjected to normal stresses σ_{11} and σ_{22} in the plane of the sheet.

ible. For this situation $\sigma_{33} = 0$, which enables the arbitrary hydrostatic pressure to be obtained in terms of known quantities. Thus

$$\sigma_{11} = 2\left\{\lambda_1^2 \frac{\partial U}{\partial I_1} - \frac{1}{\lambda_1^2} \frac{\partial U}{\partial I_2}\right\} + \mathbf{p},$$

$$\sigma_{22} = 2\left\{\lambda_2^2 \frac{\partial U}{\partial I_1} - \frac{1}{\lambda_2^2} \frac{\partial U}{\partial I_2}\right\} + \mathbf{p},$$

$$\sigma_{33} = 2\left\{\lambda_3^2 \frac{\partial U}{\partial I_1} - \frac{1}{\lambda_3^2} \frac{\partial U}{\partial I_2}\right\} + \mathbf{p} = 0.$$

Substituting for $\mathbf{p}$ we have

$$\sigma_{11} = 2\left\{\lambda_1^2 - \frac{1}{\lambda_1^2 \lambda_2^2}\right\}\left\{\frac{\partial U}{\partial I_1} + \lambda_2^2 \frac{\partial U}{\partial I_2}\right\},$$

$$\sigma_{22} = 2\left\{\lambda_2^2 - \frac{1}{\lambda_1^2 \lambda_2^2}\right\}\left\{\frac{\partial U}{\partial I_1} + \lambda_1^2 \frac{\partial U}{\partial I_2}\right\}. \tag{3.20}$$

Solving these two equations for $\partial U/\partial I_1$ and $\partial U/\partial I_2$,

$$\frac{\partial U}{\partial I_1} = \left\{\frac{\lambda_1^2 \sigma_{11}}{\lambda_1^2 - 1/\lambda_1^2 \lambda_2^2} - \frac{\lambda_2^2 \sigma_{22}}{\lambda_2^2 - 1/\lambda_1^2 \lambda_2^2}\right\} \bigg/ 2(\lambda_1^2 - \lambda_2^2)$$

and

$$\frac{\partial U}{\partial I_2} = \left\{\frac{\sigma_{11}}{\lambda_1^2 - 1/\lambda_1^2 \lambda_2^2} - \frac{\sigma_{22}}{\lambda_2^2 - 1/\lambda_1^2 \lambda_2^2}\right\} \bigg/ 2(\lambda_2^2 - \lambda_1^2). \tag{3.21}$$

If λ_1, λ_2, are varied such that I_2 remains constant, we obtain values of $\partial U/\partial I_1$ and $\partial U/\partial I_2$ for various values of I_1 and constant I_2. If on the other hand, λ_1, λ_2 are varied such that I_1 remains constant, we obtain values of $\partial U/\partial I_1$ and $\partial U/\partial I_2$ for various I_2 at constant I_1. The results obtained by Rivlin and Saunders for vulcanized rubber are shown in Figure 3.4. It can be seen that $\partial U/\partial I_1$ is approximately constant and independent of both I_1 and I_2. $\partial U/\partial I_2$, however, is independent of I_1 but decreases with increasing I_2. If we assume the polynomial form for the strain-energy function (p. 44) the terms which must be retained to satisfy these results give

$$U = C_1(I_1 - 3) + f(I_2 - 3),$$

i.e. the sum of the first-order neo-Hookean term and a function of I_2. The second term is generally small compared with the first, and decreases as I_2 increases.

In this case

$$\frac{\partial U}{\partial I_1} \sim 1.7 \times 10^5 \, \text{N m}^{-2},$$

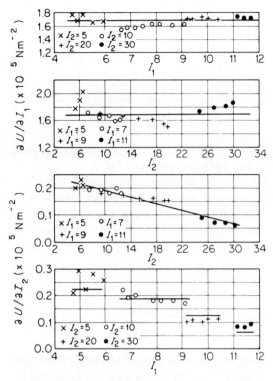

Figure 3.4. Dependence of $\partial U/\partial I_1$ and $\partial U/\partial I_2$ on I_1 and I_2. [Redrawn with permission from Rivlin and Saunders, *Phil. Trans. Roy. Soc. A*, **243**, 251 (1951).]

whereas

$$\frac{\partial U}{\partial I_2} \sim 1.5 \times 10^4 \, \text{N m}^{-2},$$

i.e. about one-tenth as large.

It is next important to consider whether this strain-energy function is consistent with results obtained for different types of deformation. Rivlin and Saunders also undertook measurements in pure shear and in simple elongation, and these will now be discussed in turn.

Pure Shear

Consider a sheet clamped between two edges AB, CD (Figure 3.5). If the width of the sheet (AB or CD) is sufficiently large compared with its length, the non-uniformity of strain arising because the outer edges AC, BD are unconstrained can be neglected. The strain parallel to the edges AB, CD can

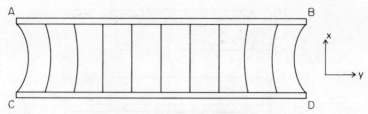

Figure 3.5. Pure shear deformation.

then be considered to remain constant in a deformation produced by moving AB and CD apart but keeping them parallel.

Shear is by definition a deformation in which the strain is zero in one direction and there is no volume change. If we make the simplifying assumption that the rubber is incompressible, the deformation which has been described (AB and CD displaced normal to their length) will be pure shear.

From the stress–strain relationship (equations (3.20)) we have

$$\sigma_{11} = 2\left\{\lambda_1^2 - \frac{1}{\lambda_1^2\lambda_2^2}\right\}\left\{\frac{\partial U}{\partial I_1} + \lambda_2^2\frac{\partial U}{\partial I_2}\right\}.$$

The pure strain experiments suggested that

$$U = C_1(I_1 - 3) + f(I_2 - 3),$$

i.e. $\partial U/\partial I_1 = C_1$. For the pure shear experiment $\lambda_2 = \text{constant} = 1$ and

$$\sigma_{11} = 2\left(\lambda_1^2 - \frac{1}{\lambda_1^2}\right)\left(C_1 + \frac{\partial U}{\partial I_2}\right).$$

Thus if λ_1^2 is varied (and hence I_2) by measuring σ_{11} we obtain values of $\partial U/\partial I_2$ as a function of I_2.

Rivlin and Saunders then did a second experiment in which pure shear was superimposed on an initial extension of the sheet along the AB/CD direction with $\lambda_2 = 0.776$.

They compared the calculated values of

$$\left(\frac{\partial U}{\partial I_1} + (0.776)^2\frac{\partial U}{\partial I_2}\right)$$

using the results from the first pure shear experiment, with these obtained from direct experiment (using equations (3.21)). Good agreement was obtained, confirming that the assumed form of the strain-energy function was a good mathematical model for the material.

Simple Elongation

A further key experiment is simple elongation. From equation (3.20) above when

$$\sigma_{22} = \sigma_{33} = 0 \quad \text{and} \quad \lambda_1 = \frac{1}{\lambda_2^2} = \lambda$$

(since $\lambda_2 = \lambda_3$ and $\lambda_1\lambda_2\lambda_3 = 1$), we have

$$\boldsymbol{\sigma}_{11} = 2\left(\lambda^2 - \frac{1}{\lambda}\right)\left(\frac{\partial U}{\partial I_1} + \frac{1}{\lambda}\frac{\partial U}{\partial I_2}\right). \tag{3.22}$$

The load f required to extend a specimen of initial cross-sectional area A is

$$f = \frac{A\boldsymbol{\sigma}_{11}}{\lambda} = 2A\left(\lambda - \frac{1}{\lambda^2}\right)\left(\frac{\partial U}{\partial I_1} + \frac{1}{\lambda}\frac{\partial U}{\partial I_2}\right).$$

The well known general shape of the load–extension curve is shown in Figure 3.6. A more revealing curve is obtained by plotting

$$\left(\frac{\partial U}{\partial I_1} + \frac{1}{\lambda}\frac{\partial U}{\partial I_2}\right) \quad \text{against} \quad \frac{1}{\lambda},$$

i.e.

$$\frac{f}{2A(\lambda - 1/\lambda^2)} \quad \text{against} \quad \frac{1}{\lambda}.$$

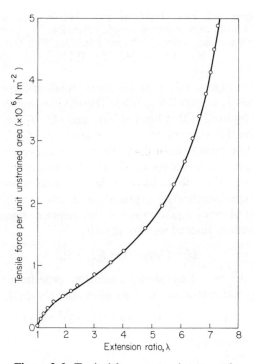

Figure 3.6. Typical force–extension curve for natural rubber. [Redrawn with permission from Treloar, *Introduction to Polymer Science*, Wykeham Publications (London) Ltd, London and Winchester 1970.

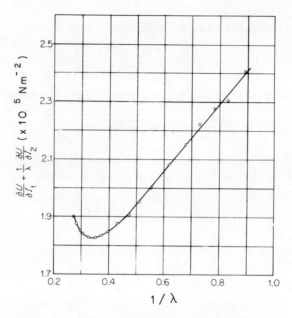

Figure 3.7. Plot of $[\partial U/\partial I_1 + (1/\lambda)\,\partial U/\partial I_2]$ against $1/\lambda$ from experiment in simple extension. [Redrawn with permission from Rivlin and Saunders, *Phil. Trans. Roy. Soc. A*, **243**, 251 (1951).]

This is shown in Figure 3.7. A nearly linear relationship is observed over the range of values $1/\lambda$ from 0.9 to 0.45. The obvious simple inference from this result taken by itself is that both $\partial U/\partial I_1$ and $\partial U/\partial I_2$ are constant in this range of extension, i.e. that the Mooney equation $U = C_1(I_1 + 3) + C_2(I_2 - 3)$ is adequate for $1/\lambda$ greater than 0.45. There is, however, a hidden snag in this simple interpretation. Values of C_1 and C_2 chosen in this way do not then agree with the values obtained from the two-dimensional extension and the pure shear experiments. The explanation of this apparent contradiction is as follows. The Mooney equation is not an adequate representation of the strain-energy function. Instead we should write

$$U = C_1(I_1 - 3) + f(I_2 - 3),$$

as indicated by the two-dimensional extension experiment. Expanding in powers of $(I_2 - 3)$ and curtailing the expansion after the cubic term,

$$U = C_1(I_1 - 3) + C_2(I_2 - 3) + C_3(I_2 - 3)^2 + C_4(I_2 - 3)^3.$$

For simple extension this gives

$$f = 2A\left(\lambda - \frac{1}{\lambda^2}\right)\left\{\left[(C_1 + 4C_3 - 36C_4) + \frac{1}{\lambda}(C_2 - 6C_3 + 27C_4)\right]\right.$$
$$\left. + \left[\left(12\lambda + \frac{12}{\lambda^2} - \frac{18}{\lambda^3} + \frac{3}{\lambda^5}\right)C_4 + \frac{2}{\lambda^3}C_3\right]\right\}.$$

Now C_1 and C_2 of the Mooney equation are replaced by $(C_1+4C_3-36C_4)$ and $(C_2-6C_3+27C_4)$ respectively. There are also additional groups of terms involving different powers of λ. If C_3 and C_4 are small, this latter group may well only give some scatter in the experimental data. But it could be that C_3 and C_4 are of different sign, e.g. C_3 could be negative as found from the two-dimensional extension experiment and C_4 positive. Close examination of the results of simple elongation and the two-dimensional extension confirms that this is the case, leading to a significant decrease in the apparent value of C_1 (if the Mooney equation is adopted) and a corresponding increase in the apparent value of C_2.

This example has been discussed in some detail because it emphasizes that any conclusions regarding the form of the strain-energy function in rubbers which are derived from a one-dimensional simple elongation experiment are necessarily suspect.

Recent work by Becker[8] has confirmed the validity of Rivlin and Saunders' work over the intermediate range of strains, but suggests that this formulation may not be adequate at low strains.

Simple Shear

The last situation to be discussed in rubber elasticity is that of simple shear. Consider an elemental cube, whose $x_3 = 0$ plane is shown in Figure 3.8, which is bounded initially by the planes $x_i = 0$, $x_i = a_i$ and subjected to simple shear

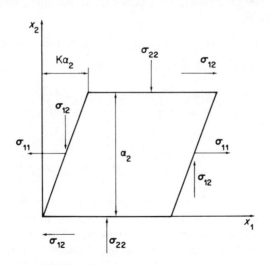

Figure 3.8. Simple shear deformation. [Reproduced with permission from Atkins, in *Progress in Solid Mechanics* (I. N. Sneddon and R. Hill, eds) North-Holland, Amsterdam, 1961, p. 28.]

in which planes parallel to the plane $x_2 = 0$ move parallel to the x_1 axis by amounts proportional to their x_2 coordinate.

For general finite deformation, points x_i suffer displacements u_i, the new coordinates being $x_i + u_i$. Put

$$x_i + u_i = x_i'$$

An element of length ds in the undeformed body changes to $\overline{\text{d}s}$ where

$$\text{d}s^2 = \text{d}x_i\,\text{d}x_i$$

and

$$(\overline{\text{d}s})^2 = \text{d}x_i'\,\text{d}x_i' = \frac{\partial x_i'}{\partial x_r}\frac{\partial x_i'}{\partial x_s}\,\text{d}x_r\,\text{d}x_s.$$

Summing over repeated suffices,

$$(\overline{\text{d}s})^2 - \text{d}s^2 = 2\mathbf{e}_{ij}\,\text{d}x_i\,\text{d}x_j,$$

where

$$\mathbf{e}_{ij} = \frac{1}{2}\left(\frac{\partial x_r'}{\partial x_i}\frac{\partial x_r'}{\partial x_j} - \delta_{ij}\right).$$

This definition of strain enables a direct evaluation of the strain components.

The coordinates x_i' for the deformation are given by

$$x_1' = x_1 + Kx_2, \qquad x_2' = x_2, \qquad x_3' = x_3,$$

where K is a constant defining the magnitude of the shear strain. This leads to strain invariants

$$I_1 = I_2 = 3 + K^2 \qquad (I_3 = 1, \text{ for incompressible solid}),$$

and it can be shown that the stresses are given by

$$\sigma_{11} = 2K^2\frac{\partial U}{\partial I_1}, \qquad \sigma_{22} = -2K^2\frac{\partial U}{\partial I_2}$$

and

$$\sigma_{12} = 2K\left(\frac{\partial U}{\partial I_1} + \frac{\partial U}{\partial I_2}\right), \qquad \sigma_{13} = \sigma_{23} = \sigma_{33} = 0.$$

To produce a finite shear, therefore, it is necessary to apply *normal components of stress* σ_{22} to the surfaces $x_2 = 0$, a_2 and σ_{11} to the surfaces initially at $x_1 = 0$, a_1 in addition to the shear stress σ_{12}. This is equivalent to the situation in viscous flow called the Weissenberg effect[9]: It is to be noted that these normal stresses are second order in K, which explains why they vanish for small strains.

The important point is that if the strain-energy function U is a linear function of the strain invariants I_1, I_2 the shear stress σ_{12} is proportional to the shear displacement K. This result was obtained by Mooney[6] and led to his proposal that the strain-energy function $U = C_1(I_1 - 3) + C_2(I_2 - 3)$. If we examine the problem of simple shear in detail, however, it turns out that for quite large amounts of shear the strain invariants I_1 and I_2 remain fairly small. Thus, if the higher order derivatives of U are fairly small compared with $\partial U/\partial I_1$ and $\partial U/\partial I_2$, which is actually the case for rubber, an approximately linear load–deformation relationship results.

The final conclusion of this discussion is that we must be very careful not to suppose that experiments on rubbers involving simple extension or simple shear only, will lead to a satisfactory understanding of the form of the strain-energy function.

3.6 RECENT DEVELOPMENTS IN THE MECHANICS OF RUBBER ELASTICITY

In this chapter we have developed two approaches to the definition of the constitutive equations for rubber elasticity, both due to Rivlin. The first approach was by analogy with the stress–strain relationships for an isotropic, incompressible material at low strains (Section 3.3). The second approach also uses the analogy with small strain elasticity and develops the strain-energy function in terms of the strain invariants (Section 3.4). This led to equation (3.12).

Rivlin adopted a formulation for the strain-energy function U which involved the squares of the extension ratios λ_i because he envisaged that negative values for these quantities were a mathematical possibility, and U must always be positive. However, as shown in Section 3.1 above, if we choose a suitable rotation of coordinates, the most general deformation can be reduced to homogeneous pure strain. The deformation can then be defined in terms of the three principal extension ratios λ_1, λ_2, λ_3 which are all essentially positive. In physical terms these extension ratios define the dimensions of the principal axes of the strain ellipsoid.

Following these ideas, we can then consider more general forms of the strain-energy function by removing the constraint that the form of U should be restricted to even-powered functions of the extension ratios. Recent developments along these lines have been made by Valanis and Landel[10] and Ogden[11-13], and by Treloar and his associates[14,15]. Treloar has also given comprehensive reviews of this work[16,17], and of further experimental studies by Obata, Kawabata and Kawai[18].

Valanis and Landel[10] noted the practical difficulties of using the Rivlin equations (3.21) above because these equations are very sensitive to experimental error for values of the invariants of less than about 5. Thus they are not suitable for comparatively small finite deformations, as indicated by the studies of Becker[8] referred to above. Valanis and Landel therefore postulated a form for the strain-energy function U which is reasonable on intuitive physical grounds and is a function, not of the strain invariants, but of the extension ratios λ_i. We have seen that there is no reason to reject a formulation of this nature and, in fact, the only constraint on the form of U is that imposed by material isotropy, which implies that $U(\lambda_1, \lambda_2, \lambda_3)$ should be a symmetric function of the extension ratios, i.e. invariant to any permutation of the indices 1, 2, 3.

Valanis and Landel proposed as a hypothesis that

$$U(\lambda_1, \lambda_2, \lambda_3) = u(\lambda_1) + u(\lambda_2) + u(\lambda_3). \tag{3.23}$$

Following equations (3.16) above we obtain analogous equations to (3.19):

$$\boldsymbol{\sigma}_{ii} = \lambda_i u'(\lambda_i) + \mathbf{p}, \tag{3.24}$$

where $u'(\lambda_i) = \partial u / \partial \lambda_i$ and $\mathbf{p}$ is an arbitrary hydrostatic pressure. The stress–strain relations are therefore of the form

$$\boldsymbol{\sigma}_{xx} - \boldsymbol{\sigma}_{yy} = \lambda_1 u'(\lambda_1) - \lambda_2 u'(\lambda_2). \tag{3.25}$$

To obtain the form of $u'(\lambda_1)$ from experiment it is convenient to apply a biaxial strain such that one of the principal extension ratios (say λ_2) is maintained constant while the other (λ_1) is varied. We then have

$$\boldsymbol{\sigma}_{xx} - \boldsymbol{\sigma}_{yy} = \lambda_1 u'(\lambda_1) - c, \tag{3.26}$$

where $c = [\lambda u'(\lambda)]_{\lambda = \lambda_2}$ is a constant.

Valanis and Landel used the data of Becker[8] to obtain the form of $u'(\lambda)$ from pure shear measurements ($\boldsymbol{\sigma}_{zz} = 0$, $\lambda_2 = 1$). Noting that the constant c does not enter into equation (3.24), which is the practical equation used, they assumed $c = 0$. As pointed out by Treloar, this is incorrect, although the error is of no significance for the derivation of the form of $u'(\lambda)$.

For the range $1 < \lambda < 2.5$, Valanis and Landel proposed that

$$u'(\lambda) = 2\mu \ln \lambda. \tag{3.27}$$

Since for small deformations equation (3.27) must reduce to

$$\sigma_{xx} - \sigma_{yy} = 2\mu e_{xx} \tag{3.28}$$

the constant μ must be the shear modulus.

Valanis and Landel then compared results from other workers by plotting

$$\frac{1}{2\mu}(\boldsymbol{\sigma}_{xx} - \boldsymbol{\sigma}_{yy}) \quad \text{versus} \quad \frac{1}{2\mu}[\lambda_1 u'(\lambda_1) - \lambda_2 u'(\lambda_2)],$$

i.e. adopting a more general fit to $u'(\lambda)$ than that of equation (3.26) but allowing for a scale adjustment according to the modulus μ in any particular case.

The results of the comparison are shown in Figure 3.9, and give excellent confirmation for the proposals of Valanis and Landel. Further support for the form of the strain-energy function (3.21) comes from later experimental studies by Obata, Kawabata and Kawai[18]. Biaxial tests, of a similar nature

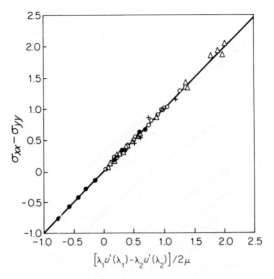

Figure 3.9. Experimental tests of equations (3.23) and (3.27) for natural rubber, using three sets of experimental data from + Becker (uniaxial) △ (biaxial), $\mu = 4.1 \times 10^5 \, \mathrm{N\,m^{-2}}$; ○ Rivlin, $\mu = 3.9 \times 10^5 \, \mathrm{N\,m^{-2}}$; ● Treloar (uniaxial compression) $\mu = 4 \times 10^5 \, \mathrm{N\,m^{-2}}$. [Redrawn with permission from Valanis and Landel, *J. Appl. Phys.*, **38**, 2997 (1967).]

to the classic experiments of Rivlin and Saunders described in Section 3.5 above, were carried out on several natural rubber vulcanizates. Plots were obtained for $\sigma_{xx} - \sigma_{yy}$ versus λ_2 at fixed values of λ_1. It was found that these curves were of identical form and were superposable by a suitable vertical displacement only, as predicted by equation (3.26). The results are shown in Figure 3.10(a) and (b).

In a slightly later development, Ogden[13] reverted to a more formal treatment, based on the idea that it is appropriate to seek an expression for U which is dependent only on the extension ratios. Following some earlier unpublished work by R. Hill, Ogden assumed a strain-energy function of the form

$$U = \sum_n \frac{\mu_n}{\alpha_n} (\lambda_1^{\alpha_n} + \lambda_2^{\alpha_n} + \lambda_3^{\alpha_n} - 3), \qquad (3.29)$$

which leads to the associated equation for the principal stresses

$$\sigma_{ii} = \sum_n \mu_n \lambda_i^{\alpha_n} - \mathbf{p}. \qquad (3.30)$$

Ogden showed that a satisfactory fit could be obtained to the experimental data of Treloar[19] for tension, pure shear and equibiaxial tension, using a

58

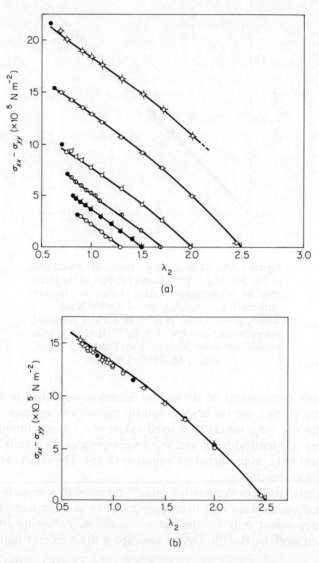

(a)

(b)

Figure 3.10. (a) Plots of $\sigma_{xx} - \sigma_{yy}$ at fixed values of λ_1 as a function of λ_2 for a natural rubber vulcanizate at 31 °C. λ_1 values are Φ, 3.0; $\ominus$, 2.5; σ, 2.0; Φ, 1.7; $\bullet$, 1.5; $\bigcirc$, 1.3. $\bullet$, uniaxial data. (b) Superposition of data of Figure 3.10(a) on to curve for pure shear ($\lambda_1 = 1$). [Redrawn with permission from Obata, Kawabata and Kawai, *J. Polymer. Sci. A2,* **8,** 903 (1970).]

three-term expression with

$$\alpha_1 = 1.3, \qquad \alpha_2 = 5.0, \qquad \alpha_3 = -2.0,$$

$$\mu_1 = 6.2 \times 10^5 \, \text{N m}^{-2}, \qquad \mu_2 = 1.2 \times 10^3 \, \text{N m}^{-2}, \qquad \mu_3 = -1 \times 10^4 \, \text{N m}^{-2}.$$

This is shown in Figure 3.11.

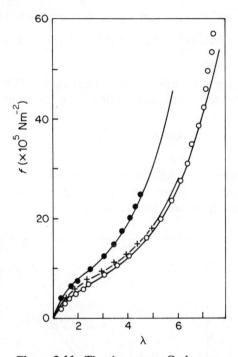

Figure 3.11. The three-term Ogden representation (equation (3.29)) compared with the Treloar data for simple tension, ○; pure shear, +; and equibiaxial tension, ●. [Redrawn with permission from Treloar, *Proc. Roy. Soc. A*, **351**, 301 (1976) and Ogden, *Proc. Roy. Soc. A*, **326**, 565 (1972).]

Treloar has also discussed the relationship between the Ogden formulation and the Valanis–Landel representation. The Ogden form of the strain-energy function in fact satisfied the Valanis–Landel representation, giving

$$u(\lambda) = \sum_n \frac{\mu_n}{\alpha_n} (\lambda^{\alpha_n} - 1) \tag{3.31}$$

and hence

$$\lambda u'(\lambda) = \sum_n \mu_n \lambda^{\alpha_n}.$$

60

It is convenient to write

$$\lambda u'(\lambda) - c = \sum_n \mu_n(\lambda^{\alpha_n} - 1), \tag{3.32}$$

where

$$c = [u'(\lambda)]_{\lambda=1} = \sum_n \mu_n.$$

Jones and Treloar[14] showed that a good fit was obtained by

$$\lambda u'(\lambda) - c = 6.9 \times 10^5(\lambda^{1.3} - 1) + 1 \times 10^4(\lambda^{4.0} - 1) - 1.22 \times 10^4(\lambda^{-2.0} - 1)$$
$$\tag{3.33}$$

in which the units are again newtons per square metre. This function is shown in Figure 3.12, together with that predicted by the simple (neo-Hookean) form for which

$$\lambda u'(\lambda) - c = G(\lambda^2 - 1), \tag{3.34}$$

where G is the shear modulus.

There are very significant differences between the two functions (3.33) and (3.34) at low strains, and these were confirmed by a detailed evaluation of the results of Obata, Kawabata and Kawai.

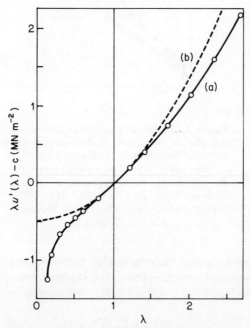

Figure 3.12. The function $\lambda u'(\lambda) - c$. (a) Continuous curve, equation (3.33), based on points experimental; (b) simple statistical theory. [Redrawn with permission from Treloar, *Proc. Roy. Soc. A*, **351**, 301 (1976).]

An important advantage of the Ogden form of representation as applied to the experimental data referred to above is that the stress–strain relations for *any* type of strain may be represented by a single algebraic function. One consequence of this is that the apparent discontinuity in passing from extension ($\lambda > 1$) to compression ($\lambda < 1$) previously obtained on the basis of the Mooney equation (3.14) is automatically eliminated. This has been fully discussed by Treloar[17].

In conclusion, it may be said that these ideas are valuable, because they suggest that if we are to seek theories which are more sophisticated than the simple statistical molecular theory which leads to the neo-Hookean equation it may be more appropriate to formulate such theories in terms of the extension ratios directly, rather than seek a physical explanation of the Mooney C_2 term or additional terms in a strain-energy function based on the strain invariants. As pointed out by Hopkins[20], the use of the extension ratios seems wholly natural in any physical theory, and this is especially true when we are translating the extension ratios to changes in the molecular conformation, i.e. directly to events at a microscopic level. We will see in the next chapter that this is an essential ingredient of the molecular theory of rubber elasticity.

REFERENCES

1. A. E. Green and W. Zerna, *Theoretical Elasticity*, Clarendon Press, Oxford, 1954.
2. A. E. Green and J. E. Adkins, *Large Elastic Deformations and Non-Linear Continuum Mechanics*, Clarendon Press, Oxford, 1960.
3. A. E. H. Love, *A Treatise on the Mathematical Theory of Elasticity*, Cambridge, Cambridge University Press, 1927.
4. H. Goldstein, *Classical Mechanics*, Harvard University Press, Cambridge, Mass., 1959, p. 119.
5. R. S. Rivlin, *Phil. Trans. Roy. Soc. A*, **240**, 459, 491 (1948) and **241**, 379 (1949).
6. M. Mooney, *J. Appl. Phys.*, **11**, 582 (1940).
7. R. S. Rivlin and D. W. Saunders, *Phil. Trans. Roy. Soc. A*, **243**, 251 (1951); *Trans. Faraday Soc.*, **48**, 200 (1952).
8. G. W. Becker, *J. Polymer Sci. C*, **16**, 2893 (1967).
9. K. Weissenberg, *Nature*, **159**, 310 (1947).
10. K. C. Valanis and R. F. Landel, *J. Appl. Phys.*, **38**, 2997 (1967).
11. R. W. Ogden, *Proc. Roy. Soc. A*, **326**, 565 (1972).
12. R. W. Ogden and P. Chadwick, *J. Mech. Phys. Solids*, **20**, 77 (1972).
13. R. W. Ogden, P. Chadwick and E. W. Haddon, *Quart. J. Mech. Appl. Math.*, **26**, 23 (1973).
14. D. F. Jones and L. R. G. Treloar, *J. Phys. D*, **8**, 1285 (1975).
15. H. Vangerko and L. R. G. Treloar, *J. Phys. D*, **11**, 1969 (1978).
16. L. R. G. Treloar, *The Physics of Rubber Elasticity*, 3rd edn, Clarendon Press, Oxford, 1975, Chapter 11.
17. L. R. G. Treloar, *Proc. Roy. Soc. A*, **351**, 301 (1976).
18. Y. Obata, S. Kawabata and H. Kawai, *J. Polymer Sci. A2*, **8**, 903 (1970).
19. L. R. G. Treloar, *Trans. Faraday Soc.*, **40**, 59 (1944).
20. H. G. Hopkins, *Proc. Roy. Soc. A*, **351**, 322 (1976).

4

The Statistical Molecular Theories
of the Rubber-like State

4.1. THERMODYNAMIC CONSIDERATIONS

The rubber-like state is the only part of polymer mechanical behaviour which can be understood in terms of a well established molecular theory. In formal mathematical terms the aim is to predict the strain-energy function U in terms of the components of finite strain (in practice the strain invariants or the extension ratios), and the appropriate molecular parameters. The theory is based on statistical mechanical considerations and processes are considered to be reversible in the thermodynamic sense. We will therefore develop our arguments within the framework of Section 3.4.1. above.

For a reversible isothermal change of state at constant volume, the work done can be equated to the change in the Helmholtz free energy A.

For simplicity we will consider the uniaxial extension under a tensile force f of an elastic solid of initial length l.

The work done on the solid in an infinitesimal displacement $\mathrm{d}l$ is

$$\mathrm{d}W = f\,\mathrm{d}l = \mathrm{d}A = \mathrm{d}U - T\,\mathrm{d}S. \tag{4.1}$$

At constant volume we have

$$f = \left(\frac{\partial A}{\partial l}\right)_T = \left(\frac{\partial U}{\partial l}\right)_T - T\left(\frac{\partial S}{\partial l}\right)_T \tag{4.2}$$

A further manipulation of the thermodynamic quantities is required to show that

$$\left(\frac{\partial U}{\partial l}\right)_T = f - T\left(\frac{\partial f}{\partial T}\right)_l \tag{4.3}$$

At constant volume

$$\mathrm{d}U = f\,\mathrm{d}l + T\,\mathrm{d}S. \tag{4.4}$$

Hence

$$\mathrm{d}A = f\,\mathrm{d}l - S\,\mathrm{d}T. \tag{4.5}$$

Then

$$\left(\frac{\partial A}{\partial l}\right)_T = f \quad \text{and} \quad \left(\frac{\partial A}{\partial T}\right)_l = -S. \tag{4.6}$$

But

$$\frac{\partial}{\partial l}\left(\frac{\partial A}{\partial T}\right)_l = \frac{\partial}{\partial T}\left(\frac{\partial A}{\partial l}\right)_T$$

Substituting we have

$$\left(\frac{\partial S}{\partial l}\right)_T = -\left(\frac{\partial f}{\partial T}\right)_l \tag{4.7}$$

Using (4.2) we then find equation (4.3).

The classic experiments of Meyer and Ferri[1] on the stress–temperature behaviour of rubber showed that the tensile force at constant length was very nearly proportional to the absolute temperature. The right-hand side of equation (4.3) is therefore close to zero, showing that the internal energy term is very small, and that the elasticity arises almost entirely from changes in entropy.

4.1.1. The Thermoelastic Inversion Effect

A typical set of results for the tensile force at constant length as a function of temperature are shown in Figure 4.1. The curves are linear at all elongations

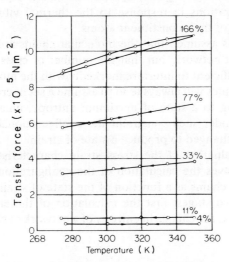

Figure 4.1. Force at constant length as a function of absolute temperature. Elongations as indicated (Meyer and Ferri[1]). [Redrawn with permission from Treloar, *The Physics of Rubber Elasticity*, 2nd edn, Oxford University Press, Oxford, 1958.]

but it is to be noted that whereas above about 10% elongation the tensile force increases with increasing temperature, below this elongation it decreases slightly. This is called the thermoelastic inversion effect. In physical terms it is caused by the thermal expansion of the rubber with increasing temperature. This increases the length in the unstrained state and hence reduces the effective elongation. It was therefore considered by Gee[2] and others that the more appropriate experiment is to measure the tensile force as a function of temperature at constant extension ratio where the expansion is corrected for. The experimental results obtained in this way by Gee[2] showed that the stress was directly proportional to absolute temperature and hence suggested that there was no internal energy change (equation (4.3)).

The conclusion is based on certain approximations relating to the difference between constant volume conditions and constant pressure conditions which have subsequently been reconsidered. (For a discussion of this see Section 4.2 below.)

4.1.2 The Statistical Theory

The kinetic or statistical theory of rubber elasticity was originally proposed by Meyer, Susich and Valko[3] and subsequently developed by Guth and Mark[4], Kuhn[5] and others[6]. It is assumed that the individual molecules of the rubber exist in the form of very long chains, each of which is capable of assuming a variety of configurations in response to the thermal vibrations of 'micro-Brownian' motion of their constituent atoms.

Furthermore it is assumed that the molecular chains are interlinked so as to form a coherent network, but that the number of cross-links is relatively small and is not sufficient to interfere markedly with the motion of the chains.

The chain molecules will always tend to assume a set of crumpled configurations corresponding to a state of maximum entropy, unless constrained by external forces. Under such constraint the configurational arrangements of the chains will be changed to produce a state of strain.

Quantitative evaluation of the stress–strain characteristics of the rubber network then involves the calculation of the configurational entropy of the whole assembly of chains as a function of the state of strain. This calculation is considered in two stages, first the calculation of the entropy of a single chain, and, secondly, the change in entropy of a network of chains as a function of strain.

The Entropy of a Single Chain

The simplest consideration of the structure of a single chain can be made in terms of a polyethylene molecule $(CH_2)_n$. Fully extended, this takes the form of the planar zig-zag shown in Figure 4.2. If we allow free rotation from one

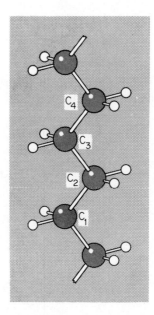

Figure 4.2. The polyethylene chain.

conformation† to another the local situation $C_1C_2C_3C_4$ can change from the planar zig-zag conformation to a variety of conformations, with the restriction that in each case the valence bond angle between carbon atoms must remain at $109\frac{1}{2}°$. It is in principle possible to calculate the number of possible configurations which correspond to a chosen end-to-end distance and hence the entropy of such molecular chain. It is, however, easier to consider instead a mathematical abstraction—the 'freely jointed' chain. The freely jointed chain, as its name implies, consists of a chain of equal links jointed without the restriction that the valence angles should remain constant, i.e. random jointing is assumed.

This problem is more tractable mathematically, the first analysis of this type being undertaken by Kuhn[5] and by Guth and Mark[4].

Consider a chain of n links each of length l, which has a configuration such that one end P is at the origin (Figure 4.3). The probability distribution for the position of the end Q is derived using approximations which are valid providing that the distance between the chain ends P and Q is much less than the extended chain length nl. The probability that Q lies within the elemental volume $dx \, dy \, dz$ at the point (x, y, z) can be shown to be

$$p(x, y, z) \, dx \, dy \, dz = \frac{b^3}{\pi^{3/2}} \exp(-b^2 r^2) \, dx \, dy \, dz \qquad (4.8)$$

---

† Conformation is used to denote differences in the immediate situation of a bond, e.g. *trans* and *gauche* conformations. Configuration is retained to refer to the arrangement of the whole molecular chain.

where

$$b^2 = \frac{3}{2nl^2}$$

and $r^2 = x^2 + y^2 + z^2$, i.e. the distribution of end-to-end vectors is defined by the Gaussian error function.

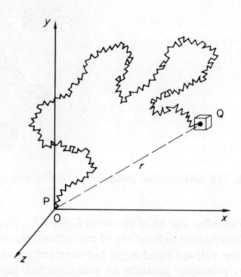

Figure 4.3. The freely jointed chain.

This shows immediately that the distribution is spherically symmetrical. It also shows that the most probable position of the end Q is at the origin. This does not mean that the most probable end-to-end distance is zero for the following reason. The probability that the chain end Q is in any given elemental volume between r and dr, irrespective of its direction, is the product of the probability distribution $p(r)$ and $4\pi r^2$ dr, the volume of a concentric shell of radius r and thickness dr.

This probability is

$$P(r)\,dr = p(r)4\pi r^2\,dr = \frac{b^3}{\pi^{3/2}}\,e^{-b^2r^2}4\pi r^2\,dr$$

$$= \left(\frac{4b^3}{\pi^{1/2}}\right)r^2\,e^{-b^2r^2}\,dr \tag{4.9}$$

and is shown in Figure 4.4.

It is seen that the most probable end-to-end distance, irrespective of direction, is not zero, but it is a function of b, i.e. of the length l of the links and the number n of links in the chain.

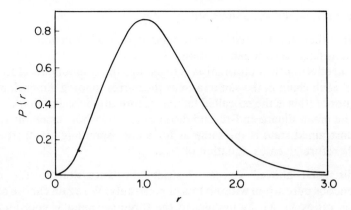

Figure 4.4. The distribution function $P(r) = \text{const } r^2 e^{-b^2 r^2}$.

Another important quantity is the root mean square chain length $(\overline{r^2})^{1/2}$

$$\overline{r^2} = \int_0^\infty r^2 P(r) \, dr.$$

Substitution of the above expression for $P(r)$ gives

$$\overline{r^2} = \tfrac{3}{2}b^2 = nl^2 \tag{4.10}$$

or that the root mean square length $(\overline{r^2})^{1/2} = l\sqrt{n}$, i.e. it is proportional to the square root of the number of links in the chain.

The entropy of the freely jointed chain s is proportional to the logarithm of the number of configurations Ω so that

$$s = k \ln \Omega,$$

where k is Boltzmann's constant. If $dx\,dy\,dz$ is constant, the number of configurations available to the chain is proportional to the probability per unit volume $p(x, y, z)$. The entropy of the chain is thus given by

$$s = c - kb^2 r^2 = c - kb^2(x^2 + y^2 + z^2), \tag{4.11}$$

where c is an arbitrary constant.

Elasticity of a Molecular Network

We now wish to calculate the strain-energy function for a molecular network, assuming that this is given by the change in entropy of a network of chains as a function of strain.

The actual network is replaced by an ideal network in which each segment of a molecule between successive points of cross-linkage is considered to be a Gaussian chain.

68

Three additional assumptions are introduced:

(1) In either the strained or unstrained state, each junction point may be regarded as fixed at its mean position.

(2) The effect of the deformation is to change the components of the vector length of each chain in the same ratio as the corresponding dimensions of the bulk material (this is the so-called 'affine' deformation assumption).

(3) The mean square end-to-end distance for the whole assembly of chains in the unstrained state is the same as for a corresponding set of free chains and is therefore given by equation (4.10).

As discussed above, we can restrict our discussion to the case of homogeneous pure strain without loss of generality. We again choose principal extension ratios λ_1, λ_2, λ_3 parallel to the three rectangular coordinate axes x, y, z. The affine deformation assumption implies that the relative displacement of the chain ends is defined by the macroscopic deformation. Thus, in Figure 4.5, we take a system of coordinates x, y, z in the undeformed body.

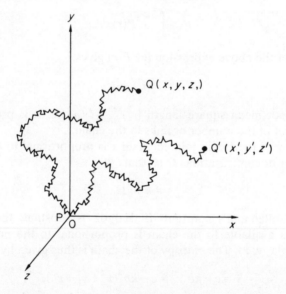

Figure 4.5. The end of the chain Q (x, y, z) is displaced to Q (x', y', z').

In this coordinate system a representative chain PQ has one end P at the origin. We refer any point in the deformed body to this system of coordinates. Thus the origin, i.e. the end of the chain P, is convected during the deformation. The other end Q (x, y, z) is displaced to the point Q' (x', y', z') and from the affine deformation assumption we have

$$x' = \lambda_1 x, \qquad y' = \lambda_2 y, \qquad z' = \lambda_3 z.$$

The entropy of the chain in the undeformed state is given by equation (4.11). On deformation the entropy changes to

$$s' = c - kb^2(\lambda_1^2 x^2 + \lambda_2^2 y + \lambda_3^2 z^2).$$ (4.12)

The entropy change on deformation is therefore

$$\Delta s = s' - s = -kb^2\{(\lambda_1^2 - 1)x^2 + (\lambda_2^2 - 1)y^2 + (\lambda_3^2 - 1)z^2\}.$$ (4.13)

The total entropy change for all chains in the network (say n per unit volume) with the same value of b (say b_p) is therefore given by

$$\Delta s_b = \sum_1^n \Delta s = -kb_p^2\left\{(\lambda_1^2 - 1)\sum_1^n x^2 + (\lambda_2^2 - 1)\sum_1^n y^2 + (\lambda_3^2 - 1)\sum_1^n z^2\right\},$$ (4.14)

where

$$\sum_1^n x^2$$

is the sum of the squares of the x components in the undeformed state of the network for these n chains. Since there will be no preferred direction for the chain vectors in the undeformed state, there will be no preference for the x, y or z directions (material isotropy).

We then have

$$\sum_1^n x^2 + \sum_1^n y^2 + \sum_1^n z^2 = \sum_1^n r^2$$

and

$$\sum_1^n x^2 = \sum_1^n y^2 = \sum_1^n z^2 = \frac{1}{3}\sum_1^n r^2.$$ (4.15)

From equation (4.10)

$$\sum_1^n r^2 = n\overline{r^2} = n(\tfrac{3}{2}b_p^2).$$ (4.16)

This gives

$$\Delta s_b = -\tfrac{1}{2}nk\{\lambda_1^2 + \lambda_2^2 + \lambda_3^2 - 3\}.$$ (4.17)

We can now add the contribution of all the chains in the network (N per unit volume), and obtain the entropy change of the network ΔS where

$$\Delta S = \sum_1^N \Delta s = -\tfrac{1}{2}Nk\{\lambda_1^2 + \lambda_2^2 + \lambda_3^2 - 3\}.$$ (4.18)

Assuming no change in internal energy on deformation this gives the change in the Helmholtz free energy.

$$\Delta A = -T\Delta S = \tfrac{1}{2}NkT(\lambda_1^2 + \lambda_2^2 + \lambda_3^2 - 3).$$

If we assume that the strain-energy function U is zero in the undeformed state this gives

$$U = \Delta A = \tfrac{1}{2}NkT(\lambda_1^2 + \lambda_2^2 + \lambda_3^2 - 3).$$

Thus we arrive at the neo-Hookean form for the strain-energy function U, in which U is a function of the strain invariant $I_1 = \lambda_1^2 + \lambda_2^2 + \lambda_3^2$ only, and

$$U = \tfrac{1}{2} NkT(I_1 - 3).$$

Consider simple elongation λ in the x direction. The incompressibility relationship gives $\lambda_1 \lambda_2 \lambda_3 = 1$. Hence, by symmetry, $\lambda_2 = \lambda_3 = \lambda^{-1/2}$ and

$$U = \tfrac{1}{2} NkT\left(\lambda^2 + \frac{2}{\lambda} - 3\right). \tag{4.22}$$

From equation (3.16) above

$$f = \frac{\partial U}{\partial \lambda} = NkT\left(\lambda - \frac{1}{\lambda^2}\right). \tag{4.23}$$

We have therefore obtained the neo-Hookean relationship of Section 3.3, with a constant NkT. For small strain we can put $\lambda = 1 + e_{xx}$ and it follows from equation (4.23) that

$$f = \sigma_{xx} = 3NkTe_{xx} = Ee_{xx},$$

where E is the Young's modulus. Since for an incompressible material $E \equiv 3G$, we see that the quantity NkT in equation (4.23) is equivalent to the shear modulus of the rubber, G. This term is sometimes written in terms of the mean molecular weight M_c of the chains, i.e. between successive points of cross-linkage. Then

$$G = NkT = \rho RT / M_c,$$

where ρ is the density of the rubber and R is the gas constant.

This simple theory of the deformation of a molecular network has formed the starting point for substantial research activity in polymer science. It is convenient to discuss further developments under three major headings. First, there are theories which consider that the strain-energy function is adequately described by a single term of the form $\tfrac{1}{2} G(I_1 - 3)$ and examine how the molecular theory can be modified to include such factors as non-Gaussian behaviour for the deformation of a single chain and non-affine deformation of the network junction points. Secondly, there is the necessity of re-examining the assumption that there is no contribution to the free energy from internal energy terms. Finally, there is the question of the significance of additional terms in the expansion of the strain-energy function. These aspects will now be discussed in turn.

4.2 MODIFICATIONS OF SIMPLE MOLECULAR THEORY

There are a number of refinements and modifications to the simple network theory which do not involve a change in the basic principles of the calculation.

(1) N has been referred to as the number of chains per unit volume. It is determined by the number of junction points in the network. 'Junction points'

can mean either chemical cross-links (as in vulcanized rubber) or physical entanglements (as in an amorphous polymer above its glass transition temperature). These considerations have led to theoretical attempts to analyse the cross-link situation in more detail[7] to take into account the fact that not all the cross-links are effective. There will be 'loose loops' where a chain folds back on itself, indicated by symbol (a) in Figure 4.6, and 'loose ends' where a chain does no contribute to the network following a cross-link point which is close to the end of a chain molecule, indicated by symbol (b) in Figure 4.6.

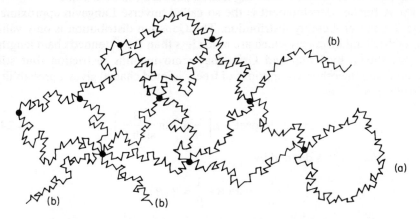

Figure 4.6. Types of network defect: (a) loose loop; (b) loose end.

A simple correction for loose ends by Flory[7] leads to the result that

$$G = \frac{\rho RT}{M_c}\left(1 - \frac{2M_c}{M}\right),$$

where M is the molecular weight of the primary molecules, before cross-linking produces the network.

(2) We have already hinted at the extension of this treatment to real chains with fixed bond angles with free or hindered rotation. This leads to the concept of the equivalent freely jointed chains[8]. It can be shown, for example, that for a paraffin-type chain with freely rotating bonds the root mean square end-to-end distance is $\sqrt{2}$ times that of a randomly jointed chain with bonds of the same length. More sophisticated treatments have been undertaken by making computer calculations based on the statistics of random walks[9]. Also, in recent years, attempts have been made to take into account the excluded volume, i.e. the fact that a chain cannot bend back to occupy its portion of space twice[10].

(3) One of the basic assumptions of the simple theory is that the junction points are assumed to remain fixed in the material body. It was soon shown by James and Guth[11] that this assumption is unnecessarily restrictive. They considered that it was adequate to assume that the cross-links move within

the range permitted by the geometry of chemical bonds, i.e. that they fluctuate around their most probable positions. Recent work[12-14] suggests that the elongation of a network can only be represented as affine in the range of small deformations. At higher deformations, it is postulated that fluctuations of the junction points reduce the $\frac{1}{2}NkT$ term by a factor A_ϕ so that $U = \frac{1}{2}A_\phi NkT$. It is envisaged that the limit of non-affine deformation would be exhibited by a 'phantom' network where the chains are portrayed as being able to transect one another with A_ϕ taking the value $A_\phi = 1 - 2/\phi$, where ϕ is the number of chains emanating from a network junction point[14].

(4) A further development is the so-called 'inverse Langevin approximation' for the probability distribution. The Gaussian distribution is only valid for end-to-end distances which are much less than the extended chain length. It was shown by Kuhn and Grün that removing this restriction (but still maintaining the other assumptions of freely jointed chains) gives a probability distribution $p(r)$ as

$$\ln p(r) = \text{const.} - n\left[\frac{r}{nl}\beta + \ln\frac{\beta}{\sinh\beta}\right]. \tag{4.24}$$

In this equation β is defined by

$$\frac{r}{nl} = \coth\beta - \frac{1}{\beta} = \mathscr{L}(\beta),$$

where $\mathscr{L}$ is the Langevin function and $\beta = \mathscr{L}^{-1}(r/nl)$ is the inverse Langevin function. This expression may be expanded to give

$$\ln p(r) = \text{const.} - n\left[\frac{3}{2}\left(\frac{r}{nl}\right)^2 + \frac{9}{20}\left(\frac{r}{nl}\right)^4 + \frac{99}{350}\left(\frac{r}{nl}\right)^6 + \ldots\right], \tag{4.25}$$

from which it can be seen that the Gaussian distribution is the first term of the series, an adequate approximation for $r \ll nl$.

The distribution functions for 25- and 100-link random chains obtained from the Gaussian and inverse Langevin approximations respectively are compared in Figure 4.7.

For the inverse Langevin approximation we have from equation (4.24)

$$S = k \ln p(r) = c - kn\left(\frac{r}{nl}\beta + \ln\frac{\beta}{\sin h\beta}\right) \tag{4.26}$$

and the tension on a chain

$$f = T\frac{\partial S}{\partial r} = \frac{kT}{l}\mathscr{L}^{-1}\left(\frac{r}{nl}\right) \tag{4.27}$$

which, following an expansion similar to equation (4.25), gives

$$f = \frac{kT}{l}\left\{3\left(\frac{r}{nl}\right) + \frac{9}{5}\left(\frac{r}{nl}\right)^3 + \frac{297}{175}\left(\frac{r}{nl}\right)^5 + \frac{1539}{875}\left(\frac{r}{nl}\right)^7 + \ldots\right\}. \tag{4.28}$$

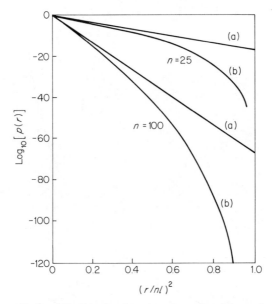

Figure 4.7. Distribution functions for 25- and 100-link random chains: (a) Gaussian approximation; (b) inverse Langevin approximation. [Redrawn with permission from Treloar, *The Physics of Rubber Elasticity*, 2nd edn, Oxford University Press, Oxford, 1958.]

The final part of the exercise is to reconsider the stress–strain relations using the inverse Langevin distribution function. This was done by James and Guth using an analogous development to that for the Gaussian distribution function.

The tensile force per unit unstrained area is

$$f = \frac{NkT}{3} n^{1/2} \left\{ \mathscr{L}^{-1}\left(\frac{\lambda}{n^{1/2}}\right) - \lambda^{-3/2} \mathscr{L}^{-1}\left(\frac{1}{\lambda^{1/2} n^{1/2}}\right) \right\}.$$

Figure 4.8 shows Treloar's fit to the experimental data for natural rubber, using this relationship and a suitable choice of the parameters N and n. The maximum extension of the network is primarily determined by the number of random links n. This result is relevant to the cold-drawing and crazing behaviour of polymers (Sections 11.6.3 and 12.3 below), where the basic deformation also involves the extension of a molecular network.

It can be seen from Figure 4.8 that there is broad agreement between the form of the force–extension curve predicted by the inverse Langevin approximation and that observed in practice.

(5) Although we have seen how the extension to non-Gaussian statistics gives rise to a very large increase in tensile stress at large extensions, in the case of natural rubber it has been proposed that the observed increase in

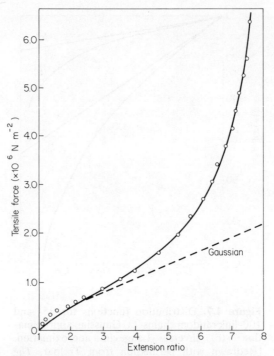

Figure 4.8. Theoretical non-Gaussian free–extension curve—obtained by fitting experimental data 0 to the James and Guth theory, with $NkT = 0.273$ MNm^{-2}, $n = 75$. [Redrawn with permission from Treloar, *The Physics of Rubber Elasticity*, 3rd edn, Oxford University Press, Oxford, 1975.]

tensile stress occurs primarily because of strain-induced crystallization.[15] The basic physical idea is that the melting point T_c of the rubber is increased due to extension. $T_c = \Delta H/\Delta S$, where ΔH and ΔS are the enthalphy and entropy of fusion respectively. Because the entropy of the extended rubber is low, the change in entropy on crystallization is reduced and T_c correspondingly increased. A higher degree of supercooling then gives rise to crystallization and the crystallites act to increase the modulus by forming additional physical crosslinks.

The precise effect of crystallization is, however, difficult to determine. Experiments on butadiene rubbers[16] which do not crystallize on extension, when directly compared with natural rubber, suggested that the influence of crystallization was relatively unimportant.

4.3 THE INTERNAL ENERGY CONTRIBUTION TO RUBBER ELASTICITY

The simple treatment of rubber elasticity given above makes two assumptions which require further consideration. First, it has been assumed that the internal energy contribution is negligible, which implies that different molecular conformations of the chains have identical internal energies. Secondly, the thermodynamic formulae which have been derived are, strictly, only applicable to measurements at constant volume, whereas most experimental results are obtained at constant pressure. These two assumptions are interrelated in the sense that the experimental work of Gee (see section 4.1.1 above) based on the approximation

$$\left(\frac{\partial f}{\partial T}\right)_{P,\lambda} = \left(\frac{\partial f}{\partial T}\right)_{V,l}$$

(where λ is the extension ratio) lead to the conclusion that the internal energy contribution was zero. Although Gee's approximation is much better than the assumption that

$$\left(\frac{\partial f}{\partial T}\right)_{P,l} = \left(\frac{\partial f}{\partial T}\right)_{V,l},$$

it is based on the assumption that the material is isotropically compressible under hydrostatic pressure even when strained, i.e. that

$$\frac{dV}{V} = 3\frac{dl}{l} \quad \text{or} \quad \left(\frac{\partial(\ln l)}{\partial(\ln V)}\right)_{f,T} = \frac{1}{3}.$$

More rigorous consideration suggests that this is not correct and that there is an anisotropic compressibility in the strained state. Gee's experimental result that the internal energy contribution is negligible, based on measurements of the change in stress with temperature at constant length, is therefore capable of another interpretation, which leads to the conclusion that the internal energy contribution may not be zero.

The internal energy component of the tensile force f is given by

$$f_e = \left(\frac{\partial U}{\partial l}\right)_{T,V}$$

(see equation (4.2) above). Volkenstein and Ptitsyn[17] showed that, if the unperturbed dimensions of an isolated chain are temperature dependent, f_e is given by

$$\frac{f_e}{f} = T\frac{\partial(\ln \overline{r_0^2})}{\partial T},$$

where $\overline{r_0^2}$ is the root mean square length (where the subscript 0 indicates a free chain unconstrained by cross-linkages).

Experimental data on dilute solutions of polymers using light scattering and viscosity measurements show that $\overline{r_0^2}$ depends in general on temperature. This implies that the energy of a chain depends on its conformation and that for a rubber, in general, f_e will differ from zero. Flory and his collaborators[18] have been particularly prominent in performing stress–temperature measurements on polymer networks, together with physicochemical measurements, to confirm these points and to obtain the energy difference between different conformations, e.g. the *trans* and *gauche* conformations of a polyethylene chain.

The value f_e can be expressed in terms of the tensile force–temperature relationship by the equation

$$\frac{f_e}{f} = -T\left[\frac{\partial\{\ln f/T\}}{\partial T}\right]_{V,l},\tag{4.29}$$

which follows by manipulating equation (4.3) above. The investigations by Allen and his coworkers involving measurements at constant volume[19] find f_e/f directly from this equation.

It is more usual, however, to make measurements at constant pressure, and in this case obtaining f_e/f is then more elaborate.

The procedure adopted by Flory was based on the theory of the Gaussian network. Flory, Ciferri and Hoeve[18] showed that if the rubber network obeys Gaussian statistics, the expression for measurement of simple extension at constant presssure is

$$\left[\frac{\partial\{\ln f/T\}}{\partial T}\right]_{P,l} + \frac{\alpha}{\lambda^3-1} = -\frac{d(\ln \overline{r_0^2})}{dT},$$

where α is the coefficient of volume expansion of the rubber at constant pressure, i.e.

$$\alpha = \frac{1}{V}\left(\frac{\partial V}{\partial T}\right)_P.$$

Treloar[20,21] has more recently considered the behaviour in torsion and shown that the corresponding relationship is

$$\left[\frac{\partial\{\ln (M/T)\}}{\partial T}\right]_{P,l,\psi} = \frac{-d(\ln \overline{r_0^2})}{dT} + \alpha,$$

where M is the torsional couple and ψ is the torsion (expressed in radians per unit length of the strained axis). M_e, the internal energy contribution to the couple at constant volume, is given by an equation similar to (4.29):

$$\frac{M_e}{M} = -T\left[\frac{\partial\{\ln (M/T)\}}{\partial T}\right]_{V,l,\psi} = T\frac{d(\ln \overline{r_0^2})}{dT}.\tag{4.30}$$

These and similar equations have been used to measure f_e or M_e and in all cases it has been found that the internal energy makes a significant contribution. In natural rubber f_e is positive and forms approximately 20% of the total force at room temperature[19,22], in polydimethyl siloxane[23] the contribution is positive, whereas in swollen and unswollen polyethylene it is negative[24,25].

Boyce and Treloar[26] showed that M_e/M for natural rubber is about 13%, which was within the range of all reported values of f_e/f, although lower than the 20% values which we have quoted here.

There have been a number of measurements, which, although indicating signficant energy contributions, show a definite decrease in f_e with increase in extension ratio[27-31]. This decrease does not, of course, agree with the theory in which f_e was shown to be a function of $\overline{r_0^2}$ only.

It is of particular interest to note Shen's conclusion[31] that the apparent dependence of f_e/f on strain could be removed by plotting f as a function of $\lambda - \lambda^{-2}$. Results for different temperatures showed a series of linear plots, so that f_e/f was independent of strain. It appears, therefore, that the apparent variation of f_e/f with strain arises from the sensitivity of this quantity to small inaccuracies in the measurement of the unstrained length of the specimen. These inaccuracies produce disproportionate errors in the values of the strain which are eliminated by Shen's modified technique.

It was also found that the value of f_e/f was not affected by the presence of a swelling liquid. This confirms that f_e/f is determined by the intramolecular energy contribution, i.e. the difference between the internal energy of different molecular conformations and not by intermolecular forces.

For a comprehensive account of recent work in this area the reader is referred to Chapter 13 in Treloar[6].

4.4 THE POSSIBLE SIGNIFICANCE OF HIGHER ORDER TERMS IN THE STRAIN–ENERGY FUNCTION

From a phenomenological viewpoint, it has often been customary to deal with the experimental data for rubbers in simple tension in terms of the Mooney–Rivlin equation (3.14). As discussed in Section 3.6 above there is, however, no particular reason to quantify deviations from the simple neo-Hookean equation in this way, rather than to adopt a more complex formulation in terms of the extension ratios directly. There must therefore be considerable reservations regarding any attempts to give molecular significance to the higher order terms in the strain–energy function, and in particular to the C_2 term in the Mooney equation. With these reservations in mind, a very brief summary of the historical situation will be given.

An extreme view, expressed in some early papers, was that the deviation from Gaussian theory at moderate extensions is entirely due to the failure to reach equilibrium. Ciferri and Flory[32] undertook a series of experiments over a wide range of conditions under which equilibrium was not necessarily achieved. If the results were fitted to the Mooney equation, it appeared that the value of C_2 increased with the hysteresis of the straining cycle and differed for different polymers. In agreement with studies by other workers[33], it was shown that C_2 decreased with the swelling of the polymer. It was also shown that C_2 decreased with the removal of physical entanglements and with the increase of the time of the experiment[34]. In the most recent papers of Flory and his associates the deviations from Gaussian theory are accepted (see p. 72) as real, and the Mooney–Rivlin equation is employed as an operational

78

procedure for extrapolating data taken in uniaxial extension to zero extension[14].

A number of workers[27,35] have associated the C_2 term either entirely or partly with internal energy effects, but their conclusions must now be regarded with some scepticism in the light of the recent developments discussed in this chapter and the previous one.

REFERENCES

1. K. H. Meyer and C. Ferri, *Helv. Chim. Acta*, **18**, 570 (1935).
2. G. Gee, *Trans. Faraday Soc.*, **42**, 585 (1946).
3. K. H. Meyer, G. Von Susich and E. Valko, *Kolloidzeitschrift*, **59**, 208 (1932).
4. E. Guth and H. Mark, *Lit. Chem.*, **65**, 93 (1934).
5. W. Kuhn, *Kolloidzeitschrift*, **68**, 2 (1934); **76**, 258 (1936).
6. L. R. G. Treloar, *The Physics of Rubber Elasticity*, 3rd edn, Clarendon Press, Oxford, 1975.
7. P. J. Flory, *Chem. Rev.*, **35**, 51 (1944).
8. W. Kuhn, *Kolloidzeitschrift*, **76**, 258 (1936); **87**, 3 (1939).
9. F. T. Wall and J. J. Erpenbeck, *J. Chem. Phys.*, **30**, 634 (1959).
10. S. F. Edwards, *Proc. Phys. Soc.*, **91**, 513 (1967); Series **21A**, 15 (1968) and **92**, 9 (1967).
11. H. M. James and E. Guth, *J. Chem. Phys.*, **11**, 455 (1943).
12. P. J. Flory, *Proc. Roy. Soc. A*, **351**, 351 (1976).
13. W. W. Graessley, *Macromolecules*, **8**, 186, 865 (1975).
14. P. J. Flory, *Polymer*, **20**, 1317 (1979).
15. J. E. Mark, *Polymer Eng. Sci.*, **19**, 254; 409 (1979).
16. W. O. S. Doherty, K. L. Lee and L. R. G. Treloar, *Brit. Polymer J.*, **12**, 19 (1980).
17. M. V. Volkenstein and O. B. Ptitsyn, *Dokl. Akad. SSR*, **91**, 1313 (1953); *Zhur. Tekh. Fiz.*, **25**, 649, 662 (1955).
18. P. J. Flory, A. Ciferri and C. A. J. Hoeve, *J. Polymer Sci.*, **45**, 235 (1960).
19. G. Allen, U. Bianchi and C. Price, *Trans. Faraday Soc.*, **59**, 2493 (1963).
20. L. R. G. Treloar, *Polymer*, **10**, 279 (1969).
21. L. R. G. Treloar, *Polymer*, **10**, 291 (1969).
22. A. Ciferri, *Makromolek. Chem.*, **43**, 152 (1961).
23. A. Ciferri, *Trans. Faraday Soc.*, **57**, 846 (1961).
24. P. J. Flory, C. A. J. Hoeve and A. Ciferri, *J. Polymer Sci.*, **34**, 337 (1959).
25. A. Ciferri, C. A. J. Hoeve and P. J. Flory, *J. Amer. Chem. Soc.*, **83**, 1015 (1961).
26. P. H. Boyce and L. R. G. Treloar, *Polymer*, **11**, 21 (1970).
27. R. J. Roe and W. R. Krigbaum, *J. Polymer Sci.*, **61**, 167 (1962).
28. G. Crespi and U. Flisi, *Makromolek. Chem.*, **60**, 191 (1963).
29. U. Bianchi and E. Pedemonte, *J. Polymer Sci. A2*, **2**, 5039 (1964).
30. M. C. Shen, D. A. McQuarrie and J. L. Jackson, *J. Appl. Phys.*, **38**, 791 (1967).
31. M. C. Shen, *Macromolecules*, **2**, 358 (1969).
32. A. Ciferri and P. J. Flory, *J. Appl. Phys.*, **30**, 1498 (1959).
33. S. M. Gumbrell, L. Mullins and R. S. Rivlin, *Trans. Faraday Soc.*, **49**, 1495 (1953).
34. G. Kraus and G. A. Moczvgemba, *J. Polymer Sci. A2*, **2**, 277 (1964).
35. M. C. Wang and E. Guth, *J. Chem. Phys.*, **20**, 1144 (1953).

5

Linear Viscoelastic Behaviour

5.1 VISCOELASTIC BEHAVIOUR

In text-books on properties of matter two particular types of ideal material are discussed, the elastic solid and the viscous liquid. The elastic solid has a definite shape and is deformed by external forces into a new equilibrium shape. On removal of these external forces it reverts exactly to its original form. The solid stores all the energy which it obtains from the work done by the external forces during deformation. This energy is then available to restore the body to its original shape when these forces are removed. A viscous liquid, on the other hand, has no definite shape and flows irreversibly under the action of external forces. Real materials have properties which are intermediate between those of an elastic solid and a viscous liquid. As discussed in Section 2.2 above, one of the most interesting features of high polymers is that a given polymer can display all the intermediate range of properties depending on temperature and the experimentally chosen time scale.

5.1.1 Linear Viscoelastic Behaviour

Newton's law of viscosity defines viscosity η by stating that stress σ is proportional to the velocity gradient in the liquid:

$$\sigma = \eta \frac{\partial V}{\partial y},$$

where V is the velocity, and y is the direction of the velocity gradient. For a velocity gradient in the xy plane,

$$\sigma_{xy} = \eta \left(\frac{\partial V_x}{\partial y} + \frac{\partial V_y}{\partial x} \right),$$

where $\partial V_x / \partial y$ and $\partial V_y / \partial x$ are the velocity gradients in the y and x directions respectively (see figure 5.1 for the case where the velocity gradient is in the y direction).

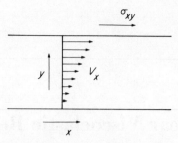

Figure 5.1. The velocity gradient.

Since $V_x = \partial u/\partial t$ and $V_y = \partial v/\partial t$ where u and v are the displacements in the x and y directions respectively, we have

$$\sigma_{xy} = \eta \left[\frac{\partial}{\partial y} \left(\frac{\partial u}{\partial t} \right) + \frac{\partial}{\partial x} \left(\frac{\partial v}{\partial t} \right) \right]$$

$$= \eta \frac{\partial}{\partial t} \left(\frac{\partial u}{\partial y} + \frac{\partial v}{\partial x} \right)$$

$$= \eta \frac{\partial e_{xy}}{\partial t}.$$

It can be seen that the shear stress σ_{xy} is directly proportional to the rate of change of shear strain with time. This formulation brings out the analogy between Hooke's law for elastic solids and Newton's law for viscous liquids. In the former the stress is linearly related to the *strain*, in the latter the stress is linearly related to the *rate of change of strain* or *strain rate*.

Hooke's law describes the behaviour of a linear *elastic* solid and Newton's law that of a linear *viscous* liquid. A simple constitutive relation for the behaviour of a linear viscoelastic solid is obtained by combining these two laws.

For elastic behaviour $(\sigma_{xy})_E = G e_{xy}$, where G is the shear modulus.

For viscous behaviour $(\sigma_{xy})_V = \eta (\partial e_{xy}/\partial t)$.

A possible formulation of linear viscoelastic behaviour combines these equations; thus

$$\sigma_{xy} = (\sigma_{xy})_E + (\sigma_{xy})_V = G e_{xy} + \eta \frac{\partial e_{xy}}{\partial t}.$$

This makes the simplest possible assumption that the shear stresses related to strain and strain rate are additive. The equation represents one of the simple models for linear viscoelastic behaviour (the Voigt or Kelvin model) and will be discussed in detail in Section 5.2.6 below.

Most of the experimental work on linear viscoelastic behaviour is confined to a single mode of deformation, usually corresponding to a measurement of the Young's modulus or the shear modulus. Our initial discussion of linear viscoelasticity will therefore be confined to the one-dimensional situation, recognizing that greater complexity will be required to describe the viscoelastic behaviour fully. For the simplest case of an isotropic polymer at least two of the modes of deformation defining two of the quantities E, G and K for an elastic solid must be examined, if the behaviour is to be completely specified.

In defining the constitutive relations for an elastic solid we have assumed that the *strains* are *small* and that there are linear relationships between stress and strain. We now ask how the principle of linearity can be extended to materials where the deformations are time-dependent. The basis of the discussion is the Boltzmann Superposition Principle[1]. This states that in linear viscoelasticity effects are simply additive, as in classical elasticity, the difference being that in linear viscoelasticity it matters at which instant an effect is created. Although the application of stress may now cause a time-dependent deformation, it can still be assumed that each increment of stress makes an independent contribution. From the present discussion it can be seen that the linear viscoelastic theory must also contain the additional assumption that the strains are small. In Chapter 9 we will deal with attempts to extend linear viscoelastic theory either to take into account non-linear effects at small strains or to deal with the situation at large strains.

5.1.2 Creep

It is convenient to introduce the discussion of linear viscoelastic behaviour with the one-dimensional situation of creep under a fixed load. For an elastic solid the following is observed at the two levels of stress σ_0 and $2\sigma_0$ (Figure 5.2(a)).

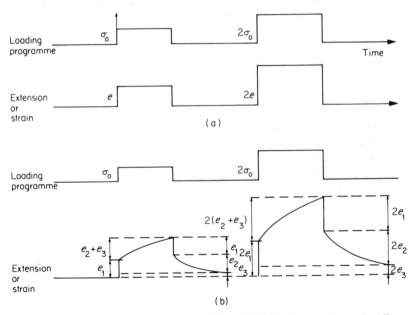

Figure 5.2. (a) Deformation of an elastic solid; (b) deformation of a linear viscoelastic solid.

The strain follows the pattern of the loading programme exactly and in exact proportionality to the magnitude of the loads applied.

The effect of applying a similar loading programme to a linear viscoelastic solid has several similarities (Figure 5.2(b)). In the most general case the total strain e is the sum of three separate parts e_1, e_2 and e_3. e_1 and e_2 are often termed the *immediate* elastic deformation and the *delayed* elastic deformation respectively. e_3 is the Newtonian flow, i.e. that part of the deformation which is identical with the deformation of a viscous liquid obeying Newton's law of viscosity.

Because the material shows linear behaviour the magnitude of the strains e_1, e_2 and e_3 are exactly proportional to the magnitude of the applied stress. Thus the simple loading experiment defines a *creep compliance* $J(t)$ which is only a function of time:

$$\frac{e(t)}{\sigma} = J(t) = J_1 + J_2 + J_3,$$

where J_1, J_2 and J_3 correspond to e_1, e_2 and e_3 respectively.

The term J_3, which defines the Newtonian flow, can be neglected for rigid polymers at ordinary temperatures, because their flow viscosities are very large. Linear amorphous polymers do show a finite J_3 at temperatures above their glass transitions, but at lower temperatures their behaviour is dominated by J_1 and J_2. Cross-linked polymers do not show a J_3 term, and this is true to a very good approximation for highly crystalline polymers as well.

This leaves J_1 and J_2. At any given temperature the separation of the compliance into terms J_1 and J_2 may involve an arbitrary division, which expresses the fact that at the shortest experimentally accessible times we will observe a limiting complicance J_1. We will assume, however, that there is a real distinction between the elastic and delayed responses. In some texts the immediate elastic response in a creep experiment is called the 'unrelaxed' response to distinguish it from the 'relaxed' response which is observed at times sufficiently long for the various relaxation mechanisms to have occurred. To emphasize that the values of such terms as J_1 are sometimes arbitrary we will enclose them in brackets.

We have already discussed in the introductory chapter how polymers can behave as glassy solids, viscoelastic solids, rubbers or viscous liquids depending on the time scale or on the temperature of the experiment. How does this fit in with our present discussion? Figure 5.3 shows the variation of compliance with time at constant temperature over a very wide time scale for an idealized amorphous polymer with only one relaxation transition. This diagram shows that for short-time experiments the observed compliance is $10^{-9}\,\mathrm{m^2\,N^{-1}}$, that for a glassy solid. It is also time independent. At very long times the observed compliance is $10^{-5}\,\mathrm{m^2\,N^{-1}}$, that for a rubbery solid, and it is again time independent. At intermediate times the compliance lies between these values and is time dependent; this is the general situation of viscoelastic behaviour.

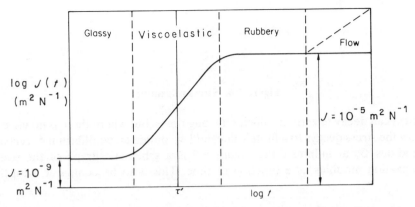

Figure 5.3. The creep compliance $J(t)$ as a function of time t. τ' is the characteristic time (the retardation time).

These considerations suggest that the observed behaviour will depend on the time scale of the experiment relative to some basic time parameter of the polymer. For creep this parameter is called the retardation time τ' and falls in the middle range of our time scale as shown in the diagram. The distinction between a rubber and a glassy plastic can then be seen as somewhat artificial in that it depends only on the value of τ' at room temperature for each polymer. Thus for a rubber τ' is very small at room temperature compared with normal experimental times which are greater than say 1 s, whereas the opposite is true for a glassy plastic. The value of this parameter τ' for a given polymer relates to its molecular constitution, as will be discussed later.

These considerations lead immediately to a qualitative understanding of the influence of temperature on polymer properties. With increasing temperature the frequency of molecular rearrangements is increased, reducing the value of τ'. Thus at very low temperatures a rubber will behave like a glassy solid, as is well known, and equally a glassy plastic will soften at high temperatures to become rubber-like.

In the diagram illustrating creep under constant load recovery curves are also displayed. We will presently show that the recovery behaviour is basically similar to the creep behaviour if we neglect the quantity e_3, the Newtonian flow. This is a direct consequence of linear viscoelastic behaviour.

5.1.3 Stress Relaxation

The counterpart of creep is stress relaxation, where the sample is subjected to constant strain e, and the decay of stress $\sigma(t)$ is observed. This is illustrated in Figure 5.4.

The assumption of linear behaviour enables us to define the *stress relaxation modulus* $G(t) = \sigma(t)/e$. In the case of stress relaxation the presence of viscous flow will affect the limiting value of the stress. Where viscous flow occurs the

84

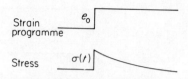

Figure 5.4. Stress relaxation.

stress can decay to zero at sufficiently long times, but where there is no viscous flow the stress decays to a finite value, and we obtain an equilibrium or relaxed modulus G_r at infinite time. Figure 5.5 is a schematic graph of the stress relaxation modulus as a function of time. This is to be compared with the

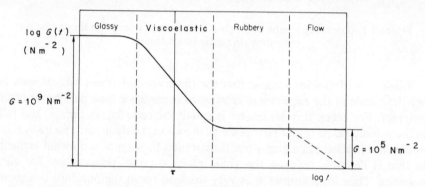

Figure 5.5. The stress relaxation modulus $G(t)$ as a function of time t. τ is the characteristic time (the relaxation time).

corresponding graph for creep (Figure 5.3). The same regions of behaviour, viz. glassy, viscoelastic, rubbery and flow, can be identified, and a transition time τ is defined which characterizes the time scale of the viscoelastic behaviour. We will shortly discuss stress relaxation and creep in detail and show that as customarily defined, the characteristic times τ and τ', although of the same order of magnitude, are not identical. Similar considerations to those discussed for creep apply to the effect of changing temperature on stress relaxation; i.e. changing temperature is equivalent to changing the time scale. Time–temperature equivalence is applicable to all linear viscoelastic behaviour in polymers and is considered fully in Chapter 7. The measurement of G_r may present difficulties as in the case of the elastic response. We will assume that there is a relaxed response to which it relates, but enclose the term involving it in brackets as for those involving the elastic response.

5.2 MATHEMATICAL TREATMENT OF LINEAR VISCOELASTIC BEHAVIOUR

The discussion must now be placed on a quantitative basis. It is a question of personal taste how this is done, as it depends on the relative merits of a formal mathematical treatment with its manipulative advantages, and a less formal treatment which provides more physical insight into viscoelastic behaviour. We will endeavour to strike a balance here by describing the several representations of linear viscoelastic behaviour in turn, and completing the presentation by indicating the formal connections between them.

5.2.1 The Boltzmann Superposition Principle and the Definition of Creep Compliance

The general pattern of creep and stress relaxation behaviour has been discussed, indicating some of the simpler consequences of assuming linear viscoelastic behaviour. The Boltzmann Superposition Principle[1] is the first mathematical statement of linear viscoelastic behaviour. Boltzmann proposed: (1) that the creep in a specimen is a function of the entire loading history, and (2) that each loading step makes an independent contribution to the final deformation and that the final deformation can be obtained by the simple addition of each contribution.

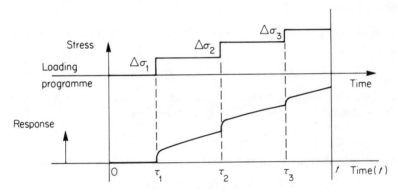

Figure 5.6. The creep behaviour of a linear viscoelastic solid.

Consider a several-stage loading programme (Figure 5.6) in which incremental stresses $\Delta\sigma_1$, $\Delta\sigma_2$, $\Delta\sigma_3$, etc., are added at times τ_1, τ_2, τ_3, etc., respectively. The total creep at time t is then given by

$$e(t) = \Delta\sigma_1 J(t-\tau_1) + \Delta\sigma_2 J(t-\tau_2) + \Delta\sigma_3 J(t-\tau_3) + \ldots, \tag{5.1}$$

where $J(t-\tau)$ is the *creep compliance function*. The contribution of each loading step is the product of the incremental stress and a general function of time, the creep compliance function, which depends only on the interval in time between the instant at which the incremental stress is applied and the instant at which the creep is measured.

Equation (5.1) can be generalized to give the integral

$$e(t) = \int_{-\infty}^{t} J(t-\tau)\, d\sigma(t) \tag{5.2}$$

and is usually rewritten as

$$e(t) = \left[\frac{\sigma}{G_u}\right] + \int_{-\infty}^{t} J(t-\tau)\frac{d\sigma(\tau)}{d\tau} d\tau, \tag{5.3}$$

where the 'immediate elastic' contribution to the compliance is included in terms of an elastic modulus G_u (G_u is the unrelaxed modulus) and the integral term is written in its more correct mathematical form. It is to be noted that the integral is taken from $-\infty$ to t. This follows from the hypothesis of the Boltzmann principle that *all* previous elements of the loading history were to be taken into account. In all actual experiment a conditioning procedure may be required to destroy the long term memory of the specimen (see Chapter 6 below).

This integral is called a *Duhamel* integral, and it is a useful illustration of the consequences of the Boltzmann superposition principle to evaluate the response for a number of simple loading programmes. Recalling the development which leads to equation (5.2) it can be seen that the Duhamel integral is most simply evaluated by treating it as the summation of a number of response terms. Consider three specific cases:

(1) Single-step loading of a stress σ_0 at time $\tau = 0$ (Figure 5.7(a)). For this case

$$J(t-\tau) = J(t) \quad \text{and} \quad e(t) = \sigma_0 J(t).$$

(2) Two-step loading, a stress σ_0 at time $\tau = 0$, followed by an *additional* stress σ_0 at time $\tau = t_1$ (Figure 5.7(b)). For this case

$$e_1 = \sigma_0 J(t), \qquad e_2 = \sigma_0 J(t-t_1)$$

give the creep deformations produced by the two loading steps, and

$$e(t) = e_1 + e_2 = \sigma_0 J(t) + \sigma_0 J(t-t_1).$$

This shows that the 'additional creep' e_c' $(t-t_1)$ produced by the second loading step is given by

$$e_c'(t-t_1) = \sigma_0 J(t) + \sigma_0 J(t-t_1) - \sigma_0 J(t) = \sigma_0 J(t-t_1).$$

This illustrates one consequence of the Boltzmann principle, viz. that the additional creep $e_c'(t-t_1)$ produced by adding the stress σ_0 is identical with the creep which would have occurred had this stress σ_0 been applied without any previous loading at the same instant in time t_1.

(3) Creep and recovery. In this case (Figure 5.7(c)) the stress σ_0 is applied at time $\tau = 0$ and removed at time $\tau = t_1$. The deformation $e(t)$ at a time $t > t_1$ is given by the addition of two terms $e_1 = \sigma_0 J(t)$ and $e_2 = -\sigma_0 J(t-t_1)$, which express the application and removal of the stress σ_0 respectively. Thus

$$e(t) = \sigma_0 J(t) - \sigma_0 J(t-t_1).$$

The *recovery* $e_r(t-t_1)$ will be defined as the difference between the anticipated

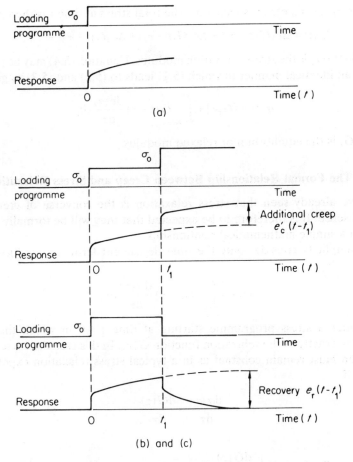

Figure 5.7. Response of a viscoelastic solid to single-step loading (a); two-step loading (b); and loading and unloading (c).

creep under the initial stress and the actual measured response. Thus

$$e_r(t - t_1) = \sigma_0 J(t) - [\sigma_0 J(t) - \sigma_0 J(t - t_1)] = \sigma_0 J(t - t_1)$$

It can be seen that this is identical with the creep response to a stress σ_0 applied at a time t_1. This demonstrates a second consequence of the Boltzmann Superposition Principle, that the creep and recovery responses are identical in magnitude.

5.2.2 The Stress Relaxation Modulus

Stress relaxation behaviour can be represented in an exactly complementary fashion using the Boltzmann Superposition Principle. Consider a stress relaxation programme in which incremental strains Δe_1, Δe_2, Δe_3, etc., are added at times τ_1, τ_2, τ_3, etc., respectively. The total stress at time t is then given by

$$\sigma(t) = \Delta e_1 G(t - \tau_1) + \Delta e_2 G(t - \tau_2) + \Delta e_3 G(t - \tau_3) + \ldots, \tag{5.4}$$

where $G(t - \tau)$ is the stress relaxation modulus. Equation (5.4) may be generalized in an identical manner in which (5.1) leads to (5.2) and (5.3) to give

$$\sigma(t) = [G_r e] + \int_{-\infty}^{t} G(t - \tau) \frac{de(\tau)}{d\tau} dt, \tag{5.5}$$

where G_r is the equilibrium or relaxed modulus.

5.2.3 The Formal Relationship Between Creep and Stress Relaxation

We have already seen that stress relaxation is the converse of creep in a general sense. It is therefore to be expected that they will be formally related through a simple mathematical relationship.

For simplicity consider only the time-dependent terms in equation (5.3). Then

$$e(t) = \int_{-\infty}^{t} J(t - \tau) \frac{d\sigma(\tau)}{d\tau} d\tau.$$

Consider a stress programme starting at time $\tau = 0$ in which the stress decreases exactly as the relaxation function $G(\tau)$. In this case the corresponding strain must remain constant as in a typical stress relaxation experiment.

Thus if

$$\frac{d\sigma(\tau)}{d\tau} = \frac{dG(\tau)}{d\tau},$$

then

$$\int_0^t \frac{dG(\tau)}{d\tau} J(t - \tau) \, d\tau = \text{constant}. \tag{5.6}$$

For simplicity we can normalize the definitions of $G(\tau)$ and $J(\tau)$ so that the constant is unity. We then have

$$\int_0^t \frac{dG(\tau)}{d\tau} J(t - \tau) \, d\tau = 1. \tag{5.7}$$

This expression is sometimes integrated to give

$$\int_0^t G(\tau) J(t - \tau) \, d\tau = t. \tag{5.8}$$

These equations provide a formal connection between the creep and stress relaxation functions. However, this approach is of greatest interest from a purely theoretical standpoint. In practice the problem of interchangeability of creep and stress relaxation data is usually dealt with via relaxation or retardation spectra, and by approximate methods.

5.2.4 Mechanical Models, Relaxation and Retardation Time Spectra

The Boltzmann Superposition Principle is one starting point for a theory of linear viscoelastic behaviour, and is sometimes called the '*integral representation* of linear viscoelasticity', because it defines an integral equation. An equally valid starting point is to relate the stress to the strain by a linear differential equation, which leads to a *differential representation* of linear viscoelasticity. In its most general form the equation has the form

$$P\sigma = Qe,$$

where P and Q are linear differential operators with respect to time. This representation has been found of particular value in obtaining solutions to specific problems in the deformation of viscoelastic solids[2].

Most generally the differential equation is

$$a_0\sigma + a_1\frac{d\sigma}{dt} + a_2\frac{d^2\sigma}{dt^2} + \ldots = b_0e + b_1\frac{de}{dt} + b_2\frac{d^2e}{dt^2} + \ldots. \tag{5.9}$$

It is often adequate to represent the experimental data obtained over a limited time scale by including only one or two terms on each side of this equation. We will now show that this is equivalent to describing the viscoelastic behaviour by mechanical models constructed of elastic springs which obey Hooke's law and viscous dashpots which obey Newton's law of viscosity.

The simplest models consist of a single spring and a single dashpot either in series or in parallel and these are known as the Maxwell model and the Kelvin or Voigt models respectively.

5.2.5 The Maxwell Model

The Maxwell model consists of a spring and dashpot in series as shown in Figure 5.8.

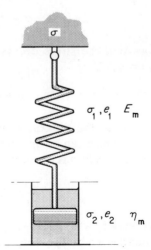

Figure 5.8. The Maxwell model.

The equations for the stress–strain relations are

$$\sigma_1 = E_m e_1 \qquad (5.10a)$$

relating the stress σ_1 and the strain e_1 in the spring and

$$\sigma_2 = \eta_m \frac{de_2}{dt} \qquad (5.10b)$$

relating the stress σ_2 and the strain e_2 in the dashopot. Now relate the total stress σ and the total strain e. We have that $\sigma = \sigma_1 = \sigma_2$ since the stress is identical for the spring and dashpot, and $e = e_1 + e_2$, the total strain being the sum of the strain in the spring and the dashpot. Equation (5.10a) can be written as

$$\frac{d\sigma}{dt} = E_m \frac{de_1}{dt}$$

and added to (5.10b), giving

$$\frac{de}{dt} = \frac{1}{E_m} \frac{d\sigma}{dt} + \frac{\sigma}{\eta_m}. \qquad (5.11)$$

The Maxwell model is of particular value in considering a stress relaxation experiment. In this case

$$\frac{de}{dt} = 0 \quad \text{and} \quad \frac{1}{E_m} \frac{d\sigma}{dt} + \frac{\sigma}{\eta_m} = 0.$$

Thus

$$\frac{d\sigma}{\sigma} = -\frac{E_m}{\eta_m} dt.$$

At time $t = 0$, $\sigma = \sigma_0$, the initial stress, and integrating we have

$$\sigma = \sigma_0 \exp \frac{-E_m}{\eta_m} t. \qquad (5.12)$$

This shows that the stress decays exponentially with a characteristic time constant $\tau = \eta_m / E_m$:

$$\sigma = \sigma_0 \exp \frac{-t}{\tau},$$

where τ is called the 'relaxation time'. There are two inadequacies of this simple model which can be understood immediately.

First, under conditions of constant stress, i.e.

$$\frac{d\sigma}{dt} = 0, \qquad \frac{de}{dt} = \frac{\sigma}{\eta_m},$$

and Newtonian flow is observed. This is clearly not generally true for visco-elastic materials where the creep behaviour is more complex.

Secondly, the stress relaxation behaviour cannot usually be represented by a single exponential decay term, nor does it necessarily decay to zero at infinite time.

5.2.6 The Kelvin or Voigt Model

The Kelvin or Voigt model consists of a spring and dashpot in parallel as shown in Figure 5.9.

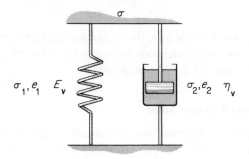

Figure 5.9. The Kelvin or Voigt model.

The stress–strain relations are

$$\sigma_1 = E_v e_1 \tag{5.13}$$

and

$$\sigma_2 = \eta_v \frac{de_2}{dt} \tag{5.14}$$

for the spring and dashpot respectively.

Again it is required to relate the total stress σ to the total strain e. In this case we have

$$e = e_1 = e_2 \quad \text{and} \quad \sigma = \sigma_1 + \sigma_2$$

Thus

$$\sigma = E_v e + \eta_v \frac{de}{dt} \tag{5.15}$$

For stress relaxation, where $de/dt = 0$, the Kelvin model gives $\sigma = E_v e$, i.e. a constant stress, implying that the material behaves as an elastic solid, which is clearly an inadequate representation for general viscoelastic behaviour.

On the other hand, the Kelvin model does represent creep behaviour to a first approximation. For creep under constant load $\sigma = \sigma_0$ it may readily be shown that

$$e = \frac{\sigma_0}{E_v} \left(1 - \exp \frac{-E_v}{\eta_v} t \right) \tag{5.16}$$

In fact, the response of a Kelvin element to constant load conditions is most readily understood by considering the recovery response, where $\sigma = 0$. Here

$$E_v e + \eta_v \frac{\mathrm{d}e}{\mathrm{d}t} = 0,$$

giving a solution for the strain

$$e = e_0 \exp \frac{-t}{\tau'},$$

where $\tau' = \eta_v / E_v$ is a characteristic time constant called the 'retardation time'. All these relationships are exactly analogous to the stress relaxation behaviour and the relaxation time of the Maxwell model.

5.2.7 The Standard Linear Solid

We have seen that the Maxwell model describes the stress relaxation of a viscoelastic solid to a first approximation, and the Kelvin model the creep behaviour, but that neither model is adequate for the general behaviour of a viscoelastic solid where it is necessary to describe both stress relaxation and creep.

Consider again the general linear differential equation which represents linear viscoelastic behaviour. From the present discussion it follows that to obtain even an approximate description of both stress relaxation and creep, at least the first two terms on each side of equation (5.9) must be retained, i.e. the simplest equation will be of the form

$$a_0 \sigma + a_1 \frac{\mathrm{d}\sigma}{\mathrm{d}t} = b_0 e + b_1 \frac{\mathrm{d}e}{\mathrm{d}t}. \tag{5.17}$$

This will be adequate to a first approximation for creep (when $\mathrm{d}\sigma/\mathrm{d}t = 0$) and for stress relaxation (when $\mathrm{d}e/\mathrm{d}t = 0$), giving an exponential response in both cases.

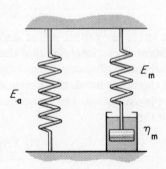

Figure 5.10. The standard linear solid.

It is very easy to show that the model shown in Figure 5.10 has this form. The stress–strain relationship is

$$\sigma + \tau \frac{d\sigma}{dt} = E_a e + (E_m + E_a)\tau \frac{de}{dt}, \quad \text{where } \tau = \frac{\eta_m}{E_m} \tag{5.18}$$

This model is known as the 'standard linear solid' and is usually attributed to Zener[3]. It provides an approximate representation to the observed behaviour of polymers in their main viscoelstic range. As has been discussed, it predicts an exponential response only. To describe the observed viscoelastic behaviour quantitatively would require the inclusion of many terms in the linear differential equation (5.9). These more complicated equations are equivalent to either a large number of Maxwell elements in parallel or a large number of Voigt elements in series (Figures 5.11(a) and (b)).

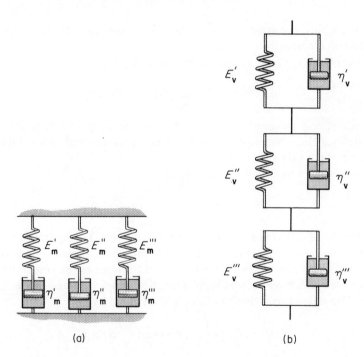

(a)   (b)

Figure 5.11. (a) Maxwell elements in parallel; (b) Voigt elements in series.

5.2.8 Relaxation Time Spectra and Retardation Time Spectra

It is next required to obtain a quantitative description of stress relaxation and creep which will help to form a link with the original mathematical description in terms of the Boltzmann integrals. It is simple and instructive to do this by development of the Maxwell and Kelvin models.

Consider first stress relaxation, described by

$$\sigma(t) = [G_r e] + \int_{-\infty}^{t} G(t - \tau) \frac{de(\tau)}{d\tau} d\tau \qquad (5.4)$$

where $G(t)$ is the stress relaxation modulus. For stress relaxation at constant strain e, equation (5.12) shows that the Maxwell model gives

$$\sigma(t) = E_m e \, \exp \frac{-t}{\tau}$$

and the stress relaxation modulus $G(t) = E_m \exp(-t/\tau)$. For a series of Maxwell elements joined in parallel, again at constant strain e, the stress is given by

$$\sigma(t) = e \sum_{n}^{n} E_n \, \exp \frac{-t}{\tau_n},$$

where E_n, τ_n are the spring constant and relaxation time respectively of the nth Maxwell element.

The summation can be written as an integral, giving

$$\sigma(t) = [G_r e] + e \int_{0}^{\infty} f(\tau) \, \exp \frac{-t}{\tau} d\tau, \qquad (5.19)$$

where the spring constant E_n is replaced by the weighting function $f(\tau) \, d\tau$ which defines the concentration of Maxwell elements with relaxation times between τ and $\tau + d\tau$.

The stress relaxation modulus is given by

$$G(t) = [G_r] + \int_{0}^{\infty} f(\tau) \, \exp \frac{-t}{\tau} d\tau. \qquad (5.20)$$

The term $f(\tau)$ is called the 'relaxation time spectrum'. In practice, it has been found more convenient to use a logarithmic time scale. A new relaxation time spectrum $H(\tau)$ is now defined, where $H(\tau) \, d(\ln \tau)$ gives the contributions to the stress relaxation associated with relaxation times between $\ln \tau$ and $\ln \tau + d(\ln \tau)$. The stress relaxation modulus is then given by

$$G(t) = [G_r] + \int_{-\infty}^{\infty} H(\tau) \, \exp \frac{-t}{\tau} d(\ln \tau) \qquad (5.21)$$

An exactly analogous treatment, using a series of the Kelvin models, leads to a similar expression for the creep compliance $J(t)$.

Thus

$$J(t) = [J_u] + \int_{-\infty}^{\infty} L(\tau) \left(1 - \exp \frac{-t}{\tau} \right) d(\ln \tau) \qquad (5.22)$$

where J_u is the instantaneous elastic compliance and $L(\tau)$ is the *retardation time spectrum*, $L(\tau) \, d(\ln \tau)$ defining the contributions to the creep compliance associated with retardation times between $\ln \tau$ and $\ln \tau + d(\ln \tau)$.

The relaxation time spectrum can be calculated exactly from the measured stress relaxation modulus using Fourier or Laplace transform methods, and similar considerations apply to the retardation time spectrum and the creep compliance. It is more convenient to consider these transformations at a later stage, when the final representation of linear viscoelasticity, that of the complex modulus and complex compliance, has been discussed.

It is important to recognize that the relaxation time spectrum and the retardation time spectrum are only mathematical descriptions of the macroscopic behaviour and do not necessarily have a simple interpretation in molecular terms. It is a quite separate exercise to correlate observed patterns in the relaxation behaviour, such as a predominant relaxation time, with a specific molecular process. It should also be emphasized, as will be apparent from the further detailed discussion, that qualitative interpretations in general molecular terms can often be obtained from the experimental data directly, without recourse to calculation of the relaxation time spectrum or the retardation time spectrum.

5.3 DYNAMICAL MECHANICAL MEASUREMENTS: THE COMPLEX MODULUS AND COMPLEX COMPLIANCE

An alternative experimental procedure to creep and stress relaxation is to subject the specimen to an alternating strain and simultaneously measure the stress. For linear viscoelastic behaviour, when equilibrium is reached, the stress and strain will both vary sinusoidally, but the strain lags behind the stress.

Thus we write

$$\text{strain } e = e_0 \sin \omega t,$$

$$\text{stress } \sigma = \sigma_0 \sin (\omega t + \delta),$$

where ω is the angular frequency, δ is the phase lag.

Expanding $\sigma = \sigma_0 \sin \omega t \cos \delta + \sigma_0 \cos \omega t \sin \delta$ we see that the stress can be considered to consist of two components: (1) of magnitude $(\sigma_0 \cos \delta)$ in phase with the strain; (2) of magnitude $(\sigma_0 \sin \delta)$ 90° out of phase with the strain.

The stress–strain relationship can therefore be defined by a quantity G_1 in phase with the strain and by a quantity G_2 which is 90° out of phase with the strain, i.e.

$$\sigma = e_0 G_1 \sin \omega t + e_0 G_2 \cos \omega t,$$

where

$$G_1 = \frac{\sigma_0}{e_0} \cos \delta \quad \text{and} \quad G_2 = \frac{\sigma_0}{e_0} \sin \delta.$$

This immediately suggests a complex representation for the modulus as shown in Figure 5.12.

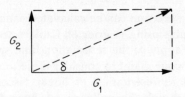

Figure 5.12. The complex modulus
$G^* = G_1 + iG_2$ and $\tan \delta = G_2/G_1$.

If we write

$$e = e_0 \exp i\omega t, \qquad \sigma = \sigma_0 \exp i(\omega t + \delta).$$

Then

$$\frac{\sigma}{e} = G^* = \frac{\sigma_0}{e_0} e^{i\delta}$$

$$= \frac{\sigma_0}{e_0} (\cos \delta + i \sin \delta)$$

$$= G_1 + iG_2.$$

The real part of the modulus G_1, which is in phase with the strain, is often called the *storage modulus* because it defines the energy stored in the specimen due to the applied strain. The imaginary part of the modulus G_2, which is out of phase with the strain, defines the dissipation of energy and is often called the *loss modulus*. The reason for this is seen by calculating the energy dissipated per cycle, $\Delta \mathscr{E}$:

$$\Delta \mathscr{E} = \int \sigma \, de = \int_0^{2\pi/\omega} \frac{\sigma \, de}{dt} \, dt.$$

Substituting for σ and e we have

$$\Delta \mathscr{E} = \omega e_0^2 \int_0^{2\pi/\omega} (G_1 \sin \omega t \cos \omega t + G_2 \cos^2 \omega t) \, dt$$

$$= \pi G_2 e_0^2$$

In most cases G_2 is small compared with G_1. $|G^*|$ is therefore approximately equal to G_1. $|G^*|$ is sometimes loosely referred to as the 'modulus' G. It is customary to define the dynamic mechanical behaviour in terms of the 'modulus' $G \doteqdot G_1$, and the phase angle δ or often $\tan \delta = G_2/G_1$. To a good approximation $\delta = \tan \delta$ when the loss modulus G_2 is small. Typical values of G_1, G_2 and $\tan \delta$ for a polymer would be $10^9 \, \mathrm{N \, m^{-2}}$, $10^7 \, \mathrm{N \, m^{-2}}$ and 0.01 respectively.

An exactly complementary treatment can be developed to define a complex compliance

$$J^* = J_1 - iJ_2$$

This is directly related to the complex modulus since

$$G^* = 1/J^*$$

We have so far ignored any question of frequency or time dependence. Here there is an exact analogy with creep and stress relaxation and it is necessary to determine G_1 and G_2 (or $\tan \delta$) or J_1 and J_2 (or $\tan \delta$) as a function of frequency if we wish to specify the viscoelastic behaviour completely.

5.3.1 Experimental Patterns for G_1, G_2, etc., as a Function of Frequency

Now consider the complex moduli and compliances as a function of frequency for a typical viscoelastic solid, in a similar manner to the creep and stress relaxation as a function of time.

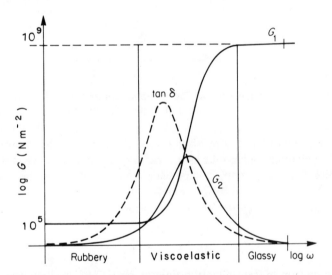

Figure 5.13. The complex modulus $G_1 + iG_2$ as a function of frequency ω.

Figure 5.13 shows the variation of G_1, G_2 and $\tan \delta$ with frequency for a polymer which shows no flow. At low frequencies the polymer is rubber-like and has a low modulus G_1 of $\sim 10^5 \, \mathrm{N \, m^{-2}}$, which is independent of frequency. At high frequencies the polymer is glassy with a modulus of $\sim 10^9 \, \mathrm{N \, m^{-2}}$, which is again independent of frequency. At intermediate frequencies the polymer behaves as a viscoelastic solid, and its modulus G_1 increases with increasing frequency.

The complementary pattern of behaviour is shown by the loss modulus G_2. At low and high frequencies G_2 is zero, the stress and strain being exactly in phase for the rubbery and glassy states. At intermediate frequencies, where the polymer is viscoelastic, G_2 rises to a maximum value, this occurring at a frequency close to that for which the storage modulus is changing most rapidly with frequency. The viscoelastic region is also characterized by a maximum in the loss factor tan δ, but this occurs at a slightly lower frequency than that in G_2, since tan $\delta = G_2/G_1$ and G_1 is also changing rapidly in this frequency range.

An analogous diagram (Figure 5.14) shows the variation of the compliances J_1 and J_2 with frequency.

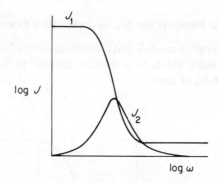

Figure 5.14. The complex compliance $J^* = J_1 - iJ_2$ as a function of frequency ω.

The next development is to obtain a mathematical representation for the dynamic mechanical behaviour as a function of frequency. As in the case of stress relaxation and creep, a very easy starting point for the argument is based on the Maxwell and Voigt models.

Using equation (5.11) for the Maxwell model,

$$\frac{de}{dt} = \frac{1}{E_m}\frac{d\sigma}{dt} + \frac{\sigma}{\eta_m},$$

and the definition of the relaxation time as $\tau = \eta_m/E_m$, we can write

$$\sigma + \tau\frac{d\sigma}{dt} = E_m\tau\frac{de}{dt}.$$

Put

$$\sigma = \sigma_0\, e^{i\omega t} = (G_1 + iG_2)e.$$

This gives

$$\sigma_0\, e^{i\omega t} + i\omega\tau\sigma_0\, e^{i\omega t} = \frac{E_m\tau i\omega\sigma_0\, e^{i\omega t}}{G_1 + iG_2},$$

from which it follows that

$$G_1 + iG_2 = \frac{E_m i\omega\tau}{1 + i\omega\tau},$$

i.e.

$$G_1 = \frac{E_m\omega^2\tau^2}{1+\omega^2\tau^2}, \qquad G_2 = \frac{E_m\omega\tau}{1+\omega^2\tau^2} \quad \text{and} \quad \tan\delta = \frac{1}{\omega\tau}. \qquad (5.23)$$

This result gives the pattern shown in Figure 5.15 for G_1, G_2 and $\tan\delta$ as a function of frequency (or $\omega\tau$, which is more convenient). It is seen that the qualitative features are correct in the case of G_1 and G_2, but not for $\tan\delta$.

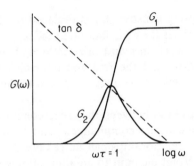

Figure 5.15. The complex modulus $G^* = G_1 + iG_2$ as a function of frequency ω.

A similar manipulation of the equation representing the Voigt model, introducing the complex compliances, leads to a comparable qualitative picture of J_1, J_2 and $\tan\delta$ as a function of frequency. Again the qualitative features are correct for J_1 and J_2 but not for $\tan\delta$, which in this case is equal to $\omega\tau'$.

The Maxwell and Voigt models are therefore inadequate to describe the dynamic mechanical behaviour of a polymer, as they do not provide an adequate representation of both the creep and stress relaxation behaviour. A good measure of qualitative improvement could be gained, as in the previous discussion of creep and stress relaxation, by using a three-parameter model, e.g. the standard linear solid, and it is an interesting exercise to show that this model gives a more realistic variation of G_1, G_2 and $\tan\delta$ with frequency.

It is, however, desirable to move directly to the general representation, using the relaxation time spectrum.

The general representation for the stress relaxation modulus (equation (5.21)),

$$G(t) = \frac{\sigma(t)}{e} = [G_r] + \int_{-\infty}^{\infty} H(\tau) \exp\frac{-t}{\tau} \, d(\ln\tau),$$

follows by generalizing the stress relaxation response from a single Maxwell element where $G(t) = E_m \exp(-t/\tau)$.

The response of a Maxwell element to an alternating strain is defined by the relationships

$$G_1 = \frac{E_m\omega^2\tau^2}{1+\omega^2\tau^2} \quad \text{and} \quad G_2 = \frac{E_m\omega\tau}{1+\omega^2\tau^2}$$

An identical generalization to the previous one then gives

$$G_1(\omega) = [G_r] + \int_{-\infty}^{\infty} \frac{H(\tau)\omega^2\tau^2}{1+\omega^2\tau^2} \, d(\ln \tau) \qquad (5.24)$$

and

$$G_2(\omega) = \int_{-\infty}^{\infty} \frac{H(\tau)\omega\tau}{1+\omega^2\tau^2} \, d(\ln \tau). \qquad (5.25)$$

As previously, the spring constant E_m is replaced by the weighting function $H(\tau) \, d(\ln \tau)$ which defines the contribution to the response of elements whose relaxation time is between $\ln \tau$ and $\ln \tau + d(\ln \tau)$. It is seen that the stress relaxation modulus $G(t)$ and the real and imaginary parts of the complex compliance G_1 and G_2 can all be directly related to the same relaxation time spectrum $H(\tau)$.

Similar relationships hold between the creep compliance $J(t)$, the real and imaginary parts of the complex compliance J_1 and J_2 and the retardation time spectrum $L(\tau)$. These relationships can be readily derived by consideration of the response of a Voigt element to an alternating stress. They will not be derived here, but the results are quoted for completeness:

$$J_1(\omega) = [J_u] + \int_{-\infty}^{\infty} \frac{L(\tau)}{1+\omega^2\tau^2} \, d(\ln \tau); \qquad (5.26)$$

$$J_2(\omega) = \int_{-\infty}^{\infty} \frac{J(\tau)\omega\tau}{1+\omega^2\tau^2} \, d(\ln \tau). \qquad (5.27)$$

5.4 THE RELATIONSHIPS BETWEEN THE COMPLEX MODULI AND THE STRESS RELAXATION MODULUS

The exact formal relationships between the various viscoelastic functions are conveniently expressed using Fourier or Laplace transform methods (cf. Section 5.4.2 below).

It is often required to combine dynamic mechanical measurement of complex moduli at high frequencies (or short times) with stress relaxation modulus measurements at long times (or low frequencies). The most usual approach is to use approximate methods to convert both types of measurement to the determination of the relaxation time spectrum.

Now

$$G(t) = [G_r] + \int_{-\infty}^{\infty} H(\tau) \exp\frac{-t}{\tau} \, d(\ln \tau). \qquad (5.21)$$

A simple approximation is to assume that $e^{-t/\tau} = 0$ up to the time $\tau = t$, and $e^{-t/\tau} = 1$ for $\tau > t$.

Then we can write

$$G(t) = [G_r] + \int_{\ln t}^{\infty} H(\tau) \, d(\ln \tau).$$

This gives the relaxation time spectrum

$$H(\tau) = -\left[\frac{dG(t)}{d \ln t}\right]_{t=\tau}, \tag{5.28}$$

which is known as the 'Alfrey approximation'[4].

The relaxation time spectrum can be expressed to a similar degree of approximation in terms of the real and imaginary parts of the complex modulus:

$$H(\tau) = \left[\frac{d\, G_1(\omega)}{d \ln \omega}\right]_{1/\omega=\tau} = \frac{2}{\pi}[G_2(\omega)]_{1/\omega=\tau}. \tag{5.29}$$

These relationships are illustrated diagrammatically for the case of a single relaxation transition in Figure 5.16(a) and (b). To obtain the complete relaxation time spectrum the longer time part of $H(\tau)$ will be found from the stress relaxation modulus data of Figure 5.16(a) and the shorter time part from the dynamic mechanical data of Figure 5.16(b).

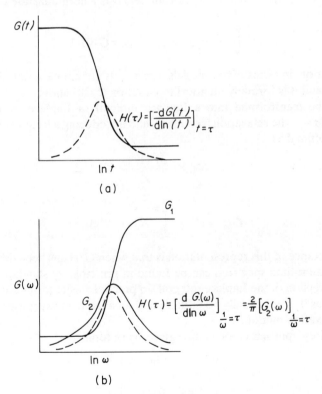

Figure 5.16. The Alfrey approximations for the relaxation time spectrum $H(\tau)$: (a) from the stress relaxation modulus $G(t)$; (b) from the real and imaginary parts G_1 and G_2 respectively of the complex modulus $G(\omega)$.

Complementary relationships can be used to obtain the retardation time spectrum in terms of the complex compliances and the creep compliance.

5.4.1 Formal Representations of the Stress Relaxation Modulus and the Complex Modulus

A complete exposition of the mathematical structure of linear viscoelasticity has been given by Gross[5]. Here we will only summarize certain parts of his argument to illustrate the use of Laplace and Fourier transforms in establishing the formal connections between various viscoelastic functions:

(1) *The stress relaxation modulus.* This modulus is a continuous, decreasing function which goes to zero at infinite time. In Gross' nomenclature it is represented in integral form as

$$G(t) = [G_r] + \int_0^\infty \bar{\beta} \bar{F}(\tau) \exp \frac{-t}{\tau} \, d\tau \tag{5.30}$$

where $\bar{F}(\tau) \, d\tau$ is the relaxation spectrum, and $\bar{\beta}$ is a normalization factor such that

$$\int_0^\infty \bar{F}(\tau) \, d\tau = 1, \qquad \bar{\beta} = G(0).$$

This equation, in terms of our models, represents an infinite series of Maxwell elements and it is formally identical to equation (5.20) above.

It can be transformed into a Laplace integral or Laplace transform by putting $1/\tau = s$, the relaxation frequency, and introducing a frequency function $\bar{N}(s) \, ds$ defined as

$$\bar{N}(s) = \frac{\bar{\beta} \bar{F}(1/s)}{s^2}.$$

Thus

$$G(t) = [G_r] + \int_0^\infty \bar{N}(s) \, e^{-ts} \, ds. \tag{5.31}$$

The importance of this representation is that when $G(t)$ has been determined, the relaxation time spectrum can be found in principle by standard methods for the inversion of the Laplace integral. In practice this requires computation methods, as it is not usually possible to find an analytical expression to fit the stress relaxation modulus.

The Alfrey approximation is now given by putting

$$e^{-ts} = 1 \quad \text{for } s \leq 1/t$$

and

$$e^{-ts} = 0 \quad \text{for } s > 1/t$$

and

$$G(t) = [G_r] + \int_0^{1/t} \bar{N}(s) \, ds. \tag{5.32}$$

(2) *The complex modulus.* The Boltzmann Superposition Principle gives us

$$\sigma(t) = [G_r e(t)] + \int_{-\infty}^{t} G(t-\tau) \frac{de(\tau)}{d\tau} d\tau$$

Put $e(\tau) = e_0 e^{i\omega\tau}$. Then

$$\sigma(t) = [G_r e(t)] + i\omega \int_{-\infty}^{t} G(t-\tau)e_0 e^{i\omega\tau} d\tau. \qquad (5.33)$$

Put $t - \tau = T$. Then

$$\sigma(t) = [G_r e(t)] + i\omega \int_{0}^{\infty} G(T) e^{-i\omega T} dT e_0 e^{i\omega t} \qquad (5.34)$$

Now $e(t) = e_0 e^{i\omega t}$. Thus

$$\frac{\sigma(t)}{e(t)} = [G_r] + i\omega \int_{0}^{\infty} G(\tau) e^{-i\omega\tau} d\tau = G^*(\omega),$$

the complex modulus, where we have changed the dummy variable from T back to τ. Therefore

$$G_1(\omega) = \omega \int_{0}^{\infty} G(\tau) \sin \omega\tau \, d\tau \qquad (5.35)$$

and

$$G_2(\omega) = \omega \int_{0}^{\infty} G(\tau) \cos \omega\tau \, d\tau, \qquad (5.36)$$

where $G^*(\omega) = (G_r + G_1) + iG_2$. Equations (5.35) and (5.36) are one-sided Fourier transforms.

Inversion gives the stress relaxation modulus

$$G(t) = \frac{2}{\pi} \int_{0}^{\infty} \frac{G_1(\omega)}{\omega} \sin \omega t \, d\omega \qquad (5.37)$$

and

$$G(t) = \frac{2}{\pi} \int_{0}^{\infty} \frac{G_2(\omega)}{\omega} \cos \omega t \, d\omega. \qquad (5.38)$$

These equations imply a relationship between $G_1(\omega)$ and $G_2(\omega)$, the dispersion or compatibility relations, which are the viscoelastic analogue of the Kramers–Krönig relations for optical dispersion and magnetic relaxation.

5.4.2 Formal Representations of the Creep Compliance and the Complex Compliance

Similar relationships hold for the creep compliance and complex compliance to those derived for the stress relaxation modulus and the complex modulus.

The details of the derivations will not be given, but the results are quoted for completeness.

(1) *Creep compliance.* In this case the *rate of change* of creep compliance is expressed as a Laplace integral. Thus

$$\frac{\mathrm{d}J(t)}{\mathrm{d}t} = \int_0^\infty sN(s)\,\mathrm{e}^{-ts}\,\mathrm{d}s, \tag{5.39}$$

where

$$N(s) = \frac{F(1/s)}{s^2}, \qquad s = \frac{1}{\tau}$$

and $F(\tau)\,\mathrm{d}\tau$ is the distribution of retardation times. Note that $N(s) \neq \bar{N}(s)$, and that $F(\tau) \neq \bar{F}(\tau)$, i.e. the retardation time spectrum is not identical to the relaxation time spectrum.

(2) *Complex compliance.* Here it is found that

$$J_1(\omega) = \int_0^\infty \frac{\mathrm{d}J(\tau)}{\mathrm{d}\tau} \cos \omega\tau\,\mathrm{d}\tau \tag{5.40}$$

and

$$J_2(\omega) = -\int_0^\infty \frac{\mathrm{d}J(\tau)}{\mathrm{d}\tau} \sin \omega\tau\,\mathrm{d}\tau. \tag{5.41}$$

Again both $J_1(\omega)$ and $J_2(\omega)$ are Fourier transforms, which may be inverted to give the creep compliance in terms of the components of the complex compliance. The inversion formulae both give the creep compliance, implying a relationship between the real and imaginary parts of the complex compliance, as in the case of the complex modulus.

5.4.3 The Formal Structure of Linear Viscoelasticity

Gross[5] has discussed the formal structure of the theory of linear viscoelasticity. A summary of his treatment will be presented here, as a suitable conclusion to our discussion.

There are two groups of experiments:

Group 1: Experiments which take place under a given stress, either fixed or alternating. These define the creep compliance or the complex compliance.

Group 2: Experiments which take place under a given strain, either fixed or alternating. These define the stress relaxation modulus or the complex modulus.

Within each group the viscoelastic functions exist in three levels:

(a) Top level   Complex compliance   (Group 1)
                  Complex modulus   (Group 2)

(b) Medium level	Creep function	(Group 1)
	Relaxation function	(Group 2)
(c) Bottom level	Retardation spectrum	(Group 1)
	Relaxation spectrum	(Group 2)

To go *up* a level, one applies either a Laplace transform or a one-sided complex Fourier transform.

To go *down* a level, one applies either an inverse Laplace transform or an inverse Fourier transform.

The relationships between the groups vary in complexity. At the top level, the complex compliance is merely the inverse of the complex modulus. The relationships between the creep function and the relaxation function and between the retardation spectrum and the relaxation spectrum involve integral equations and integral transforms respectively.

5.5 THE RELAXATION STRENGTH

A concept which is of value in considering the relationship of viscoelastic behaviour to physical and chemical structure is that of 'relaxation strength'. In a stress relaxation experiment the modulus relaxes from a value G_u at very short times to G_r at very long times (Figure 5.17(b)). Similarly in a dynamic

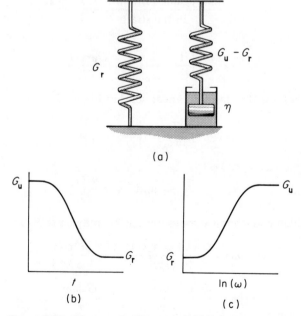

Figure 5.17. The standard linear solid (a) gives a response in a stress relaxation test shown in (b) and in a dynamic mechanical test in (c). G_u is the unrelaxed modulus and G_r the relaxed modulus.

mechanical experiment the modulus changes from G_r at low frequencies to G_u at very high frequencies. G_u is the unrelaxed modulus and G_r the relaxed modulus (Figure 5.17(c)),

This behaviour is shown by the standard linear solid of Figure 5.17(a). Consider in turn the behaviour in the initial unrelaxed state, and in the final relaxed state.

(1) A total applied stress σ is given in terms of the initial unrelaxed strain e_1 by adding the stresses in both springs. Thus

$$\sigma = G_r e_1 + (G_u - G_r)e_1 = G_u e_1$$

and the initial unrelaxed strain

$$e_1 = \frac{\sigma}{G_u} = \frac{\sigma}{\text{Unrelaxed modulus}}.$$

(2) The final relaxed strain e_2 is given in terms of an applied stress σ by

$$e_2 = \frac{\sigma}{G_r} = \frac{\sigma}{\text{Relaxed modulus}}$$

because the spring $(G_u - G_r)$ is now ineffective. The relaxation strength is conventionally defined as

$$\frac{\text{Final strain—Initial strain}}{\text{Initial strain}}.$$

This is

$$\frac{e_2 - e_1}{e_1} = \left\{\frac{1}{G_r} - \frac{1}{G_u}\right\}G_u = \frac{G_u - G_r}{G_r}. \tag{5.42}$$

The equation for the standard linear solid in Figure 5.17(a) is

$$\sigma + \tau_1 \frac{d\sigma}{dt} = G_r\left\{e + \tau_2 \frac{de}{dt}\right\}, \tag{5.43}$$

where

$$\tau_1 = \frac{\eta}{G_u - G_r} \quad \text{and} \quad \tau_2 = \frac{\tau_1 G_u}{G_r}.$$

Then for dynamic mechanical measurements it may be shown that

$$G_1(\omega) = \frac{G_r(1 + \omega^2 \tau_1 \tau_2)}{1 + \omega^2 \tau_1^2} = \frac{G_r + \omega^2 \tau^2 G_u}{1 + \omega^2 \tau^2}, \tag{5.44}$$

$$G_2(\omega) = \frac{G_r(\tau_2 - \tau_1)\omega}{1 + \omega^2 \tau_1^2} = \frac{(G_u - G_r)\omega\tau}{1 + \omega^2 \tau^2} \tag{5.45}$$

and

$$\tan\delta = \frac{(\tau_2 - \tau_1)\omega}{1 + \omega^2 \tau_1 \tau_2} = \frac{(G_u - G_r)\omega\tau}{G_r + \omega^2 \tau^2 G_u}, \tag{5.46}$$

where we have put

$$\tau = \tau_1 = \frac{\eta}{G_u - G_r}.$$

The following relationships then hold:

$$\tan \delta_{max}(\omega^2 \tau^2 = G_r/G_u) = \frac{G_u - G_r}{2\sqrt{G_u G_r}}, \tag{5.47}$$

$$G_{2_{max}}(\omega^2 \tau^2 = 1) = \frac{G_u - G_r}{2}, \tag{5.48}$$

$$\int_{-\infty}^{\infty} G_2(\omega)\, d(\ln \omega) = \frac{\pi}{2}(G_u - G_r), \tag{5.49}$$

$$\int_{-\infty}^{\infty} \tan d(\ln \omega) = \frac{\pi}{2} \frac{(G_u - G_r)}{\sqrt{G_u G_r}}. \tag{5.50}$$

All the relationships defined above are proportional to $G_u - G_r$ and hence to the relaxation strength. $\tan \delta_{max}$ and $\int_{-\infty}^{\infty} \tan \delta\, d(\ln \omega)$ are closest to our original definition in being normalized to a dimensionless quantity. This provides some formal justification for the use of $\tan \delta$ rather than G_2 for estimating the relaxation strength to correlate with structural parameters[6].

REFERENCES

1. L. Boltzmann, *Pogg. Ann. Phys. Chem.*, **7**, 624 (1876).
2. E. H. Lee, in *Proceedings of the First Symposium on Naval Structural Mechanics*, Pergamon Press, Oxford, 1960, p. 456.
3. C. Zener, *Elasticity and Anelasticity of Metals*, Chicago University Press, Chicago, 1948.
4. T. Alfrey, *Mechanical Behaviour of High Polymers*, Interscience Publishers, New York, 1948.
5. B. Gross, *Mathematical Structure of the Theories of Viscoelasticity*, Hermann, Paris, 1953.
6. R. W. Gray and N. G. McCrum, *J. Polymer Sci. B*, **6**, 691 (1968).

FURTHER READING

J. J. Aklonis, W. J. MacKnight and M. Shen, *Introduction to Polymer Viscoelasticity*, Wiley-Interscience, New York, 1972.

6

The Measurement of Viscoelastic Behaviour

Experimental studies of viscoelasticity in polymers are extremely extensive, and a very large number of techniques have been developed. In this chapter we will only attempt to indicate the types of methods which are available, together with a few representative examples of actual experimental arrangements. The reader is referred to standard texts on viscoelastic behaviour[1,2] and review articles[3-5] for more detailed expositions of the subject.

To obtain a satisfactory understanding of the viscoelastic behaviour, data are required over a wide range of frequency (or time) and temperature. In Chapter 5, the equivalence of creep, stress relaxation and dynamic mechanical data has been described. In Chapter 7 the equivalence of time and temperature as variables will be discussed. Although this equivalence can sometimes reduce the required range of experimental data, it is desirable in principle to be able

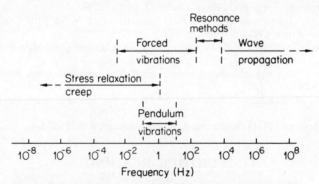

Figure 6.1. Approximate frequency scales for different experimental techniques. [Redrawn with permission from Becker, *Mater. Plast. Elast.*, **35**, 1387 (1969).]

to cover wide ranges of both time and temperature. This can only be done by combining a wide variety of techniques, the approximate time scales of which are shown in Figure 6.1. The techniques fall into five main classes:

(1) Transient measurements: creep and stress relaxation.
(2) Low frequency vibrations: free oscillation methods.

(3) High frequency vibrations: resonance methods.
(4) Forced vibration non-resonance methods.
(5) Wave propagation methods.

These will now be discussed in turn.

6.1 CREEP AND STRESS RELAXATION

The transient methods of measuring viscoelastic behaviour, creep and stress relaxation, cover the frequency range from ~1 Hz to very low frequencies. These methods are also most effective in revealing the essential nature of the non-linear viscoelasticity which is typical of most polymers at other than very low strain levels.

6.1.1 Creep: Conditioning

To undertake steady state quantitative measurements of creep and recovery under various loading conditions (i.e. measurements which are equivalent to dynamic mechanical measurements), a conditioning procedure is essential, as emphasized by Leaderman[6]. The specimen should be subjected to successive creep and recovery cycles, each cycle consisting of application of the maximum load for the maximum period of loading, followed by a recovery period after unloading of about 10 times the loading period. For creep tests over a range of temperatures, this conditioning should be carried out at the highest temperature of measurement.

The conditioning procedure has two major effects on the creep and recovery behaviour. First, subsequent creep and recovery responses under a given load are then identical, i.e. the sample has lost its 'long term' memory and now only remembers loads applied in its immediate past history. Secondly, after the conditioning procedure the deformation produced by any loading programme is almost completely recoverable provided that the recovery period is about 10 times the period during which loads are applied. For tensile creep measurements over a wide range of temperature greater elaboration is required.

6.1.2 Extensional Creep

For simplicity of interpretation, creep measurements require a dead-loading procedure. The most elementary procedure is to use a cathetometer or travelling microscope to measure the creep between two ink marks on the sample, which eliminates end effects at the clamps.

For accuracy at other than low strain levels, there should be a device for reducing the load in a manner proportional to the decrease in cross-sectional area, in order to maintain a constant stress. The system adopted by Leaderman[7] is shown in Figure 6.2.

110

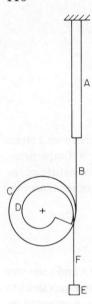

Figure 6.2. Cam arrangement for creep under constant stress. [Redrawn with permission from Leaderman, *Trans. Soc. Rheol.*, **6**, 361 (1962).]

The cylindrical drum C and the specially shaped cam D are attached to a shaft supported in cone bearings. The upper end of the specimen A is fixed, and the lower end attached to the drum by means of a thin flexible steel tape B. One end of a similar tape F is attached to the cam; the other end of this tape supports a constant weight E. A balance arm is used to set the centre of gravity of the rotating system at the axis of the shaft. As the specimen extends under load, the moment of the applied weight decreases according to the cam profile.

6.1.3 Extensometers

A more satisfactory technique for high accuracy is to use an extensometer attached directly to the specimen, the strain being converted either into rotation of mirrors or into an electrical signal by a displacement transducer. Dunn, Mills and Turner[8] describe two types of extensometer which have been used for polymers of high and medium stiffness respectively. For a polymer of high stiffness, the modified Lamb extensometer shown in Figure 6.3 was found satisfactory. A long rectangular prism of polymer (cross-section 5.5 mm × 3.2 mm) is gripped between two pairs of knife edges. The knife edges are attached to four main members and movement takes place by rotation of two steel balls at the top and by rotation of two rollers at the bottom. Mirrors attached to the rollers form part of an optical system so that extension of the specimen is converted by rotation of the rollers into movement of a light beam. A typical deflection of the light beam is 0.26 radians for a

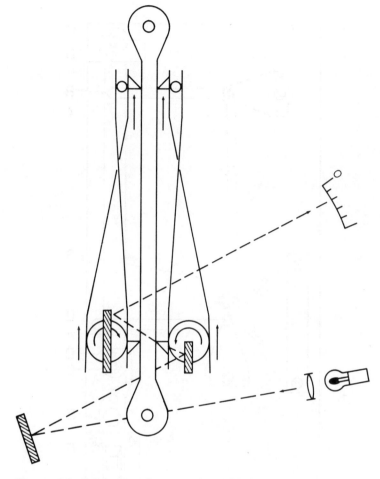

Figure 6.3. Schematic diagram of modified Lamb extensometer. [Redrawn with permission from Dunn, Mills and Turner, *Brit. Plast.*, **37**, 386 (1964).]

1% strain, i.e. ~130 mm displacement of a light spot on a scale at 0.5 m distance.

For specimens of lower stiffness, the Lamb extensometer may distort the specimen at the knife edges. The weight of the extensometer may also cause extension of the specimen. For these reasons an optical lever extensometer was developed, as shown in Figure 6.4. The two forked arms B possess steel pins which pass through holes of 1.19 mm diameter in the specimen. Mirrors attached to these forked arms form part of an optical system so that the extension of the specimen is again converted into the movement of an optical beam. Sensitivities are quoted of ~60 mm deflection of the beam for 1% strain.

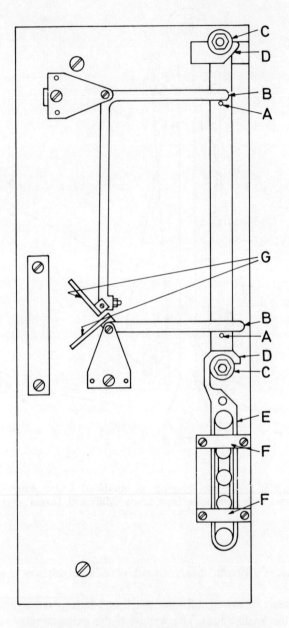

Figure 6.4. Optical lever extensometer. A, steel pins; B, forked arms; C, coaxial discs; D, slotted hooks; E, guide bar; F, guide bar support slot; G, mirrors. [Redrawn with permission from Dunn, Mills and Turner, *Brit. Plast.*, **37**, 386 (1964).]

6.1.4 High Temperature Creep

An apparatus designed to measure creep with reasonable accuracy[9] over a wide temperature range (20 – 200 °C) is shown in Figure 6.5. A heavy monofilament of 2.54 mm diameter is suspended from a micrometer head inside a copper tube which is surrounded by an oil bath, the temperature of which is thermostatically controlled to better than +0.1 °C. The monofilament is attached to long clamps (~0.3 m) at its extremities so that its entire length is well within the controlled temperature volume. The clamp at the lower end of the monofilament is attached via the slug of a differential transformer to the load W. The total weight of slug and clamp is about 20 g, and at each temperature at least 20 h are allowed for the sample to creep under this small load. The subsequent creep under this residual load is thus small compared with that due to the larger applied load. The creep is measured by continuously adjusting the micrometer so as to position the slug in the centre of the differential transformer. This gives a minimum output voltage from the transformer (which is observed on the oscilloscope). The input to the transformer comes from a 2 kHz oscillator via suitable matching impedances. The monofilament is typically about 0.3 m in length and the micrometer can be read to ±0.0125 mm.

This apparatus was used to undertake creep measurements at low strain levels. There was therefore no attempt to correct for the change in cross-sectional area under load.

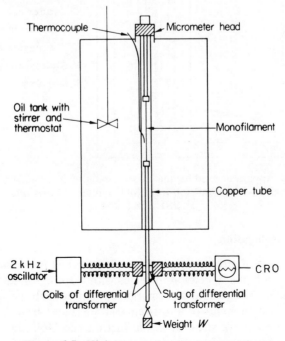

Figure 6.5. High temperature creep apparatus.

6.1.5 Torsional Creep

The measurement of creep in torsion can be made with very great accuracy. This is because the deformation can be used to cause rotation of a mirror directly and thus give large deflection of a light beam. McCrum and his collaborators, for example, have used a torsional creep apparatus of the type shown in Figure 6.6 to make very accurate torsional creep measurements on polyethylene[10].

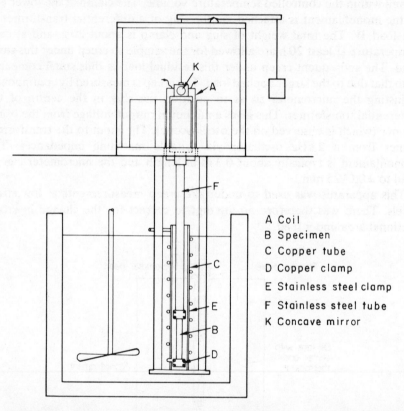

A Coil

B Specimen

C Copper tube

D Copper clamp

E Stainless steel clamp

F Stainless steel tube

K Concave mirror

Figure 6.6. Apparatus used for the measurement of creep and stress relaxation. [Redrawn with permission from McCrum and Morris, *Proc. Roy. Soc. A*, **281**, 258 (1964).]

6.1.6 Stress Relaxation

The measurement of stress relaxation can in principle most simply be made by placing the polymer specimen in series with a spring of sufficiently great stiffness to undergo negligible deformation compared with the specimen. The spring may be the elements of a resistance strain-gauge transducer, enabling a direct measurement of stress as in the Instron and Hounsfield tensometer, or it may be incorporated in a differential transformer so that its displacement again records the stress. For mathematical simplicity it is desirable to use a step function of strain, which requires very rapid application of strain if measurements are to be made at short times. An apparatus for stress relaxation measurements[11] is shown in Figure 6.7.

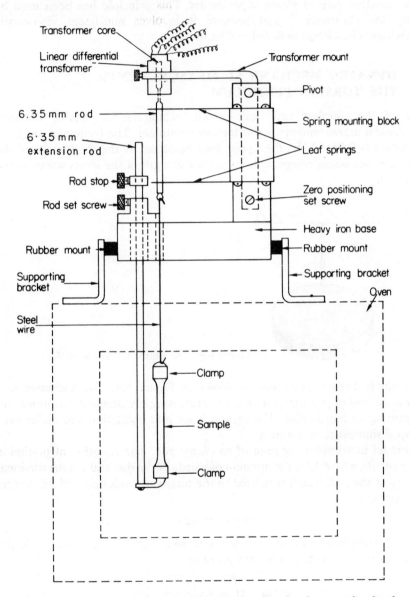

Figure 6.7. Apparatus for measuring stress relaxation in extension by the changes in position of a spring whose stiffness is much greater than that of the sample. [Redrawn with permission from McLoughlin, *Rev. Sci. Instr.*, **23**, 459 (1952).]

A more elaborate technique is to maintain the specimen at constant length by continuously adjusting the required force as the stress relaxes. This has been done by Stein and Schaevitz[12] using an automatic servomechanism.

It is also worth noting that the stress relaxation modulus can be obtained from constant rate of strain experiments. This principle has been used by Smith for elastomers[13] and because it involves non-linear viscoelastic behaviour is described in detail in Chapter 9.

6.2 DYNAMIC MECHANICAL MEASUREMENTS: THE TORSION PENDULUM

One of the simplest and best known techniques for making dynamic mechanical measurements is the torsion pendulum. The frequency range of operation is 0.01–50 Hz, the upper limit being set by the dimensions of the specimen becoming comparable to the wavelength of the stress waves in the specimen.

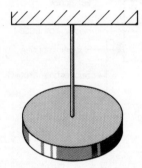

Figure 6.8. The simple torsion pendulum.

A simple torsion pendulum is shown in Figure 6.8. The specimen is a cylindrical rod of polymer, one end of which is rigidly clamped, the other end supporting an inertia disc. The system is set into oscillation and undertakes damped sinusoidal oscillations.

First let us consider the case of an *elastic* rod. The equation of motion is $M\ddot{\theta} + \tau\theta = 0$, where M is the moment of inertia of the disc and τ is the torsional rigidity of the rod, which is related to the torsional modulus G of the rod by the equation

$$\tau = G\pi r^4/2l,$$

where l = length of rod and r = radius of rod. The system executes simple harmonic motion with a frequency given by

$$\omega = \sqrt{\frac{\tau}{M}} = \sqrt{\frac{\pi r^4 G}{2lM}}.$$

Thus G can be found directly from the frequency of oscillation, M being determined separately.

The effect of the viscoelastic behaviour of the polymer is to introduce a damping term proportional to $\dot{\theta}$ into the equation of motion, giving an equation of the form

$$a\ddot{\theta} + b\dot{\theta} + c\theta = 0.$$

This has the general solution

$$\theta = A \exp\left(\frac{-b+\sqrt{b^2-4ac}}{2a}t\right) + B \exp\left(\frac{-b-\sqrt{b^2-4ac}}{2a}t\right).$$

For small damping $b^2 < 4ac$ and the motion is oscillatory. A solution is found by taking $\theta = 0$ and $\dot{\theta} = \text{constant}$ when $t = 0$. Then

$$\theta = \theta_0 \exp\left(\frac{-bt}{2a}\right) \sin \omega_1 t, \tag{6.1}$$

where

$$\omega_1 = \frac{b^2 - 4ac}{2a}.$$

This solution represents a damped oscillatory motion. It is customary to describe the damping by what is termed the 'logarithmic decrement' Λ. This is the natural logarithm of the ratio of the amplitude of successive oscillations, and equation (6.1) is then written as

$$\theta = \theta_0 \exp\left(\frac{-\omega_1 \Lambda}{2\pi}\right) \sin \omega_1 t. \tag{6.2}$$

For a linear viscoelastic solid the torsional modulus is a complex quantity and may be written as $G^* = G_1 + iG_2$. The equation of motion for the torsion pendulum may then be written as

$$M\ddot{\theta} + \frac{\pi r^4}{2l}(G_1 + iG_2)\theta = 0. \tag{6.3}$$

Assuming a solution of the form

$$\theta = \theta_0 \exp\left(\frac{-\omega \Lambda}{2\pi}t\right) \sin \omega t,$$

substituting and equating real and imaginary parts we have

$$\frac{\pi r^4}{2l} \frac{G_1}{M} = \omega^2 \tag{6.4}$$

for small damping and

$$\frac{\pi r^4}{2l} \frac{G_2}{M} = \omega^2 \frac{\Lambda}{\pi} \tag{6.5}$$

or

$$\frac{G_2}{G_1} = \frac{\Lambda}{\pi} = \tan \delta. \tag{6.6}$$

Thus the frequency of the oscillation gives the real part of the modulus G_1 and measurement of the logarithmic decrement gives tan δ and G_2.

A simple torsion pendulum with an inertia disc mounted directly on the end of the specimen was used by Schmieder and Wolf[14]. In a slightly more elaborate arrangement (Figure 6.9) the specimen is supported by a fine wire or ribbon (or by a counterbalancing weight[15]). This enables measurements to be made in temperature ranges where the weight of the inertia disc would cause extensional creep. The equation of motion for the compound pendulum is a simple extension of equation (6.3), adding a term $(\tau'\theta = (\pi r'^4/2l')G'\theta$ for a circular wire) from the supporting wire or ribbon. The additional term will not contribute to the damping as the supporting wire can be considered to be elastic.

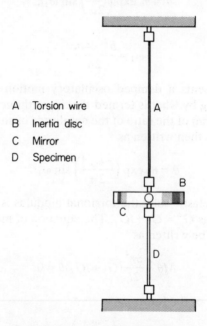

A   Torsion wire
B   Inertia disc
C   Mirror
D   Specimen

Figure 6.9. Apparatus for measuring torsional rigidity at low frequencies.

The polymer specimen is often in the form of a rectangular block or thin sheet. The term $(\pi r^4/2l)G^*$ has then to be replaced by the St Venant formula for the torsion of rectangular beams (see Chapter 10).

Figure 6.9 shows a typical experimental scheme for measuring the behaviour of a fibre monofilament. The frequency and logarithmic decrement are determined by observing the motion of a light beam reflected from the mirror.

Figure 6.10 shows the apparatus with heating and cooling facilities. The image of the light source falls on the photodyne recorder which plots out the damped sinusoidal oscillations.

Figure 6.10. The free vibration torsion pendulum.

6.3 RESONANCE METHODS

At higher frequencies the wavelength of the stress waves decreases until it becomes comparable to the dimensions of the specimen. Assuming a Young's modulus of $10^7 \, \text{N m}^{-2}$ and a density of $10^3 \, \text{kg m}^{-3}$ gives a longitudinal wave velocity of $10^2 \, \text{m s}^{-1}$. At a frequency of $10^3 \, \text{Hz}$ this gives a wavelength of 0.1 m. Thus at higher frequencies the specimens become vibrating systems of standing waves with resonances, and from the variation in amplitude of vibration with frequency we can determine the real and imaginary parts of the complex modulus. The frequency range of these methods is clearly somewhat restricted.

6.3.1 The Vibrating Reed Technique

The vibrating reed or cantilever is a very popular technique which in many industrial laboratories provides a standard method for polymer evaluation[16-18]. The small specimen (Figure 6.11) is clamped at one end in a gramophone cutter head which is driven from a variable frequency oscillator

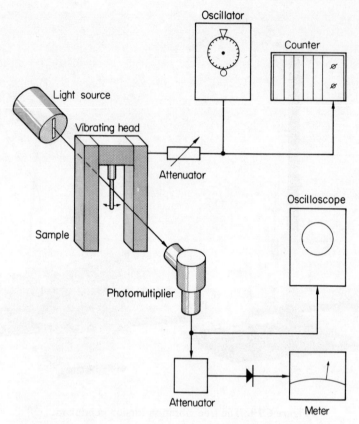

Figure 6.11. Schematic diagram of the vibrating reed apparatus.

(at a frequency typically in the range 200–1500 Hz). The light source illuminates a rectangular slit, the image of which is focused on the tip of the specimen. The motion of the specimen modulates the intensity of the light falling on the photocell and must be arranged so that the specimen is never totally in nor totally out of the illuminated region. The signal from the photocell is amplified and monitored on an oscilloscope for correct alignment of the optics, and its amplitude is measured by a meter. The amplitude of vibration,

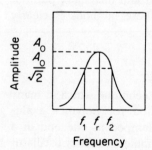

Figure 6.12. Resonance curve for the vibrating reed. $(f_2 - f_1 = \Delta f$; see text.)

i.e. the resonance curve, is measured as a function of frequency, at fixed temperatures and a curve of the form shown in Figure 6.12 is obtained.

A complete treatment of the viscoelastic behaviour of the vibrating reed has been given by Bland and Lee[19]. It is, however, satisfactory for our purposes to assume that the losses are small (in practice this means $\tan \delta < 0.2$). The 'modulus' ($|E|$) can then be obtained from the solution to the *elastic* problem and $\tan \delta$ from the resonance curve as follows.

The equation of motion of an *elastic* cantilever is (for a detailed discussion see Reference 20)

$$\frac{\partial^2 y}{\partial t^2} + \frac{E}{\rho} k^2 \frac{\partial^4 y}{\partial x^4} = 0, \tag{6.7}$$

where E is the Young's modulus of the cantilever, ρ is the density, k^2 is the radius of gyration $= h^2/12$ where h is the thickness of the cantilever, and y is the displacement from its equilibrium position of a point on the reed at a distance x from the fixed end (Figure 6.13).

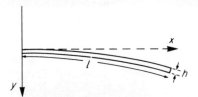

Figure 6.13. The vibrating reed.

The general solution to equation (6.7) is

$$y = \{A \cosh mx + B \sinh mx + C \cos mx + D \sin mx\} \cos \omega t, \tag{6.8}$$

where

$$m = \left\{\frac{\rho \omega^2}{Ek^2}\right\}^{1/4}$$

and ω is the angular frequency of vibration in radians per second.

The boundary conditions of the vibrating reed are $y = dy/dx = 0$ for all t at the clamped end where $x = 0$ and $d^2y/dx^2 = d^3y/dx^3 = 0$ for all t at the free end. (No couple and no shearing force.)

Substituting these boundary conditions in turn into the general solution (equation (6.8)) gives two simultaneous equations which are compatible if

$$\cosh ml \cos ml = -1, \tag{6.9}$$

where l is the length of the cantilever. The first mode solution of equation (6.9) is $ml = 1.875$. The angular frequency at resonance ω_r is given by

$$\omega_r = \frac{(1.875)^2}{l^2} \sqrt{\frac{Ek^2}{\rho}}. \tag{6.10}$$

Figure 6.14. The vibrating reed apparatus.

The resonant frequency f_r is then

$$f_r = \frac{1}{2\pi} \frac{(1.875)^2}{l^2} \sqrt{\frac{E}{\rho}} \frac{h}{2\sqrt{3}}, \tag{6.11}$$

giving

$$E = 38.24 \frac{l^4 \rho}{h^2} f_r^2. \tag{6.12}$$

It may also be shown, mostly readily on the basis of the analogy between the viscoelastic reed and an equivalent electrical circuit[16], that

$$\tan \delta = \Delta f / f_r, \tag{6.13}$$

where Δf is the band width, or the difference between those frequencies for which the amplitude of vibration has $1/\sqrt{2}$ times its maximum value.

The vibrating reed method is most often used for very small polymer samples moulded to give an isotropic specimen. It is, however, equally applicable to oriented specimens, for example, by cutting thin strips from oriented sheets.

Figure 6.14 shows a typical experimental set-up, where the reed is enclosed in a cryostat and measurements can be carried out over a wide temperature range (-150 to $+200\,°C$).

6.4 FORCED VIBRATION NON-RESONANCE METHODS

The free vibration and resonance methods are simplest and most accurate for determining the dynamic mechanical behaviour over a wide range of temperatures. They suffer from the considerable disadvantage that the frequency of measurement depends on the stiffness of the specimen, and as this changes with temperature so does the effective frequency. Thus to determine the frequency and temperature dependence of viscoelastic behaviour, forced vibration non-resonance methods are preferable.

6.4.1 Measurement of Dynamic Extensional Modulus

The principle of this technique is to apply a sinusoidal extensional strain to the specimen which takes the form of a thin strip of film, a fibre monofilament or a multifilament yarn, and simultaneously measure the stress. The viscoelastic behaviour is specified from the relative amplitudes of the stress and the strain, and from the phase shift between these (see Section 5.3.1 above).

There are two limitations to design and measurement respectively which are of some importance. First, the length of the sample must be short enough for there to be no appreciable variation of stress along the sample length, i.e. the length of the fibre must be short compared with the wavelength of the stress waves. Assuming that the lowest value of the modulus to be measured is $10^7\,N\,m^{-2}$ and that the specimen density is $10^3\,kg\,m^{-3}$, the longitudinal wave velocity is $10^2\,m\,s^{-1}$. At a frequency of $100\,Hz$ the wavelength of the stress waves is then 1 m. This suggests an upper limit of approximately 0.1 m length on specimens to be measured at 100 Hz.

Secondly, there is the limit imposed by the stress relaxation time of the material, it being clear that the stress developed in the sample must never vanish.

Several types of apparatus of this general form have been designed[5,21–24].

A block diagram of that due to Takayanagi[22] is shown in Figure 6.15. The specimen, which is usually from 0.1 to 0.3 mm in width and from 40 to 60 mm in length, undergoes an alternating extensional strain due to the electromagnetic vibrator in the frequency range 3.5–100 Hz. The strain and stress are measured by unbonded strain-gauge transducers, the signals from which are fed via additional amplifiers to a phase meter. This provides a direct reading of the relative amplitudes and phase difference of the signals from the two transducers and hence values of the modulus and tan δ. The specimen is enclosed in a thermostatted enclosure so that measurements can be made over the temperature range -170 to $+150\,^{\circ}$C.

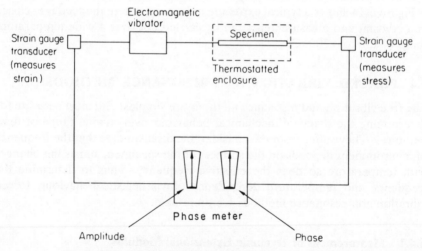

Figure 6.15. Block diagram of forced-vibration apparatus.

The range of measurement has been extended to 10^{-3} Hz by an apparatus which is identical in principle but uses a mechanical drive to produce the alternating strain (range $10–10^{-3}$ Hz) and a Servomex transfer function analyser (TFA) to make simultaneous comparison of stress and strain[24]. A schematic diagram of this apparatus is shown in Figure 6.16.

The servo-controlled motor, in addition to providing the sinusoidal strain by driving the Scotch-yoke mechanism, rotates a loop in a high frequency (2 kHz) alternating magnetic field. Since the loop rotates at a very low frequency (ω_m) this produces a carrier wave modulated at the rotational speed, which can be fed to the system under test. In this application the mechanical

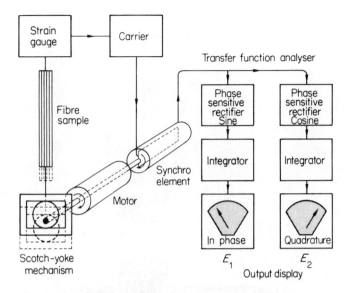

Figure 6.16. Block diagram of TFA dynamic mechanical apparatus. [Redrawn with permission from Pinnock and Ward, *Polymer*, **7**, 255 (1966.]

signal is used, and consequently the return signal presented to the measuring system by the transducer which measures the stress will be of the form $K_1 \sin(\omega_m t + \delta) + N$, where δ is the phase shift (stress leads strain) and N represents unwanted elements such as harmonics, random noise, etc. At this stage a carrier is introduced and the signal is then multiplied simultaneously but separately by $\sin \omega_m t$ and $\cos \omega_m t$. These two signals are subsequently demodulated and integrated over a complete cycle to produce terms which are proportional in magnitude and polarity to the real and imaginary parts of the system output at the fundamental frequency, i.e. to E_1 and E_2.

A photograph of this apparatus is shown in Figure 6.17. The sample is heated or cooled by means of a flow of air or dry nitrogen respectively (temperature range -150 to $+200$ °C). To isolate the sample thermally it is connected to the Scotch yoke and the resistance strain gauge at its two extremities by German silver rods and surrounded by a cylindrical copper can which fits tightly into a thermos tube. The temperature is determined by a series of thermocouples. The air is heated by passing through a hot wire spiral and the nitrogen is cooled by use of a German silver heat-exchanger cryostat. With these arrangements temperatures can be controlled to about ± 1 K.

A fixed alternating strain of 0.5% (± 0.25 mm on a sample length of 100 mm) is typical, being close to the highest strain level at which linear viscoelastic behaviour is observed. The sample is also subjected to a static strain greater than 0.25% to prevent it becoming slack.

126

Figure 6.17. Apparatus for measurement of the dynamic extensional modulus.

6.5 WAVE PROPAGATION METHODS

In the frequency range 10^3–10^4 Hz, the wavelength of the stress waves is the same order of magnitude as the length of sample. A typical stress-wave propagation method[25,26] is shown diagrammatically in Figure 6.18. The polymer is in the form of a monofilament (diameter ~ 0.254 mm) and is attached to a stiff massive diaphragm (say a loudspeaker) with a quartz piezo electric crystal which detects the signal amplitude and phase at a variable distance along its length.

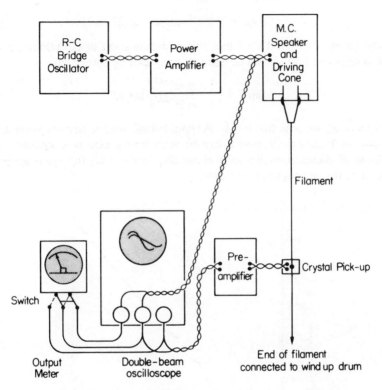

Figure 6.18. The sonic modulus apparatus of Hillier and Kolsky. [Redrawn with permission from Hillier and Kolsky, *Proc. Phys. Soc. B*, **62**, 111 (1949).]

Let the displacement of the diaphragm $u_s = A \sin \omega t$ and the displacement of the pick-up be $u_p = B \sin (\omega t + \theta)$.

There are two quantities to measure; first the ratio B/A and secondly the phase difference θ. In the first instance these can be related to the position of the pick-up l, the reflection coefficient at the pick-up m, the propagation constant $k = \omega/c$ (c = longitudinal wave velocity along the filament), and the damping or attenuation coefficient α.

In the steady state the displacement at any point can be represented by the summation of a single progressive wave travelling down the filament and a single reflected wave from the pick-up travelling back. It is important to note that the filament is assumed to be sufficiently long to ignore reflections from its far end.

The total displacement at any point x is then given by

$$u_T = u_1 + u_2$$

$$= a\ e^{-\alpha x} \sin(\omega t - kx)$$

$$+ ma\ e^{-\alpha(2l-x)} \sin[\omega t - k(2l-x)]. \tag{6.14}$$

Substituting $x = 0$ and $x = l$ into this equation gives us equations for θ and A/B in terms of l, m, k and α, from which it can be shown that

$$\tan\theta = \left\{\frac{1 + m\ e^{-2\alpha l}}{1 - m\ e^{-2\alpha l}}\right\} \tan kl. \tag{6.15}$$

For large αl, we note that $\theta = kl$. A typical result for low density polyethylene is shown in Figure 6.19, and it can be seen that a plot of θ against l takes the form of damped oscillations about the line $\theta = kl$, the slope giving the value of k, the propagation constant.

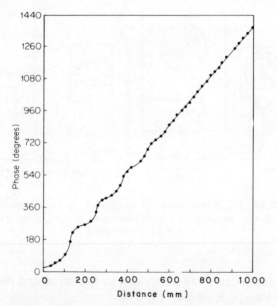

Figure 6.19. The variation of phase angle with distance along a polyethylene monofilament for transmission of sound waves at 3000 Hz. [Redrawn with permission from Hillier and Kolsky, *Proc. Phys. Soc. B*, **62**, 111 (1949).]

For small α the amplitude–distance relation is of the form shown in Figure 6.20. It can be shown[4] that to a good approximation $V_{max}/V_{min} = \tanh(\alpha l + \beta)$, where V is the amplitude of the signal received. Thus a plot of $\tanh^{-1}(V_{max}/V_{min})$ against l gives us a line of slope α, the attenuation constant.

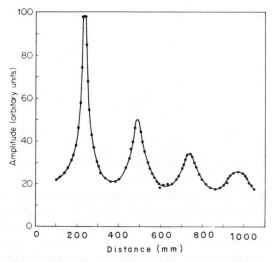

Figure 6.20. The amplitude–distance relationship for transmission of sound waves along a polyethylene monofilament at 3000 Hz. [Redrawn with permission from Hillier and Kolsky, *Proc. Phys. Soc. B*, **62**, 111 (1949).]

The second part of the exercise is to relate the propagation constant k and the attenuation coefficient α to E_1, E_2 and $\tan \delta$. It is instructive physically to introduce the phase velocity of propagation c. Then any displacement can be written as

$$u = u_0 \exp\left[-\alpha x + i\omega\left(t - \frac{x}{c}\right)\right] \qquad (6.16)$$

and strain

$$e = \frac{\partial u}{\partial x} = -\left[\alpha + \frac{i\omega}{c}\right]u. \qquad (6.17)$$

But stress is given by

$$\sigma = (E_1 + iE_2)e, \qquad (6.18)$$

where

$$E_1 = \frac{\sigma_0}{e_0}\cos\delta, \qquad E_2 = \frac{\sigma_0}{e_0}\sin\delta, \qquad \tan\delta = E_2/E_1.$$

Newton's Second Law applied to the extensional motion of a rod parallel to the x axis gives

$$\rho\frac{\partial^2 u}{\partial t^2} = \frac{\partial\sigma}{\partial x} \qquad (6.19)$$

Equation (6.18) is substituted into (6.19) and then (6.16) and (6.17) are used.

130

Equating real and imaginary parts and assuming α is small it is found that

$$E_1 = c^2\rho = \frac{\omega^2}{k^2}\rho \tag{6.20}$$

and

$$\alpha = \frac{\omega}{2c}\frac{E_2}{E_1} = \frac{\omega}{2c}\tan\delta \tag{6.21}$$

or

$$E_2 = \frac{2\alpha c}{\omega}E_1 = \frac{2\alpha c^3\rho}{\omega} \tag{6.22}$$

Figure 6.21 shows the propagation and attenuation constants for polyethylene as a function of frequency at room temperature. The velocity is approximately constant; this shows that E_1 is independent of frequency (equation (6.20)). The attenuation, however, rises approximately linearly with frequency; this shows that $\tan\delta$ is independent of frequency to the same degree of approximation (equation (6.21)).

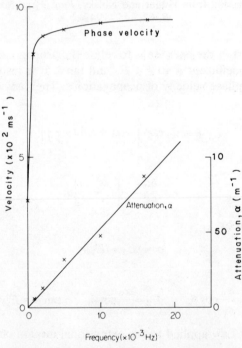

Figure 6.21. Experimental measurements of attenuation and velocity in polyethylene filaments at 10 °C. [Redrawn with permission from Kolsky, *Phil. Mag.*, Eighth Series **1**, 643 (1956).]

There have been many other wave propagation experiments, and the reader is referred to review articles[27-29] for further information.

NOTE ON COMMERCIAL INSTRUMENTS

There have been a number of commercial instruments developed using the methods outlined above. Notable among these are the following:

(1) Torsional Braid Analysis (J. K. Gillham, Department of Chemical Engineering, Princeton University, Princeton, New Jersey, USA).

(2) The Autovibron (Toyo Baldwin Co. Ltd, Tokyo, Japan).

(3) Du Pont 981 Mechanical Analyzer (Du Pont Company, Scientific and Process Instruments Division, Wilmington, Delaware, USA).

(4) Polymer Laboratories Dynamic Mechanical Thermal Analyzer (Polymer Laboratories Ltd, Church Stretton, Salop, UK).

(5) The Dynastat (S. S. Sternstein, Materials Engineering Department, Rennsselaer Polytechnic Institute, Troy, New York, USA).

REFERENCES

1. J. D. Ferry, *Viscoelastic Properties of Polymers*, 2nd edn, Wiley, New York, 1970.
2. N. G. McCrum, B. E. Read and G. Williams, *Anelastic and Dielectric Effects in Polymeric Solids*, Wiley, New York, 1967.
3. H. Kolsky, in *Mechanics and Chemistry of Solid Propellants, Proceedings of the Fourth Symposium on Naval Structural Mechanics*, Pergamon Press, New York, 1966, p. 357.
4. K. W. Hillier, in *Progress in Solid Mechanics* (I. N. Sneddon and R. Hill, eds), North-Holland Publishing Company, Amsterdam, 1961, Chapter 5.
5. G. W. Becker, *Mater. Plast. Elast.*, **35**, 1387 (1969).
6. H. Leaderman, *Elastic and Creep Properties of Filamentous Materials and Other High Polymers*, Textile Foundation, Washington D.C., 1943.
7. H. Leaderman, *Trans. Soc. Rheol.*, **6**, 361 (1962).
8. C. M. R. Dunn, W. H. Mills and S. Turner, *Brit. Plast.*, **37**, 386 (1964).
9. I. M. Ward, *Polymer*, **5**, 59 (1964).
10. N. G. McCrum and E. L. Morris, *Proc. Roy. Soc. A*, **281**, 258 (1964).
11. J. R. McLoughlin, *Rev. Sci. Instr.*, **23**, 459 (1952).
12. R. S. Stein and H. Schaevitz, *Rev. Sci. Instr.*, **19**, 835 (1948).
13. T. L. Smith, *Trans. Soc. Rheol.*, **6**, 61 (1962).
14. K. Schmieder and K. Wolf, *Kolloidzeitschrift*, **127**, 65 (1952).
15. J. Koppelman, *Kolloidzeitschrift*, **144**, 12 (1955).
16. A. W. Nolle, *J. Appl. Phys.*, **19**, 753 (1948).
17. M. Horio and S. Onogi, *J. Appl. Phys.*, **22**, 977 (1951).
18. D. W. Robinson, *J. Sci. Instr.*, **2**, 32 (1955).
19. D. R. Bland and E. H. Lee, *J. Appl. Phys.*, **26**, 1497 (1955).
20. K. W. Hillier, *Proc. Phys. Soc. B*, **64**, 998 (1951).
21. A. B. Thompson and D. W. Woods, *Trans. Faraday Soc.*, **52**, 1383 (1956).
22. M. Takayanagi, in *Proceedings of Fourth International Congress on Rheology*, Part 1, Interscience Publishers, New York, 1965, p. 161.
23. P. R. Pinnock and I. M. Ward, *Proc. Phys. Soc.*, **81**, 261 (1963).
24. P. R. Pinnock and I. M. Ward, *Polymer*, **7**, 255 (1966).

132

25. K. W. Hillier and H. Kolsky, *Proc. Phys. Soc. B*, **62**, 111 (1949).
26. J. W. Ballou and J. C. Smith, *J. Appl. Phys.*, **20**, 493 (1949).
27. H. Kolsky, *Appl. Mech. Rev.*, **11** (9) (1958).
28. H. Kolsky, in *International Symposium on Stress Wave Propagation in Materials*, Interscience Publishers, New York, 1960, p. 59.
29. H. Kolsky, *Structural Mechanics*, Pergamon Press, Oxford, 1960, p. 233.

7

Experimental Studies of the Linear Viscoelastic Behaviour of Polymers

7.1 GENERAL INTRODUCTION

An introduction to the extensive experimental studies of linear viscoelastic behaviour in polymers falls conveniently into three parts, in which amorphous polymers, temperature dependence and crystalline polymers are discussed in turn.

7.1.1 Amorphous Polymers

Most of the earlier investigations of linear viscoelastic behaviour in polymers were confined to studies of amorphous polymers. This is because amorphous polymers show more distinct changes in viscoelastic behaviour with frequency (and as we will shortly discuss, temperature) than crystalline polymers. The most extensive attempt to obtain a complete range of viscoelastic data on an amorphous polymer was in the case of polyisobutylene

$$\left[-CH_2-\underset{\underset{CH_3}{|}}{\overset{\overset{CH_3}{|}}{C}}- \right]_n.$$

R. S. Marvin of the National Bureau of Standards, Washington, collected and analysed results from many laboratories to obtain the complex shear modulus and complex shear compliance of a high molecular weight sample of this polymer over a very wide range of frequencies[1]. The results are shown in Figure 7.1(a) and (b), where in addition to the experimental points, calculated curves are shown based on the phenomenological theories of Marvin and Oser[2].

These figures show clearly the four characteristic regions for amorphous high polymers, i.e. the glassy, viscoelastic, rubbery and flow regions. It can be seen that the viscoelastic behaviour is not very dissimilar from that of the standard linear solid. The complex modulus at high frequencies is approximately constant at a value of about $10^9 \, N \, m^{-2}$, and decreases through the viscoelastic range to a value of $10^5 \, N \, m^{-2}$. In this high molecular weight sample it remains approximately constant at this value of $10^5 \, N \, m^{-2}$ over the

134

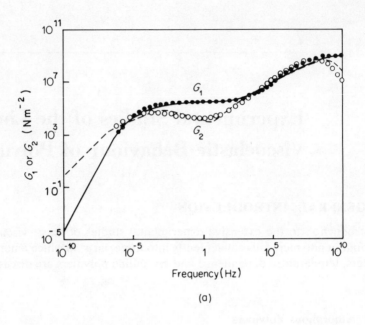

(a)

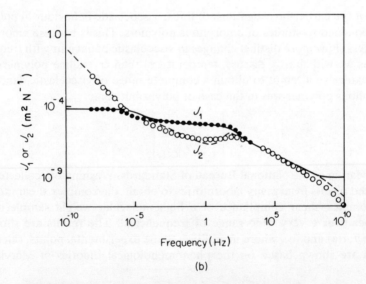

(b)

Figure 7.1. (a) Complex shear modulus, and (b) complex shear compliance for 'standard' polyisobutylene reduced to 25 °C. Points from averaged experimental measurements; curves from a theoretical model for viscoelastic behaviour. [Redrawn with permission from Marvin and Oser, *J. Res. Natl Bur. Stand. B,* **66,** 171 (1962).]

frequency range 10^2–10^{-2} Hz, and molecular flow only sets in appreciably at frequencies below 10^{-5} Hz.

The relaxation spectrum and retardation spectrum for the polymer are shown in Figure 7.2. The relaxation spectrum has the typical 'wedge and box' form to which it approximates in the manner indicated by the dotted lines. This approximation has been found useful for purposes of numerical evaluation[3].

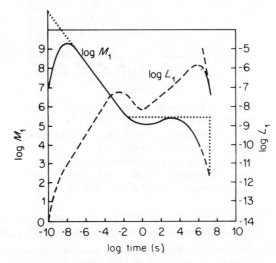

Figure 7.2. Approximate distribution functions of relaxation (M_1) and retardation (L_1) times for polyisobutylene. [Redrawn with permission from Marvin, in *Proceedings of the 2nd International Congress of Rheology*, Butterworth, London, 1954.]

For amorphous polymers large changes in viscoelastic behaviour may be brought about by the presence or absence of chemical cross-links or by changing the molecular weight which controls the degree of molecular entanglement or physical cross-linking.

The influence of chemical or physical cross-links is twofold. First, chemical cross-links prevent irreversible molecular flow at low frequencies (or, as will be discussed shortly, high temperatures) and thereby produce the rubbery plateau region of modulus or compliance. Physical cross-links due to entanglements will restrict molecular flow by causing the formation of temporary networks. At long times such physical entanglements are usually labile and lead to some irreversible flow.

Secondly, the value of the modulus in the plateau region is directly related to the number of effective cross-links per unit volume; this follows from the molecular theory of rubber elasticity (Section 4.1.2 above).

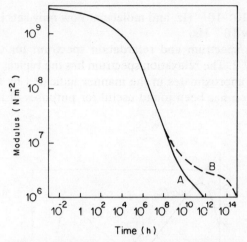

Figure 7.3. Master stress–relaxation curves for low molecular weight (molecular weight 1.5×10^5 daltons), curve A, and high molecular weight (molecular weight 3.6×10^6 daltons), curve B, polymethyl methacrylate. [Redrawn with permission from McLoughlin and Tobolsky, *J. Colloid Sci.*, **7**, 555 (1952).]

The influence of molecular entanglements is illustrated by Figure 7.3, which shows the stress relaxation behaviour for two samples of polymethyl methacrylate. It is seen that the lower molecular weight sample does not show a rubbery plateau region of modulus but passes directly from the viscoelastic region to the region of permanent flow.

7.1.2 Temperature Dependence of Viscoelastic Behaviour

Previously we have only referred indirectly to the effect of temperature on viscoelastic behaviour. From a practical viewpoint, however, the temperature dependence of polymer properties is of paramount importance because plastics and rubbers show very large changes in properties with changing temperature.

In purely scientific terms, the temperature dependence has two primary points of interest. In the first place, as we have seen in Chapter 6, it is not possible to obtain from a single experimental technique a complete range of measuring frequencies to evaluate the relaxation spectrum at a single temperature. It is therefore a matter of considerable experimental convenience to change the temperature of the experiment, and so bring the relaxation processes of interest within a time scale which is readily available. This procedure, of course, assumes that a simple interrelation exists between time scale and temperature, and we will discuss shortly the extent to which this assumption is justified.

Secondly, there is the question of obtaining a molecular interpretation of the viscoelastic behaviour. In most general terms polymers change from glass-like to rubber-like behaviour as either the temperature is raised or the time scale of the experiment is increased. In the glassy state at low temperatures we would expect the stiffness to relate to changes in the stored elastic energy on deformation which are associated with small displacements of the molecules from their equilibrium positions. In the rubbery state at high temperatures, on the other hand, the molecular chains have considerable flexibility, so that in the undeformed state they can adopt conformations which lead to maximum entropy (or more strictly, minimum free energy). The rubber-like elastic deformations are then associated with changes in the molecular conformations.

The molecular physicist is interested in understanding how this conformational freedom is achieved in terms of molecular motions, e.g. to establish which bonds in the structure become able to rotate as the temperature is raised. One approach, which has proved successful to some degree, has been to compare the viscoelastic behaviour with dielectric relaxation behaviour and more particularly with nuclear magnetic resonance behaviour.

We have tacitly assumed that there is only one viscoelastic transition, corresponding to the change from the glassy low temperature state to the rubbery state. In practice there are several relaxation transitions. For a typical

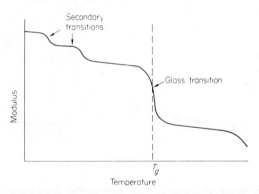

Figure 7.4. Temperature dependence of modulus in a typical polymer.

amorphous polymer the situation is summarized in Figure 7.4. At low temperatures there are usually several secondary transitions involving comparatively small changes in modulus. These transitions are attributable to such features as side-group motions, e.g. methyl ($-CH_3$) groups in polypropylene

$$\left[\begin{array}{c} CH_3 \\ | \\ CH_2-CH- \end{array} \right]_n .$$

In addition, there is one primary transition which involves a large change in modulus. This is conveniently called the 'glass transition', and the temperature at which it occurs is commonly denoted by T_g.

138

7.1.3 Crystalline Polymers

The viscoelastic behaviour of crystalline polymers is markedly different from that of amorphous polymers. The four characteristic regions, although they still exist, are not so clearly defined. This is illustrated by the data obtained by Schmieder and Wolf[4] for polychlorotrifluorethylene $[CClF-CF_2-]_n$ and polyvinyl fluoride $[-CH_2CHF-]_n$ which are shown in Figure 7.5.

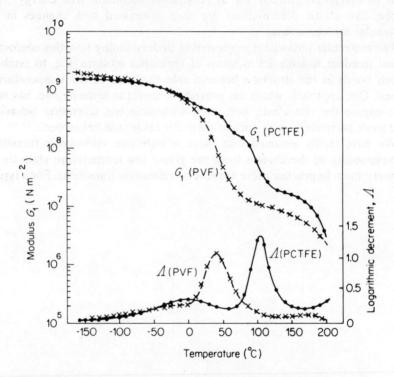

Figure 7.5. Shear modulus G_1 and logarithmic decrement Λ of poly-chlorotrifluoroethylene (PCTFE) and polyvinyl fluoride (PVF), as a function of temperature at ~3 Hz. [Redrawn with permission from Schmieder and Wolf, *Kolloidzeitschrift*, **134**, 149 (1953).]

For crystalline polymers, the fall in modulus over the glass transition region is much smaller, generally only involving a fall from $10^9 \, N \, m^{-2}$ to 10^8 or $10^7 \, N \, m^{-2}$. Also, the change in modulus or loss factor with frequency or temperature is much more gradual, giving a much broader relaxation time spectrum.

At high temperatures the molecular mobility is severely curtailed by the crystalline regions so that it is no longer correct to regard the polymer at low frequencies and/or high temperatures as simply rubber-like. These differences

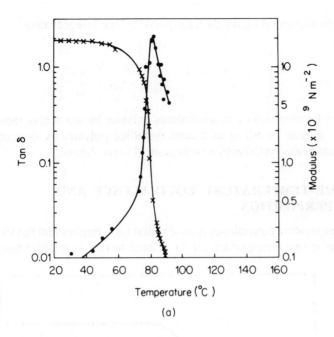

(a)

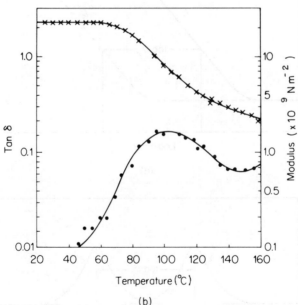

(b)

Figure 7.6. Tensile modulus and loss factor tan δ for (a)
unoriented amorphous polyethylene terephthalate and (b)
unoriented crystalline polyethylene terephthalate as a function of temperature at ~1.2 Hz. [Redrawn with permission
from Thompson and Woods, *Trans. Faraday Soc.*, **52**, 1383
(1956).]

are clearly illustrated by the data for polyethylene terephthalate[5]

$$\left[-O-CH_2-CH_2-O-C-\underset{O}{\overset{O}{\parallel}}-\underset{}{\bigcirc}-C-\right]_n$$

which can be obtained as an amorphous polymer by quenching rapidly from the melt (Figure 7.6(a)) or as a semicrystalline polymer by slow cooling or subsequent heat-crystallization treatments (Figure 7.6(b)).

7.2 TIME–TEMPERATURE EQUIVALENCE AND SUPERPOSITION

Time–temperature equivalence in its simplest form implies that the viscoelastic behaviour at one temperature can be related to that at another temperature

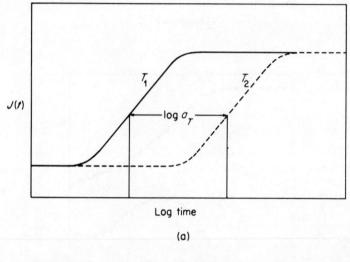

(a)

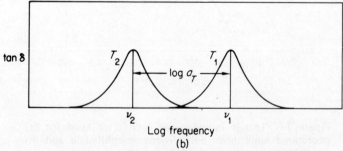

(b)

Figure 7.7. Schematic diagrams illustrating the simplest form of time–temperature equivalence for (a) compliance $J(t)$ and (b) loss factor tan δ.

by a change in the time scale only. More sophisticated schemes to be described shortly also allow for changes in the magnitude of the response scale (e.g. compliance in a creep experiment) with temperature.

To fix our ideas, consider the creep compliance curves of an idealized polymer at two temperatures T_1 and T_2, the time scale being logarithmic (Figure 7.7(a)). On the simplest scheme for time–temperature equivalence these two creep compliance curves can be superimposed exactly by a horizontal displacement $\log a_T$. This defines a shift factor a_T.

Consider also the dynamic mechanical behaviour of the same polymer, shown by the determination of $\tan \delta$ at temperatures T_1 and T_2 (Figure 7.7(b)). The simplest proposition is that these can also be superimposed by the same shift factor a_T.

In many studies of the temperature dependence of viscoelastic behaviour in polymers, particularly those involving dynamic mechanical measurements, the shift factors have been calculated on this simple basis. It has, however, been pointed out (notably by McCrum and his coworkers[6]) that it may be necessary to take into account changes in the relaxed and unrelaxed compliances with temperature. The situation is illustrated schematically in Figure 7.8. When we compare the creep compliance curves at the two temperatures T_1 and T_2 we see that the relaxed and unrelaxed compliances are both changing with temperature. McCrum and Morris[6] propose a scaling procedure for obtaining a modified or 'reduced' compliance curve at the temperature T_1, to give the dashed curve $J_\rho^{T_1}(t)$ in Figure 7.8. The shift factor is now obtained by a horizontal shift of $J_\rho^{T_1}(t)$ to superimpose it on $J^{T_2}(t)$.

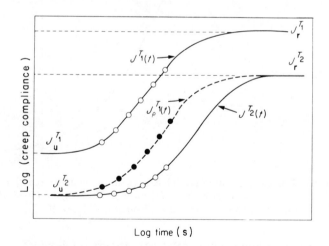

Figure 7.8. Schematic diagram illustrating McCrum's reduction procedure for superposition of creep data. $J_r^{T_1}$ and $J_u^{T_1}$ are the relaxed and unrelaxed compliances respectively at the temperature T_1; $J_r^{T_2}$ and $J_u^{T_2}$ are the corresponding quantities at the temperature T_2.

In many cases it is adequate to assume that $J_r^{T_1}/J_r^{T_2} = J_u^{T_1}/J_u^{T_2}$. This is true for the glass transition relaxation of an amorphous polymer (see Section 7.4.1 below) and is a very good approximation for some other relaxations. The correction for changes in the relaxed and unrelaxed compliances with temperature is then a vertical shift of the compliance curves (plotted on a logarithmic scale).

7.3 TRANSITION STATE THEORIES

The simplest theories which attempt to deal with the temperature dependence of viscoelastic behaviour are the transition state or barrier theories. The transition state theory of time-dependent processes stems from the theory of chemical reactions and is associated with the names of Eyring, Glasstone and others[7]. The basic idea is that for two molecules to react they must first form an *activated complex* or *transition state*, which then decomposes to give the final products of the reaction.

The potential energy diagram for two reacting molecules is closely analogous to that for the internal rotation of molecules discussed in Section 1.2.1 above. Figure 7.9 shows the change in potential energy for the reaction between an atom A and a diatomic molecule BC which results in the formation of a diatomic molecule AB and an atom C. The intermediate step is the formation of the activated complex A–B–C.

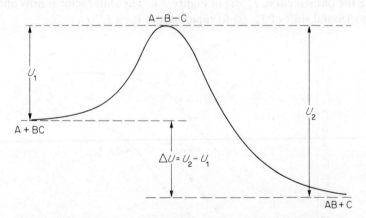

Figure 7.9. Change of potential energy in a chemical reaction. [Redrawn with permission from Glasstone, *Textbook of Physical Chemistry*, 2nd edn, Macmillan, London, 1953, p. 1092.]

The theory of absolute reaction rates now argues as follows. The activated complex can be treated by statistical methods as a normal molecule, except that in addition to having three translational degrees of freedom, it has a fourth degree of freedom of movement along what is termed the reaction coordinate. The reaction coordinate is the direction leading to the lower potential energy of the final reactants. The theory shows that the rate of reaction is the product of two quantities, the probability of forming the activated complex and the effective rate of crossing the energy barrier by the activated complexes. It can be shown that the effective rate of crossing the energy barrier, which is the low frequency vibration of the activated complex in the direction of the reaction coordinate, is equal to kT/h. This is a universal frequency whose value is dependent only on the temperature and is independent of the nature of the reactants and the type of reaction (k is Boltzmann's constant and h is Planck's constant). Because there is equilibrium between the initial reacting species and the transition state, the probability of forming the activated complex is determined in absolute terms by the Boltzmann factor $e^{-\Delta G/RT}$, where ΔG is the free energy difference per mole between the system when the reactants are relatively far from each other and when they form the activated complex. For reactions occurring under constant pressure conditions ΔG is the Gibbs free energy difference per mole.

We now argue that, by analogy, the frequency of molecular jumps between two rotational isomeric states of a molecule (Section 1.2.1 above) is given by

$$\nu = \frac{kT}{h} e^{-\Delta G/RT}, \tag{7.1}$$

where ΔG is the Gibbs free energy barrier height per mole.

This equation states that the frequency of molecular conformational changes depends on the *barrier height* and not on the free energy difference between the equilibrium sites. Equation (7.1) may be written as

$$\nu = \frac{kT}{h} e^{\Delta S/R} e^{-\Delta H/RT} = \nu_0 e^{-\Delta H/RT}. \tag{7.2}$$

This form of equation (7.1) emphasizes the way in which temperature affects ν primarily through the activation energy ΔH. To a good approximation the activation energy for the process (actually an enthalpy) is thus given by

$$\Delta H = -R \left[\frac{\partial (\ln \nu)}{\partial (1/T)} \right]_P \tag{7.3}$$

Equation (7.2) is known as the 'Arrhenius equation', because it was first shown by Arrhenius[9] that it describes the influence of temperature on the velocity of chemical reactions.

We now take the intuitive step (to be justified below by the site model theory) that the viscoelastic behaviour can be directly related to a controlling molecular rate process with a constant activation energy.

Consider the tan δ curves of Figure 7.7(b). At the temperatures T_1 and T_2 the peak value of tan δ occurs at frequencies ν_1 and ν_2 respectively. The assumption is that ν_1 and ν_2 are related by the equation

$$\frac{\nu_1}{\nu_2} = \frac{e^{-\Delta H/RT_1}}{e^{-\Delta H/RT_2}},$$

144

i.e.

$$\log \frac{\nu_1}{\nu_2} = \log a_T = \frac{\Delta H}{R} \left\{ \frac{1}{T_2} - \frac{1}{T_1} \right\}. \tag{7.4}$$

The activation energy for the process can therefore be obtained from a plot of $\log a_T$ against the reciprocal of the absolute temperature. For large values of ΔH, changes in temperature give very large changes in frequency. Dynamic mechanical data on polymers are often dealt with in terms of the Arrhenius equation and a constant activation energy. In some cases this can be regarded as only an approximate treatment due to the limited range of experimental frequencies available. In general it has been found that the temperature dependence of the glass transition relaxation behaviour of amorphous and crystalline polymers does not fit a constant activation energy, in contrast to more localized molecular relaxations.

7.3.1 The Site Model Theory

The site model theory is based on transition state theory, and although first developed to explain the dielectric behaviour of crystalline solids[10,11] has also been applied to mechanical relaxations in polymers[12].

In its simplest form there are two sites, separated by an equilibrium free energy difference $\Delta G_1 - \Delta G_2$, the barrier heights being ΔG_1 and ΔG_2 per mole, respectively (Figure 7.10).

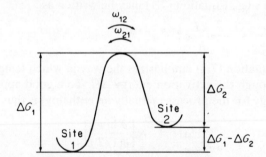

Figure 7.10. The two-site model.

The transition probability for a jump from site 1 to site 2 is given by

$$\omega_{12}^0 = A' e^{-\Delta G_1/RT} \tag{7.5}$$

and for a jump from site 2 to site 1 by

$$\omega_{21}^0 = A' e^{-\Delta G_2/RT}, \tag{7.6}$$

where A' is a constant.

(In some treatments the change in molecular conformation is imagined to be a simple rotation of 180° around one bond. It is then considered that the transition probability is $2\omega_{12}^0$ where ω_{12}^0 is the probability for a jump in either a clockwise or anticlockwise direction.)

To give rise to a mechanical relaxation process, the energy difference between the two sites must be changed by the application of the applied stress. There is then a change in the populations of site 1 and site 2, and it is assumed that this relates directly to the strain. It is not difficult to imagine how this might arise at a molecular level if, for example, the uncoiling of a molecular chain involved internal rotations. Locally, the chain conformations could be changing from crumpled *gauche* conformations to extended *trans* conformations (see Section 1.2.1).

Assume that the applied stress σ causes a small linear shift in the free energies of the sites such that

$$\delta G_1' = \lambda_1 \sigma \tag{7.7}$$

and

$$\delta G_2' = \lambda_2 \sigma \tag{7.8}$$

for sites 1 and 2 respectively, where λ_1 and λ_2 are constants with the dimensions of volume. The transition probabilities ω_{12} and ω_{21} in the presence of the applied stress are then given by

$$\omega_{12} \simeq \omega_{12}^0 \left[1 - \frac{\delta G_1'}{RT}\right] = \omega_{12}^0 \left[1 - \frac{\lambda_1 \sigma}{RT}\right], \tag{7.9}$$

where ω_{12}^0 is the transition probability in the absence of the stress. Similarly

$$\omega_{21} \simeq \omega_{21}^0 \left[1 - \frac{\lambda_2 \sigma}{RT}\right]. \tag{7.10}$$

The rate equations for sites 1 and 2 are then

$$\frac{dN_1}{dt} = -N_1 \omega_{12} + N_2 \omega_{21}, \tag{7.11}$$

$$\frac{dN_2}{dt} = -N_2 \omega_{21} + N_1 \omega_{12}, \tag{7.12}$$

where we can write the occupation number N_1 of state 1 as $N_1 = N_1^0 + n$ and similarly $N_2 = N_2^0 - n$, where N_1^0 and N_2^0 are the occupation numbers at zero stress, $N_1^0 + N_2^0 = N_1 + N_2 = N$.

Combining these equations and making suitable approximations gives a rate equation

$$\frac{dn}{dt} + n(\omega_{12}^0 + \omega_{21}^0) = N_1^0 \omega_{12}^0 \left[\frac{\lambda_1 - \lambda_2}{RT}\right] \sigma, \tag{7.13}$$

which describes the change in the site population n as a function of time. Assuming that this change in site population is directly related to the observed strain e, e is given by

$$e = e_u + n\bar{e}. \tag{7.14}$$

In this equation e_u is the instantaneous or unrelaxed elastic deformation and it is considered that each change in site population produces a proportionate change in strain by an amount $\bar{e}$.

Equation (7.13) can then be seen to have the form

$$\frac{de}{dt} + Be = C,$$

where B and C are constants. This is formally identical to the equation of a Voigt element, with a characteristic retardation time given by $\tau' = 1/B$ which is

$$\tau' = \frac{1}{(\omega_{12}^0 + \omega_{21}^0)} = \frac{e^{\Delta G_2/RT}}{A'[\exp\{-(\Delta G_1 - \Delta G_2)/RT\} + 1]}. \qquad (7.15)$$

Since RT is usually small compared with the equilibrium free energy difference we may approximate to

$$\tau' = \frac{1}{A'} e^{\Delta G_2/RT}. \qquad (7.16)$$

Equation (7.16) is formally equivalent to equation (7.1).

It follows that the time–temperature behaviour of the relaxation process is governed by the unperturbed transition probabilities, and to a good approximation by ΔG_2, a free energy of activation. The *magnitude* of the relaxation[12,13], on the other hand, is proportional to

$$p\left[\frac{\exp\left[-(\Delta G_1 - \Delta G_2)/RT\right]}{(1 + \exp\left[-(\Delta G_1 - \Delta G_2)/RT\right])^2}\right]\frac{(\lambda_1 - \lambda_2)^2}{RT},$$

where p is the number of species per unit volume. Thus the intensity of the relaxation on this model is low at both high and low temperatures and passes through a maximum when the free energy difference $(\Delta G_1 - \Delta G_2)$ and RT are of the same order of magnitude.

The site model is applicable to relaxation processes showing a constant activation energy, e.g. those associated with localized motions in the crystalline regions of semicrystalline polymers.

7.4. THE TIME–TEMPERATURE EQUIVALENCE OF THE GLASS TRANSITION VISCOELASTIC BEHAVIOUR IN AMORPHOUS POLYMERS AND THE WILLIAMS, LANDEL AND FERRY (WLF) EQUATION

In considering time–temperature equivalence of the glass transition behaviour in amorphous polymers, we will follow a treatment very close to that given by Ferry[14]. To fix our ideas, consider the storage compliance J_1 of an amorphous polymer (poly-n-octyl methacrylate) as a function of temperature and frequency (Figure 7.11). It can be seen that there is an overall change in the

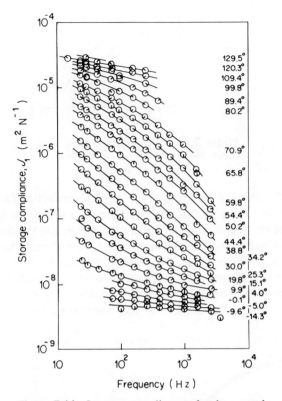

Figure 7.11. Storage compliance of poly-*n*-octyl methacrylate in the glass transition region plotted against frequency at 24 temperatures as indicated. [Redrawn from Ferry, *Viscoelastic Properties of Polymers* 1st Edn, Wiley, New York, 1961, Chap. 11.]

shape of the compliance–frequency curve as the temperature changes. At high temperatures there is an approximately constant high compliance, the rubbery compliance. At low temperatures the compliance is again approximately constant but at a low value, the glassy compliance. At intermediate temperatures there is the frequency-dependent viscoelastic compliance.

The simplest way of applying time–temperature equivalence is to produce a 'master compliance curve' by choosing one particular temperature and applying only a horizontal shift on a logarithmic time scale to make the compliance curves for other temperatures join as smoothly as possible to the curve at this particular temperature. This simple procedure is very nearly, but not quite, the procedure adopted by Ferry and his coworkers. The molecular theories of viscoelasticity suggest that there should be an additional small vertical shift factor $T_0\rho_0/T\rho$ in changing from the actual temperature T in kelvins (at a density ρ) to the reference temperature T_0 in kelvins (at a

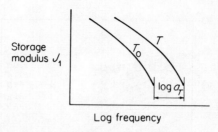

Figure 7.12. Diagram illustrating shift factor $\log a_T$ for change in temperature T to T_0.

density ρ_0). The physical meaning of this vertical correction factor is that the molecular theories suggest that the equilibrium modulus changes with temperature in the transition range in a manner according to the theory of rubber elasticity (see Chapter 4). This is quite distinct from changes in the molecular relaxation times, which affect the measured modulus at a given time or frequency due to affecting the viscoelastic behaviour. In practice, the correction factor has a very small effect in the viscoelastic range of temperatures compared with the large changes in the viscoelastic behaviour. Thus it is

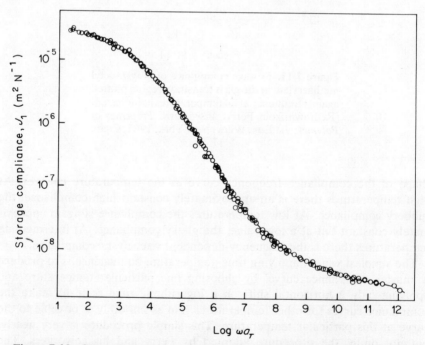

Figure 7.13. Composite curve obtained by plotting the data of Figure 7.11 with suitable shift factors, giving the behaviours over an extended frequency scale at temperature T_0. [Redrawn from Ferry, *Viscoelastic Properties of Polymers*, 1st Edn, Wiley, New York, 1961, Chap. 11.]

usually adequate to apply a simple horizontal shift on the time scale only (see Figure 7.12).

This procedure gives the storage compliance as a function of frequency over a very wide range of frequencies, as shown in Figure 7.13. Thus it is now possible to calculate the retardation time spectrum, and compare this with any theoretical models which may be proposed.

We may also consider the significance of the horizontal shift on the logarithmic time scale, as shown in Figure 7.14.

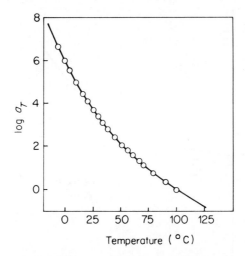

Figure 7.14. Temperature dependence of the shift factor a_T used in plotting Figure 7.13. Points, chosen empirically; curve is WLF equations with a suitable choice of T_g (or T_s). [Redrawn from Ferry, *Viscoelastic Properties of Polymers*, 1st Edn, Wiley, New York, 1961, Chap. 11.]

The remarkable observation, which was established largely by the work of Williams, Landel and Ferry, is that for all amorphous polymers this shift factor–temperature relationship is *approximately* identical.

It was found that the relationship

$$\log a_T = \frac{C_1(T - T_s)}{C_2 + (T - T_s)},$$

where C_1 and C_2 are constants, and T_s is a reference temperature peculiar to a particular polymer, holds extremely well over the temperature range $T = T_s \pm 50\,°C$ for all amorphous polymers. This equation, known as the 'WLF equation'[15] (we shall see that there are other forms for the WLF equation), was originally considered to be only an empirical equation and the constants

C_1 and C_2 were originally determined by arbitrarily choosing $T_s = 243$ K for polyisobutylene.

Following this empirical discovery, there was naturally some speculation as to whether the WLF equation has a more fundamental interpretation. This brings us to considerations of the dilatometric glass transition and to discussion of the use of the concept of free volume.

The glass transition can be defined on the basis of dilatometric measurements. As shown in Figure 7.15, if the specific volume of the polymer is

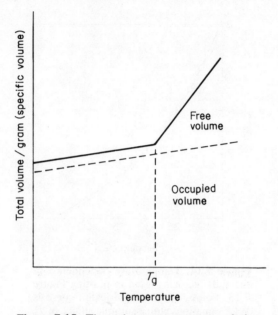

Figure 7.15. The volume–temperature relationship for a typical amorphous polymer.

measured against temperature, a change of slope is observed at a characteristic temperature, which we may call T_g. In the first place this change in slope may be somewhat less sharp than this diagram suggests. Secondly, it is known that if dilatometric measurements are carried out at very slow rates of temperature change, one approaches a roughly constant value for the glass transition temperature T_g. The value of T_g will vary by only 2–3 K when the heating rate is decreased from 1 K min^{-1} to 1 K per day. Thus it appears possible to define a rate-independent value of T_g to at least a very good approximation.

It has subsequently been shown that the original WLF equation can be rewritten in terms of this dilatometric transition temperature such that

$$\log a_T = \frac{C_1^g (T - T_g)}{C_2^g + (T - T_g)},$$

where C_1^g and C_2^g are new constants and $T_g = T_s - 50\,°\text{C}$.

Moreover it is now possible to give a plausible theoretical basis to the WLF equation in terms of the concept of free volume[15].

In liquids, the concept of free volume has proved useful in discussing transport properties such as viscosity and diffusion. These properties are considered to relate to the difference $v_f = v - v_0$, where v is the total macroscopic volume, v_0 is the actual molecular volume of the liquid molecules, the 'occupied volume', and v_f is the proportion of holes or voids, the 'free volume'.

Figure 7.15 shows the schematic division of the total volume of the polymer into both occupied and free volumes. It is argued that the occupied volume increases uniformly with temperature. The discontinuity in the expansion coefficient at T_g then corresponds to a sudden onset of expansion in the free volume. This suggests that certain molecular processes which control the viscoelastic behaviour commence at T_g, and not merely that T_g is the temperature when their time scale becomes comparable with that of the measuring time scale. This would seem to imply that T_g is a genuine thermodynamic temperature. This point is not, however, completely resolved, and it has been shown by Kovacs[16] that the T_g measured dilatometrically is still sensibly dependent on the time scale, i.e. the rate of heating. However, as already mentioned, this time dependence is small. Thus to a good approximation it can be assumed that the free volume is constant up to T_g and then increases linearly with increasing temperature.

The fractional free volume $f = v_f/v$ can therefore be written as

$$f = f_g + \alpha_f(T - T_g), \tag{7.17}$$

where f_g is the fractional free volume at the glass transition T_g and α_f is the coefficient of expansion of the free volume.

The WLF equation can now be obtained in a simple manner. The model representations of linear viscoelastic behaviour all show that the relaxation times are given by expressions of the form $\tau = \eta/E$ (see the Maxwell model in Section 5.2.5 above), where η is the viscosity of a dashpot and E the modulus of a spring.

If we ignore the changes in the modulus E with temperature compared with changes in the viscosity η, this suggests that the shift factor a_T for changing temperature from T_g to T will be given by

$$a_T = \frac{\eta_T}{\eta_{T_g}} \tag{7.18}$$

At this juncture, we introduce Doolittle's viscosity equation[17], which relates the viscosity to the free volume. This equation is based on experimental data for monomeric liquids and gives

$$\eta = a \exp(bv/v_f), \tag{7.19}$$

where a and b are constants. Using (7.18) and (7.19) it can be shown that the Doolittle equation becomes

$$\ln a_T = b\left\{\frac{1}{f} - \frac{1}{f_g}\right\}. \tag{7.20}$$

Substituting $f = f_g + \alpha_f(T - T_g)$ we have

$$\log a_T = -\frac{(b/2.303 f_g)(T - T_g)}{f_g/\alpha_f + T - T_g}, \tag{7.21}$$

which is the WLF equation.

Ferry and his coworkers have given further consideration to the exact form of the WLF equation. It can be shown that a better fit to data for different polymers can be obtained by changing the constants C_1^g and C_2^g; and that the actual values obtained for C_1^g and C_2^g yield values for f_g and α_f which are plausible on physical grounds. The reader is referred to Ferry's book[14] for detailed discussion of these points. We will, however, note here that the fractional free volume at the glass transition temperature f_g is 0.025 ± 0.003 for most amorphous polymers. The thermal coefficient of expansion of free volume α_f is a more variable quantity, but has the physically reasonable 'universal' average value of $4.8 \times 10^{-4} \, \text{K}^{-1}$.

It is of some interest to complete our discussion of the WLF equation by indicating the lines of its derivation by Bueche[18] using a transition state model.

It is possible to develop the transition state theory on the basis of free volume by expressing the frequency ν of the controlling molecular process by the equation

$$\nu = A \int_{f_c}^{\infty} \phi(f) \, df.$$

It is assumed that the required unit of structure can move when the local fractional free volume f exceeds some critical value f_c.

Bueche evaluated $\phi(f)$, and showed that with some approximations

$$\nu = \nu_g \exp\left\{-Nf_c\left[\frac{1}{f} - \frac{1}{f_g}\right]\right\}, \tag{7.22}$$

where ν_g is the frequency at T_g. If $f = f_g + \alpha_f(T - T_g)$ it may be shown that

$$\ln \frac{\nu}{\nu_g} = \frac{(Nf_c/f_g)(T - T_g)}{T - T_g + f_g/f}. \tag{7.23}$$

Assuming that there is a direct link between the shift factor a_T and the ratio of the frequencies of the controlling molecular process, equation (7.23) is identical in form to equation (7.21). Note also that equation (7.22) is Bueche's analogy to the Doolittle equation, equation (7.20).

In conclusion we observe that for time–temperature equivalence to be exact, a necessary simplicity is implied. At a molecular level, the individual relaxation times for molecular processes must shift uniformly with temperature. In phenomenological terms the spectrum of relaxation times must shift as a unit on a logarithmic time scale to shorter times with increasing temperature.

Staverman and Schwarzl[19] call these materials thermorheologically simple, and Lee and his collaborators[20] have worked out the theoretical consequences of this assumption, so that complex problems concerning the deformation of viscoelastic solids in variable temperature situations can be solved.

7.4.1 The Williams, Landel and Ferry Equation, the Free Volume Theory and Other Related Theories

The WLF equation gives the shift factor for time–temperature superposition as

$$\log a_T = \frac{C_1^g (T - T_g)}{C_2^g + (T - T_g)}.$$

We have seen that this relationship can be regarded as describing the change in the internal viscosity of the polymer as we change the temperature from the glass transition temperature T_g to the test temperature T (equation (7.18)).

We may therefore write the WLF equation as

$$\log \eta_T = \log \eta_{T_g} + \frac{C_1^g (T - T_g)}{C_2^g + (T - T_g)},$$

where η_T, η_{T_g} is the viscosity of the polymer at temperatures T, T_g respectively. In this form the equation implies that at a temperature $T = T_g - C_2^g$ (i.e. $T = T_g - 51.6$ for the WLF equation in its universal form) the viscosity of the polymer is infinite.

This has led to the view that the WLF equation should be related at molecular level to the temperature $T = T_g - 51.6$, which we will call T_2, rather than to the dilatometric glass transition T_g.

There have been two basic approaches along these lines:

(1) The free volume theory is modified so that the changes in free volume with temperature relate to a discontinuity which occurs at T_2 rather than T_g. This is discussed in Section 7.4.2.

(2) It is considered that T_2 represents a true thermodynamic transition temperature. A modified transition state theory is developed in which the frequency of molecular jumps relates to the cooperative movement of a group of segments of the chain. The number of segments acting cooperatively is then calculated from statistical thermodynamic considerations. This is the theory of Adam and Gibbs[23], which is described in Section 7.4.3 below.

7.4.2 The Free Volume Theory of Cohen and Turnbull

Cohen and Turnbull[24] have proposed that the free volume v_f corresponds to that part of the excess volume $v - v_0$ (v = total measured specific volume, v_0

= occupied volume as in Section 7.4) which can be redistributed without a change in energy. It is then assumed on the basis of arguments concerning the nature of the 'cage' formed round a molecule by its neighbours that the redistribution can take place without a change in energy at temperatures above a critical temperature, which is to be identified with T_2, where the cage reaches a critical size.

Thus

$$v_f = 0 \quad \text{for } T < T_2$$

and

$$v_f = \alpha \bar{v}_m (T - T_2) \quad \text{for } T \geq T_2$$

where α is the average expansion coefficient and $\bar{v}_m$ the average value of the molecular volume v_0 in the temperature range T_2 to T.

For a viscosity equation of the form

$$\eta = a \exp (bv/v_f) \tag{7.19}$$

this gives

$$\eta = a \exp \frac{B'}{T - T_2},$$

where B' is a constant, and correspondingly for the average relaxation time

$$\tau = \tau_0 \exp \frac{B'}{T - T_2}.$$

There is much experimental evidence from dielectric relaxation for the validity of this equation for amorphous polymers. As we have discussed, putting $T_2 = T_g - 51.6$ gives us the WLF equation and the relaxation time will become infinitely long as we approach T_2 due to the disappearance of free volume.

7.4.3 The Statistical Thermodynamic Theory of Adam and Gibbs

Gibbs and Di Marzio[21,22] proposed that the dilatometric T_g is a manifestation of a true equilibrium second-order transition at the temperature T_2. In a further development, Adams and Gibbs[23] have shown how the WLF equation can then be derived. On their theory the frequency of molecular jumps is given by

$$v_c = A \exp - \frac{n \, \Delta G^*}{kT}, \tag{7.24}$$

where A is a constant ($A = kT/h$ on the transition state theory), ΔG^* is the free energy difference hindering rearrangement *per segment* (the barrier height) and n is the number of segments acting cooperatively as a unit to make a configurational rearrangement.

The essence of the Adam and Gibbs theory is that n can be calculated on thermodynamic equilibrium grounds as follows:

If S is the configurational entropy of the system, i.e. the entropy for a mole of segments,

$$S = \frac{N_A}{n} s_n, \tag{7.25}$$

where N_A is Avogadro's number and s_n is the entropy of a unit of n segments. Thus

$$n = \frac{N_A s_n}{S} \quad \text{and} \quad \nu_c = A \exp \frac{-N_A s_n \, \Delta G^*}{SkT}. \tag{7.26}$$

It is assumed that s_n is independent of temperature and that S, the configurational entropy of the system, can be calculated directly for any temperature from the specific heat at constant pressure.

A further assumption is that $S = 0$ at the thermodynamic transition temperature T_2. In molecular terms n becomes infinite and there are no configurations available into which the system may rearrange. We may note that although the entropy S is assumed to be zero at T_2 this is not necessarily (or ever, in practice) a state of complete order.

This gives the entropy $S(T)$ at a temperature T as

$$S(T) = \Delta C_p \ln \frac{T}{T_2}, \tag{7.27}$$

where ΔC_p is the difference in specific heat between the supercooled liquid and the glass at T_g and is assumed to be constant over the temperature range considered.

Substituting for $S(T)$ and approximating somewhat we find that

$$v_c = A \exp - \frac{N_A \, \Delta G^* s_n}{k \, \Delta C_p (T - T_2)}. \tag{7.28}$$

This gives a relaxation time equation of the form

$$\tau = \tau_0 \exp \{B/(T - T_2)\},$$

which as we have seen reduces to the WLF equation if we put

$$T_2 = T_g - 51.6.$$

7.4.4 An Objection to Free Volume Theories

Hoffman, Williams and Passaglia[12] have raised a serious objection to the free volume ideas. Williams[25] showed that the β-relaxations of polymethyl acrylate and polypropylene oxide behave somewhat similarly under constant pressure and constant volume conditions. It would be expected, however, on the free volume concept, that because the occupied volume v_0 would increase with temperature the results for these two conditions would be very different. Williams concluded that the dielectric relaxation time was not a unique function of volume. He suggested that this implied that the free volume did not remain constant for constant *total* volume while temperature and pressure are varied. An alternative view is that relationships of the form

$$\eta = \eta_0 \exp U_\eta / R(T - T_2) \quad \text{and} \quad \tau = C \exp U_\tau / R(T - T_2)$$

should indeed be based on concepts which are more fundamental than those of free volume.

7.5 NORMAL MODE THEORIES BASED ON MOTION OF ISOLATED FLEXIBLE CHAINS

We have so far discussed two types of theories, those based on the site model, and those based on the WLF equation and its ramifications, which deal with time–temperature equivalence. The site model theories predict constant activation energies and are more applicable to relaxation transitions originating from localized chain motions, whereas the WLF equation theories deal with the glass transition behaviour in amorphous polymers.

In the introductory section on amorphous polymers (Section 7.1.1) we considered the relaxation spectrum of amorphous polymers and noted that it was quite complex. The normal mode theories, now to be discussed, attempt to predict the relaxation spectrum for amorphous polymers, as well as the time–temperature equivalence.

These theories are associated with the names of Rouse, Bueche and Zimm[27-29] and are based on the idea of representing the motion of polymer chains in a viscous liquid by a series of linear differential equations. They are essentially *dilute solution* theories, but we shall see that, rather unexpectedly

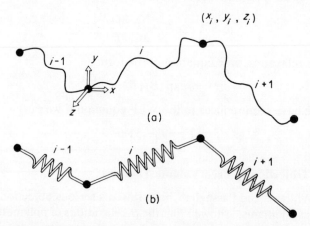

Figure 7.16. The Rouse model: (a) the network of chains; (b) the representation of the network as a combination of springs and beads.

perhaps, they can be extended to predict the behaviour of the pure polymer. Because of its simplicity we will give an account of the theory due to Rouse[27].

Each polymer chain is considered to consist of a number of submolecules. This is similar to the composition of a rubber network where molecular chains join the cross-link points (see Figure 7.16(a)). We can then represent the polymer molecules as a system of beads connected by springs whose behaviour is that of a freely jointed chain on the Gaussian theory of rubber elasticity (Figure 7.16(b)). The molecular chains between the beads are all of equal length, this portion of the polymer chain being long enough for the separation of its ends to approximate to a Gaussian probability distribution. It is assumed that only the beads interact directly with the solvent molecules. If a bead is displaced from its equilibrium position there are two types of forces acting on it; first, the forces due to this viscous interaction with the solvent molecules and secondly, the forces due to the tendency of the molecular chains to return to a state of maximum entropy by Brownian diffusional movements.

Consider the motion of the bead situated at the point $(x_i y_i z_i)$ between the ith and $(i+1)$th submolecules. The origin of coordinates is the bead between the $(i-1)$th and ith submolecules, i.e. at the other end of the ith submolecule.

For a Gaussian distribution of links in the submolecule the probability that this bead will lie at the point $x_i y_i z_i$ in the volume element $dx_i \, dy_i \, dz_i$ is

$$p_i(x_i y_i z_i) \, dx_i \, dy_i \, dz_i = \frac{b^3}{\pi^{3/2}} \exp\{-b^2(x_i^2 + y_i^2 + z_i^2)\} \, dx_i \, dy_i \, dz_i,$$

where

$$b^2 = 3/2zl^2$$

with l the length of each link, n the total number of links in the molecular chain, m the number of submolecules, giving $z = n/m =$ number of links in a submolecule.

The conformational probability of the entire chain can be represented by a point in $3m$-dimensional space. The probability that this point lies at the point $x_1 y_1 \ldots z_m$ in the volume element $dx_1 \, dy_1 \ldots dz_m$ is given by

$$P_m \, dx_1 \ldots dz_m = \prod_{i=1}^{m} p_i(x_i y_i z_i) \, dx_i \, dy_i \, dz_i$$

$$= \left(\frac{b^3}{\pi^{3/2}}\right)^m \exp\left\{-b^2\left[\sum_{i=1}^{m} x_i^2 + y_i^2 + z_i^2\right]\right\} dx_1 \ldots dz_m.$$

At equilibrium, the most probable values of the x_i, y_i and z_i coordinates are zero, i.e. each submolecule is in a coiled-up configuration. Any change from the equilibrium position will result in a decrease of entropy ΔS, or an increase in Helmholtz free energy $\Delta A = -T \, \Delta S$. (All conformations are assumed to have the same internal energy.)

Consider the change of x_i due to displacements of the ith submolecule from the equilibrium situation $(0, 0, 0)$, i.e. the change of x_i is referred to a private coordinate system with its origin at the bead between the $(i-1)$th and ith submolecules.

There will be a restoring force

$$T\left(\frac{\partial S_m}{\partial x_i} - \frac{\partial S_m}{\partial x_{i-1}}\right)$$

due to displacements of the bead between the submolecules $i-1$ and i and a restoring force

$$T\left(\frac{\partial S_m}{\partial x_{i+1}} - \frac{\partial S_m}{\partial x_i}\right)$$

due to displacements of the bead between the submolecules i and $i+1$.

The total equation of motion is

$$\eta \dot{x}_i = T\left(2\frac{\partial S_m}{\partial x_i} - \frac{\partial S_m}{\partial x_{i-1}} - \frac{\partial S_m}{\partial x_{i+1}}\right) \tag{7.29}$$

η is the coefficient of friction defining the viscous interaction between the beads and the solvent. S_m is the entropy of a molecule of conformation $x_1 y_1 \ldots z_m$ and is given by

$$S_m = k \ln P_m. \tag{7.30}$$

Combining equations (7.29) and (7.30) we have

$$\eta \dot{x}_i + \frac{3kT}{zl^2} (2x_i - x_{i-1} - x_{i+1}) = 0 \tag{7.31}$$

with $3m$ equations for coordinates $x_1 y_1 \ldots z_m$. If we make the intuitive connection between displacement and strain, we can see that these equations of motion for the chain molecules are directly equivalent to the equation of a Voigt element which has the form $\eta \dot{e} + Ee = 0$.

It therefore follows that these equations can be regarded as defining a set of creep compliances and stress relaxation moduli, or complex compliances and moduli.

The mathematical problem is to uncouple the $3m$ equations using a normal coordinate transformation. This involves obtaining the eigenfunctions which are linear combinations of the positions of the submolecules. Each eigenfunction then describes a configuration which decays with a time constant given by an associated eigenvalue, i.e. a single viscoelastic element with characteristic time-dependent properties.

For stress relaxation and dynamic mechanical experiments respectively it can be shown that the stress relaxation modulus $G(t)$ and the real part of the complex modulus $G_1(\omega)$ are given by

$$G(t) = NkT \sum_{p=1}^{m} e^{-t/\tau_p} \tag{7.32}$$

and

$$G_1(\omega) = NkT \sum_{p=1}^{m} \frac{\omega^2 \tau_p^2}{1 + \omega^2 \tau_p^2}, \tag{7.33}$$

where N is the number of molecules per cubic centimetre and τ_p, the relaxation time of the pth mode, is given by

$$\tau_p = zl^2 \eta [24kT \sin^2 \{p\pi/2(m+1)\}]^{-1}, \qquad p = 1, 2, \ldots, m. \tag{7.34}$$

These equations predict that $G(t)$ and $G_1(\omega)$ are determined by a discrete spectrum of relaxation times, each of which characterizes a given normal mode of motion. These normal modes are shown schematically in Figure 7.17. In the first mode, corresponding to $p = 1$, the ends of the molecule move whilst the centre of the molecule remains stationary. In the second mode, there are two nodes in the molecule. The general case of the pth mode has p nodes, with motion of the molecule occurring in $p + 1$ segments.

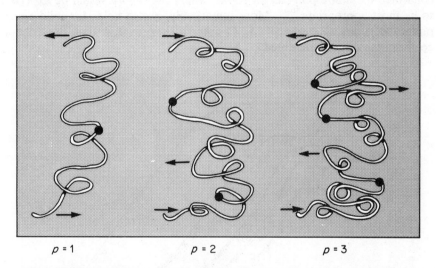

$p = 1$ $p = 2$ $p = 3$

Figure 7.17. Illustration of the first three normal modes of a chain molecule.

On this model the submolecule is the shortest length of chain which can undergo relaxation and the motion of segments *within* the submolecules are ignored. But such motions contribute to the relaxation spectrum for values of $m \geqslant 5$. Thus we would only expect the Rouse theory to be applicable for $m \gg 1$ where the equation for τ_p reduces to

$$\tau_p = \frac{m^2 zl^2 \eta}{6\pi^2 p^2 kT} = \frac{n^2 l^2 \eta_0}{6\pi^2 p^2 kT}, \tag{7.35}$$

where $\eta_0 = \eta/z$ is the friction coefficient per random link. The relaxation times depend on temperature directly through the factor $1/T$, through the

160

quantity nl^2 which defines the equilibrium mean square separation of the chain ends and may change due to differences in the energy of different chain conformations, and through changes in the friction coefficient η_0. η_0 changes rapidly with temperature and is primarily responsible for changes in τ_p. The fact that each τ_p has the same temperature dependence on this molecular theory, shows that it satisfies the requirements of thermorheological simplicity and gives theoretical justification for time–temperature equivalence.

Rouse's theory is the simplest molecular theory of polymer relaxation. A later theory of Zimm[29] does not assume that the velocity of the liquid solvent is unaffected by the movement of the polymer molecules (the 'free draining' approximation). The hydrodynamic interaction between the moving submolecules is taken into account and this gives a modified relaxation spectrum.

The Rouse, Zimm and Bueche theories are satisfactory for the longer relaxation times which involve movement of submolecules. This has been confirmed for dilute polymer solutions, where the theory would be expected to be most appropriate[30,31]. More remarkably, it also holds for solid amorphous polymers (Reference 14, Chapter 12), provided that the friction coefficient is suitably modified.

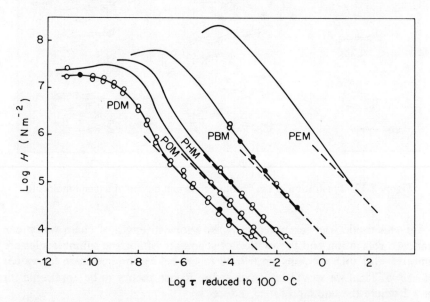

Figure 7.18. Relaxation time spectra $H(\ln \tau)$ for poly-n-dodecyl methacrylate (PDM), poly-n-octyl methacrylate (POM), poly-n-hexyl methacrylate (PHM), poly-n-butyl methacrylate (PBM), polyethyl methacrylate (PEM). Dashed lines are a slope of $-\frac{1}{2}$ predicted by the Rouse theory. [Redrawn with permission from Ferry, *Viscoelastic Properties of Polymers*, 1st Edn, Wiley, New York, 1961, Chap. 11.]

Ferry has shown that if the three longest relaxation times are ignored the distribution of relaxation times $H(\ln \tau)$, is given by

$$H(\ln \tau)\, d(\ln \tau) = -NkT\left(\frac{dp}{d\tau}\right) d\tau \qquad (7.36)$$

and from equation (7.35)

$$H(\ln \tau) = \left(\frac{Nnl}{2\pi}\right)\left(\frac{kT\eta_0}{6}\right)^{1/2} \tau^{-1/2}. \qquad (7.37)$$

This equation predicts that the plot of $\log H(\ln \tau)$ against $\log \tau$ should have a slope of $-\frac{1}{2}$. The results for five methacrylate polymers summarized in Figure 7.18 confirm this prediction for long relaxation times. The Zimm theory predicts a slope of $-\frac{2}{3}$ and is perhaps a better fit at shorter relaxation times. At very short relaxation times the theory fails completely, as we have anticipated, because the movement of short segments is involved. Another way of looking at this, suggested by Williams[32], is that a theory based essentially on the Gaussian statistics of polymer chains can only hold for low values of the 'modulus', i.e. for values less than $\sim 10^7 \, N\,m^{-2}$.

7.6 THE DYNAMICS OF HIGHLY ENTANGLED POLYMERS: THEORIES OF DE GENNES AND DOI AND EDWARDS

Our understanding of the molecular dynamics of polymer chains has advanced significantly in recent years due to the ideas of de Gennes[33] and Doi and Edwards[34].

The new approach which has been introduced stems from the idea that in a concentrated polymer solution, a melt or a solid polymer, the chains cannot pass through each other. This constraint effectively confines each chain inside a tube, the centre line of which defines the overall path of the chain in space, and has been called by Edwards the primitive chain (Figure 7.19). Each chain sees its environment as a tube, because although all the other chains are moving, there are so many entanglements that at any one time the tube is well defined. In a classic paper, de Gennes[33] discussed the possible motions of a polymer chain subject to confinement in a tube. He described these motions as worm-like and gave them the generic intuitively appealing name 'reptation'. de Gennes considered that the motions were essentially of two kinds. First, there are the comparatively short term wriggling motions which correspond to the migration of a 'defect' (a molecular kink) along the chain. The relaxation times associated with such motions take the form

$$t_p = \text{const.}/p^2,$$

where p is the mode number.

The longest relaxation time $T_d = t_{p=1}$ and it was shown that $T_d \propto M^2$, where M is the molecular weight of the chain.

162

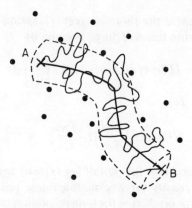

Figure 7.19. Chain segment AB in dense rubber. The points A and B denote the cross-linked points, and the dots represent other chains which, in this drawing, are assumed to be perpendicular to the paper. Due to entanglements the chain is confined to the tube-like region denoted by the broken line. The bold line shows the primitive path. [Redrawn with permission from Doi and Edwards, *J. Chem. Soc. Faraday Trans.*, **74**, 1802 (1978).]

Secondly, there is the much longer time associated with the movement of the chain as a whole through the polymer. This corresponds to the overall movement of the centre of gravity of the chain and has a characteristic time $T_R \propto M^3$. Provided therefore that the chains are long enough, T_R and T_d will be sufficiently separated to represent distinctly different relaxation times.

These ideas were extended by Doi and Edwards in a series of three related publications[34]. The theories of de Gennes for the free motions of a polymer chain were developed in more complete mathematical terms, and the problem of stress relaxation following a sudden deformation which remains constant was considered. Doi and Edwards introduced the slip-link network model which is equivalent to the tube model, where the slip-links are small rings through which the chains can pass freely and serve to define the primitive path. Figure 7.20 is their schematic illustration of the chain relaxation process. On this model the first relaxation process involves the stretched primitive chain shrinking to some finite length by sliding through the slip links with the comparatively short term snake-like motions proposed by de Gennes. In the second relaxation process the primitive chain disengages from the deformed slip link with a characteristic relaxation time which is identical to de Gennes' long relaxation time T_R. Doi and Edwards agree with de Gennes that $T_R \propto M^3$

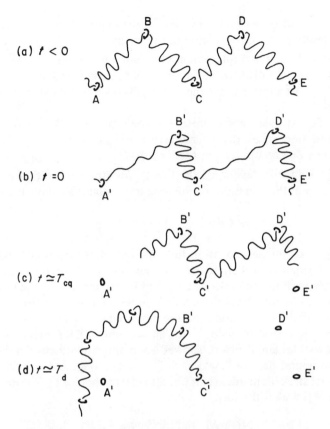

Figure 7.20. Schematic representation of the relaxation after sudden deformation: (a) initial equilibrium state; (b) immediately after the deformation—each part of the chain is stretched or compressed; (c) after the first relaxation process—the primitive chain recovers its equilibrium arc length, but the conformation of the primitive chain is still in a non-equilibrium state; (d) the second relaxation process—the chain disengages from the deformed slip-links and returns to the final equilibrium state. [Redrawn with permission from Doi and Edwards, *J. Chem. Soc. Faraday Trans.*, **74**, 1802 (1978).]

and that the diffusion coefficient D_R which defines the diffusion of the centre of mass of the chain has the form $D_R \propto M^{-2}$, a result which has been verified experimentally by Klein[35].

It is convenient to consider the chain motions in terms of the mean square displacement of a single monomer on a chain:

$$\langle r^2(t) \rangle = \langle [r_n(t) - r_n(0)]^2 \rangle,$$

where $r_n(t)$ is the position of the nth monomer at time t. Doi and Edwards[34], de Gennes[36] and Evans and Edwards[37,38] distinguish three basic regimes for the characteristic modes of motion of the chain.

(1) $t > T_R$. Here the displacement of a point on the chain will be equivalent to the displacement of the centre of mass of the chain and $\langle r^2(t) \rangle \propto t$.

(2) $t < T_R$. At shorter times the chain diffuses like a particle in a tube and $\langle r^2(t) \rangle \propto t^{1/2}$.

(3) $t \ll T_R$. At even shorter times the chain behaves like a Rouse chain in a tube and undergoes the characteristic wriggling or snake-like motions envisaged by de Gennes. In this case $\langle r^2(t) \rangle \propto t^{1/4}$.

Doi and Edwards developed the theory for stress relaxation and showed that the form of the equation for the long term relaxation gave the stress σ as

$$\sigma \propto F(e) \sum_{p \, \mathrm{odd}} \frac{8}{p^2 \pi^2} \exp\left(-\frac{t_p^2}{T_R}\right),$$

where T_R is again the relaxation time for complete renewal of the original chain conformation (called by de Gennes the disengagement time) which corresponds to diffusion of the chain out of its original tube, cage or slip-links. It is important to note that this equation shows separability between the strain and time variables, a feature of the non-linear single integral theory of Bernstein, Zapas and Kearsley[39] (for an account of BKZ theory see Section 9.3.4 below). Doi and Edwards showed good agreement between their theory and experimental data of Osaki et al.[40] and Tschoegl and coworkers[41] with regard to the strain dependence. It was also shown that for $t < T_R$ the relaxation spectrum $H(\tau)$ takes the form

$$H(\tau) \propto \int (\tau/T_R)^{1/2} \quad \text{for } \tau < T_R, \tag{7.38}$$

which contrasts with equation (7.37) where $H(\tau) \propto (\tau/T_R)^{-1/2}$ for Rouse theory.

There is some indication from experimental data[42] that equation (7.38) may be correct in the long term region. Graessley[43] has reviewed the Doi–Edwards theory in the light of the available experimental data for diffusion coefficients, plateau moduli, zero shear viscosities and the terminal relaxation times T_R. He concluded that the agreement between theory and experiment is very good, with the exception of the effects of molecular weight distribution. This is probably because chains of all lengths are compelled to diffuse by reptation whereas the cage life times may be very small compared with the disengagement time of the longest chains in the system. It may well be that the motions of the longest chains are thus modelled better by the Rouse model where tube constraints are absent.

REFERENCES

1. R. S. Marvin, in *Proceedings of the Second International Congress of Rheology*, Butterworth, London, 1954.

165

2. R. S. Marvin and H. Oser, *J. Res. Natl Bur. Stand. B*, **66**, 171 (1962).
3. R. S. Marvin and J. T. Berger, *Viscoelasticity: Phenomenological Aspects*, Academic Press, New York, 1960, p. 27.
4. K. Schmieder and K. Wolf, *Kolloidzeitschrift*, **134**, 149 (1953).
5. A. B. Thompson and D. W. Woods, *Trans. Faraday Soc.*, **52**, 1383 (1956).
6. N. G. McCrum and E. L. Morris, *Proc. Roy. Soc. A*, **281**, 258 (1964).
7. S. Glasstone, K. J. Laidler and H. Eyring, *The Theory of Rate Processes*, McGraw-Hill, New York, 1941.
8. S. Glasstone, *Textbook of Physical Chemistry*, 2nd edn, Macmillan, London, 1953.
9. Z. Arrhenius, *J. Phys. Chem.*, **4**, 226 (1889).
10. P. Debye, *Polar Molecules*, Dover Publications, New York, 1945.
11. H. Fröhlich, *Theory of Dielectrics*, Oxford University Press, Oxford, 1949.
12. J. D. Hoffman, G. Williams and E. Passaglia, *J. Polymer Sci. C*, **14**, 173 (1966).
13. J. B. Wachtman, *Phys. Rev.*, **131**, 517 (1963).
14. J. D. Ferry, *Viscoelastic Properties of Polymers*, Wiley, New York, 1961, Chapter 11.
15. M. L. Williams, R. F. Landel and J. D. Ferry, *J. Amer. Chem. Soc.*, **77**, 3701 (1955).
16. A. Kovacs, *J. Polymer Sci.*, **30**, 131 (1958).
17. A. K. Doolittle, *J. Applied Phys.*, **22**, 1471 (1951).
18. F. Bueche, *J. Chem. Phys.*, **21**, 1850 (1953).
19. A. J. Staverman and F. Schwarzl, *Die Physik der Hochpolymeren*, Springer-Verlag, Berlin, 1956, Chapter 1.
20. E. H. Lee, in *Proceedings of the First Symposium on Naval Structural Mechanics* (*London*), Pergamon Press, Oxford, 1960, p. 456.
21. J. H. Gibbs and E. A. di Marzio, *J. Chem. Phys.*, **28**, 373 (1958).
22. J. H. Gibbs and E. A. di Marzio, *J. Chem. Phys.*, **28**, 807 (1958).
23. G. Adam and J. H. Gibbs, *J. Chem. Phys.*, **43**, 139 (1965).
24. M. H. Cohen and D. Turnbull, *J. Chem. Phys.*, **31**, 1164, (1959).
25. G. Williams, *Trans. Faraday Soc.*, **60**, 1556 (1964).
26. G. Williams, *Trans. Faraday Soc.*, **61**, 1564 (1965).
27. P. E. Rouse, *J. Chem. Phys.*, **21**, 1272 (1953).
28. F. Bueche, *J. Chem. Phys.*, **22**, 603, (1953).
29. B. H. Zimm, *J. Chem. Phys.*, **24**, 269 (1956).
30. N. W. Tschoegl and J. D. Ferry, *Kolloidzeitschrift*, **189**, 37 (1963).
31. J. Lamb and A. J. Matheson, *Proc. Roy. Soc. A*, **281**, 207 (1964).
32. G. Williams, *J. Polymer Sci.*, **62**, 87 (1962).
33. P. G. de Gennes, *J. Chem. Phys.*, **55**, 572 (1971).
34. M. Doi and S. F. Edwards, *J. Chem. Soc. Faraday Trans.*, **74**, 1789, 1802, 1818 (1978).
35. J. Klein, *Nature*, **271**, 143 (1978).
36. P. G. de Gennes, *J. Chem. Phys.*, **72**, 109 (1981).
37. K. E. Evans and S. F. Edwards, *J. Chem. Soc. Faraday Trans.*, **77**, 1891, 1913, 1929 (1981).
38. K. E. Evans, *J. Chem. Soc. Faraday Trans.*, **77**, 2385 (1981).
39. B. Bernstein, E. A. Kearsley and L. J. Zapas, *Trans. Soc. Rheol.*, **7**, 91 (1963).
40. K. Osaki, S. Ohta, M. Fukada and M. Kurata, *J. Polymer Sci.*, **14**, 1701 (1976).
41. W. V. Chang, R. Black and N. W. Tschoegl, *J. Polymer Sci.*, **15**, 923 (1977).
42. J. D. Ferry, *Viscoelastic Properties of Polymers*, Wiley, New York, 1970, Chapter 12.
43. W. W. Graessley, *J. Polymer Sci.*, *Polymer Phys. Edn*, **18**, 27 (1980).

8

Relaxation Transitions and Their Relationship to Molecular Structure

In Chapter 7 we considered the main features of viscoelastic behaviour, with particular emphasis on time–temperature equivalence. The treatment was a general one and the ideas of rate-dependent processes, free volume and the dynamics of long-chain molecules are applicable to all polymers. We will now be much more specific and discuss the assignment of the viscoelastic relaxations in a molecular sense to different chemical groups in the molecule, and in a physical sense to different parts of the structure, e.g. to the motion of molecules in the crystalline or amorphous regions.

Because there are fewer structural features in amorphous polymers, discussions of the viscoelastic behaviour are less extensive and will precede those for crystalline polymers.

8.1 RELAXATION TRANSITIONS IN AMORPHOUS POLYMERS: SOME GENERAL FEATURES

Polymethyl methacrylate

$$\left[\begin{array}{c} H_3C-O-C=O \\ \quad\quad | \\ \quad -C-CH_2- \\ \quad\quad | \\ \quad CH_3 \end{array} \right]_n$$

shows four relaxation transitions as a function of temperature, three of which are shown in Figure 8.1. It is customary to label relaxation transitions in polymers in alphabetical order α, β, γ, δ, etc., with decreasing temperature, irrespective of their molecular origin. The highest temperature relaxation, the α-relaxation, is the glass transition and is associated with a large change in modulus. The β-relaxation has been shown by a combination of comparative studies on similar polymers, NMR and dielectric measurements[1-5], to be associated with side-chain motions of the ester group. The γ- and δ-relaxations involve motion of the methyl groups attached to the main chain and to the side chain, respectively.

The viscoelastic relaxations of similar polymers (such as polyethylmethacrylate) can be identified with molecular motions in an analogous manner.

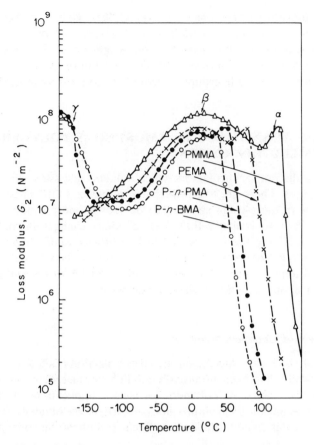

Figure 8.1. Temperature dependence of loss modulus G_2 for polymethyl methacrylate (PMMA), polyethyl methacrylate (PEMA), poly-*n*-propyl methacrylate (P-*n*-PMA) and poly-*n*-butyl methacrylate (P-*n*-BMA). [Redrawn with permission from Heijboer, in *The Physics of Non-crystalline Solids*, North-Holland, Amsterdam, 1965, p. 231.]

The principles described for polymethylmethacrylate thus form a good basis for the interpretation of the relaxation transitions in a number of amorphous polymers. There is a high temperature transition, the glass transition, associated with the onset of main-chain segmental motion, and secondary transitions which can be assigned to (1) motion of side groups, which is usually well authenticated; (2) restricted motion of the main chain or end-group motions, which are usually less well authenticated.

It should be emphasized that in many cases the assignment of a relaxation process is tentative and by no means as straightforward as in the methacrylate and acrylate polymers. For example, in amorphous polyethylene terephthalate two major relaxation transitions are observed, the high temperature transition being the glass transition and associated with general motion of the molecular chains. The secondary relaxation has been assigned only tentatively either to motion of chain ends or to some restricted motion of the main chain[5-8]. In this case, there are no side groups to provide an alternative assignment.

8.2 GLASS TRANSITIONS IN AMORPHOUS POLYMERS: DETAILED DISCUSSION

We will now discuss various factors which influence glass transitions in amorphous polymers. There are two principal approaches to the interpretation of the influence of main-chain structure, side groups and plasticizers, etc. The first approach is to consider that these produce changes in the molecular flexibility, modifying the ease with which conformational changes can take place. The second approach is to relate all these effects to changes in free volume, and in particular to the temperature at which this reaches its critical value which is, on this scheme, the glass transition.

8.2.1 Effect of Chemical Structure

The effect of chemical structure on the glass transition is a feature of polymer behaviour which has been intensively studied because of its importance in influencing the choice of polymers for useful applications. Much of our knowledge is of an empirical nature, due primarily to the difficulty of separating the intramolecular and intermolecular effects. In spite of this some generalities can be cited:

Main-chain Structure

The presence of flexible groups such as an ether link will make the main chain more flexible and reduce the glass transition temperature, whereas the introduction of an inflexible group, e.g. a terephthalate residue, will increase the glass transition temperature.

Influence of Side Groups[10]

It is generally true that bulky, inflexible side groups increase the temperature of the glass transition. This is illustrated by Table 8.1 for a series of substituted poly-α-olefins

$$\left[-CH_2-\underset{R}{\overset{|}{CH}}-\right]_n$$

Table 8.1. Glass transition of some vinyl polymers. [Reprinted with permission from Vincent, in *The Physics of Plastics* (P. D. Ritchie, ed.), 1965.]

Polymer	R	Transition temperature in °C at ~1 Hz
Polypropylene	CH_3	0
Polystyrene	C_6H_5	116
Poly-*N*-vinylcarbazole		211

There is a difference between the effect of rigid and flexible side groups, which is shown by the series of polyvinyl butyl ethers

$$\left[-CH_2-\underset{\underset{OR_1}{|}}{CH}-\right]_n$$

(Table 8.2). All these polymers have the same atoms in the side group OR_1 (R_1 represents the butyl isomeric form) but the more compact arrangements reduce the flexibility of the molecule and the transition temperature is markedly increased.

Table 8.2. Glass transition of some isomeric polyvinyl butyl ethers. [Reprinted with permission from Vincent, in *The Physics of Plastics* (P. D. Ritchie, ed.), 1965.]

Polymer	R_1	Transition temperature in °C at ~1 Hz
Polyvinyl *n*-butyl ether	$CH_2CH_2CH_2CH_3$	−32
Polyvinyl isobutyl ether	$CH_2CH(CH_3)_2$	−1
Polyvinyl *t*-butyl ether	$C(CH_3)_3$	+83

In an analogous manner, increasing the length of the flexible side groups reduces the temperature of the main transition. This can also be understood as increasing the free volume at any temperature. Data showing this are given in Table 8.3 for a series of polyvinyl *n*-alkyl ethers,

$$\left[-CH_2-\underset{\underset{OR_2}{|}}{CH}-\right]_n$$

where R_2 represents the *n*-alkyl group.

Effect of Main-chain Polarity

In Figure 8.2, the temperature of the glass transition is plotted against the number of successive $-CH_2$ or $-CH_3$ groups in the side groups, for five

Table 8.3. Glass transition of some polyvinyl n-alkyl ethers. [Reprinted with permission from Vincent, in *The Physics of Plastics* (P. D. Ritchie, ed.), 1965.]

Polymer	R_2	Transition temperature in °C at ~1 Hz
Polyvinyl methyl ether	CH_3	-10
Polyvinyl ethyl ether	CH_2CH_3	-17
Polyvinyl n-propyl ether	$CH_2CH_2CH_3$	-27
Polyvinyl n-butyl ether	$CH_2CH_2CH_2CH_3$	-32

polymer series. Each series consists of polymers of similar main-chain composition, and it can be seen that the glass transition temperature increases with increasing main-chain polarity. The associated reduction in the main-chain mobility is presumed to be due to the increase in intermolecular forces. In particular it is suggested that the higher curve for the polychloracrylic esters is due to the increased valence forces associated with the chlorine molecules.

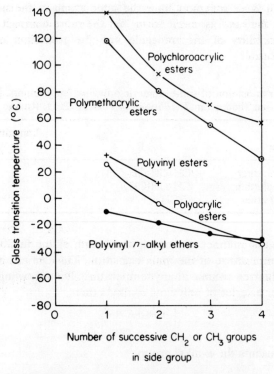

Figure 8.2. The effect of polarity on the position of the glass transition temperature for five polymer series. [Redrawn with permission from Vincent, in *The Physics of Plastics* (P. D. Ritchie, ed.), 1965.]

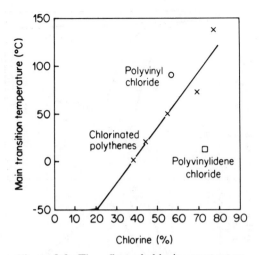

Figure 8.3. The effect of chlorine content on the softening point of polymers with varying degrees of chlorination. [Redrawn with permission from Schmieder and Wolf, *Kolloidzeitschrift*, **134**, 149 (1953).]

An even more striking case is the effect of chlorinating polyethylene[11] which is shown in Figure 8.3.

8.2.2 Effect of Molecular Weight and Cross-linking

Molecular weight does not affect the dynamic mechanical properties of polymers in the glassy low temperature state, although at low molecular weights, the glass transition temperature T_g is affected by molecular weight. This is usually explained[12,13] on the basis that the chain ends, by reducing the closeness of molecular packing, introduce extra free volume, and hence lower T_g.

As has already been discussed (Section 7.1.1 above), molecular weight has a large effect in the glass transition range, transforming the behaviour from viscous flow to a plateau range of rubber-like behaviour with increasing molecular weight. This is shown in Figure 7.3. The explanation of this behaviour is that chain entanglements prevent irreversible flow. The effects of introducing chemical cross-links are shown in Figure 8.4, where pheno–formaldehyde resin has been cross-linked with hexamethylene tetramine to concentrations of 2, 4 and 10% respectively. Chemical cross-linking raises the temperature of the glass transition and broadens the transition region[14]. It can be seen that in very highly cross-linked materials there is no glass transition.

This behaviour can again be interpreted on the basis of changes in free volume. Chemical cross-linking, by bringing adjacent chains close together, reduces the free volume and hence raises T_g.

172

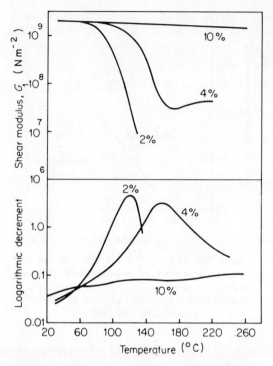

Figure 8.4. Shear modulus G_1 and logarithmic decrement of a phenol–formaldehyde resin cross-linked with hexamethylene tetramine at stated concentrations. [Redrawn with permission from Drumm, Dodge and Nielsen, *Ind. Eng. Chem.*, **48**, 76 (1956).]

8.2.3 Blends, Grafts and Copolymers

The mechanical properties of blends and graft polymers are determined primarily by the mutual solubility of the two homopolymers. If two polymers are completely soluble in one another the properties of the mixture are nearly the same as those of a random copolymer of the same composition. Figure 8.5 illustrates this point by showing that a 50:50 mixture of polyvinyl acetate and polymethyl acrylate has very similar properties to those of a copolymer of vinyl acetate and methyl acrylate[14]. It is to be noted that the damping peak for the mixture and the copolymer occurs at 30 °C while the peaks for polymethylacrylate and polyvinyl acetate occur at about 15 and 45 °C respectively.

A theoretical interpretation of the glass transition temperatures of copolymers has been given on the basis of the ideas of free volume. We have seen that the glass transition can be considered to occur at a constant value of the free volume. Gordon and Taylor[15] assume that in an ideal copolymer the

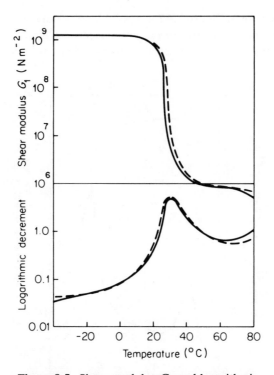

Figure 8.5. Shear modulus G_1 and logarithmic decrement for a miscible blend of polyvinyl acetate and polymethyl acrylate (——) and a copolymer of vinyl acetate and methyl acrylate (– – –). [Redrawn with permission from Nielsen, *Mechanical Properties of Polymers*, Van Nostrand–Reinhold, New York, 1962.]

partial specific volumes of the two components are constant and equal to the specific volumes of the two homopolymers. It is further assumed that the specific volume–temperature coefficients for the two components in the rubbery and glassy states remain the same in the copolymers as in the homopolymers, and are independent of temperature. It can then be shown that the glass transition temperature T_g for the copolymer is given by[16,17]

$$\frac{1}{T_g} = \frac{1}{(w_1 + Bw_2)}\left[\frac{w_1}{T_{g1}} + \frac{Bw_2}{T_{g2}}\right],$$

where w_1 and w_2 are the weight fractions of the two monomers whose homopolymers have transitions at temperatures T_{g1} and T_{g2} respectively and B is a constant which is close to unity.

If the two polymers in a mixture are insoluble they exist as two separate phases and two glass transitions are observed instead of one. This is illustrated

174

by the results shown in Figure 8.6 for a polyblend of polystyrene and styrene–butadiene rubber[14]. Two loss peaks are observed which are very close to those in pure polystyrene and pure styrene–butadiene rubber.

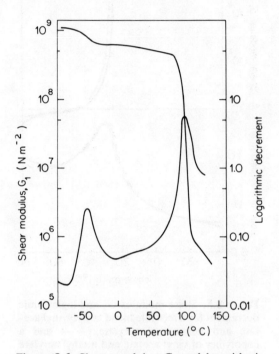

Figure 8.6. Shear modulus G_1 and logarithmic decrement for an immiscible polyblend of polystyrene and a styrene–butadiene copolymer. [Redrawn with permission from Nielsen, *Mechanical Properties of Solids*, Van Nostrand–Reinhold, New York, 1962.]

8.2.4 Effect of Plasticizers

Plasticizers are low molecular weight materials which are added to rigid polymers to soften them. Plasticizers must be soluble in the polymer and usually they dissolve it completely at high temperatures. Figure 8.7 shows the change in the loss peak associated with the glass transition of polyvinyl chloride, when plasticized with various amounts of di(ethylhexyl)phthalate[18]. The major effect of plasticizer is to lower the temperature of the glass transition; essentially plasticizers make it easier for changes in molecular conformation to occur.

Plasticizers also broaden the loss peak, and the degree of broadening depends on the nature of the interaction between the polymer and the

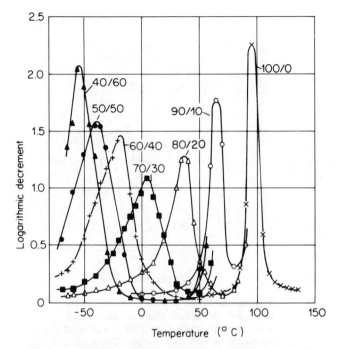

Figure 8.7. The logarithmic decrement of polyvinyl chloride plasticized with various amounts of di(ethylhexyl) phthalate. [Redrawn with permission from Wolf, *Kunststoffe*, **41**, 89 (1951).]

plasticizer. If the plasticizer has a limited solubility in the polymer, or if the plasticizer tends to associate in the presence of the polymer, a broad damping peak is found. Thus the width of the damping peak increases as the plasticizer becomes a poorer solvent. This is shown in Figure 8.8 for plasticized polyvinyl chloride[14]. Diethyl phthalate is a relatively good solvent, dibutyl phthalate is a poorer solvent, and dioctyl phthalate is a very poor solvent.

8.3 THE CRANKSHAFT MECHANISM FOR SECONDARY RELAXATIONS

In a review of relaxation transitions Willbourn[19] suggested that the γ-relaxation in both amorphous and crystalline polymers could in many cases be attributed to a restricted motion of the main chain which required at least four $-CH_2$ groups in succession on a linear part of the chain. This proposal has led to the so-called 'crankshaft' mechanisms of Shatzki[20] and Boyer[21] (Figure 8.9). Shatzki's mechanism involves the simultaneous rotation about the two bonds 1 and 7, such that the intervening carbon bonds move as a crankshaft. It is to be noted that bonds 1 and 7 are collinear, which means

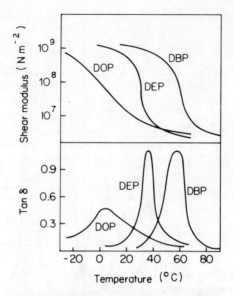

Figure 8.8. Shear modulus and loss factor tan δ for polyvinyl chloride plasticized with diethyl phthalate (DEP), dibutyl phthalate (DBP) and *n*-dioctyl phthalate (DOP). [Redrawn with permission from Nielsen, Buchdahl and Levreault, *J. Appl. Phys.*, **21**, 607 (1950).]

that bonds on either side may remain unaffected so that only a relatively small volume is required for the movement.

It has been proposed that this mechanism is relevant to the γ-relaxation in polyethylene, polyamides, polyesters, polyoxymethylene and polypropylene oxide.

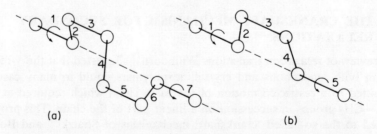

Figure 8.9. The crankshaft mechanisms of (a) Shatzki and (b) Boyer. [Redrawn with permission from McCrum, Read and Williams, *Anelastic and Dielectric Effects in Polymer Solids*, Wiley, London, 1967.]

8.4 RELAXATION TRANSITIONS IN CRYSTALLINE POLYMERS

8.4.1 General Discussion

The interpretation of the viscoelastic behaviour of crystalline polymers is still at a very speculative stage with regard to a detailed understanding. Attempts at specific interpretations are very much influenced by the opinions of the individual investigator on the structure of crystalline polymers.

The simplest starting point for discussions of viscoelastic relaxations in crystalline polymers is the two-phase fringed micelle model (see Section 1.2.2 above). On this model we would expect to identify some transitions with the crystalline regions and some with the amorphous regions. This is supported by empirical correlations between the magnitude of loss processes in crystalline polymers and the crystalline–amorphous ratio as determined by X-ray or density methods.

Two good examples of this come from the work of McCrum[22] on poly-tetrafluorethylene (PTFE) and that of Illers and Breuer[9] on polyethylene terephthalate (PET).

In this discussion we will tacitly assume that tan δ can be used as a measure of the relaxation strength. Although this is plausible in the light of previous considerations (Section 5.5 above), it is essentially at the hypothesis stage and complete justification for the use of tan δ or some other measure will only be obtained when more sophisticated structural models have been given a quantitative treatment (for further discussion see Reference 5, p. 139).

Figure 8.10 shows the temperature dependence of the logarithmic decrement for three transitions in PTFE as a function of the degree of crystallinity. The lowest temperature relaxation decreases in magnitude with increasing crystallinity in a very clear manner and on the two-phase model is identified with a transition in the amorphous regions of the polymer. The β-relaxation, on the other hand, increases in magnitude with increasing crystallinity and is therefore associated with the crystalline regions. The analysis of the α-relaxation is somewhat dependent on the method of resolving the loss peaks and determining the strength of a relaxation process, but the consensus of opinion (see Reference 5, Chapter 11) would seem to confirm that it decreases in magnitude with increasing crystallinity and is therefore associated with the amorphous regions.

In PET the effect of crystallinity on the β-relaxation is very small and has led to a very complex interpretation in terms of this loss peak being composed of several relaxation processes. No clear distinction can be drawn between relaxation processes occurring in the crystalline and the amorphous regions.

The effect of crystallinity on the α-relaxation is very striking (Figure 8.11) and more complex than in the case of the γ-relaxation of PTFE. The height of the loss peak decreases with increasing crystallinity, the peak broadens, becomes very asymmetrical and moves to higher temperatures. This suggests

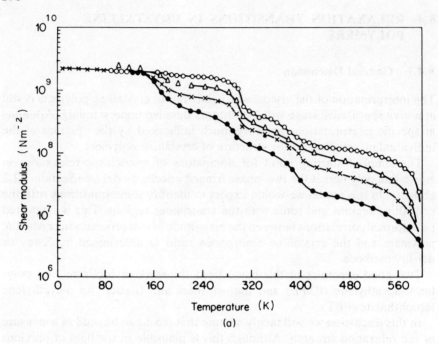

(a)

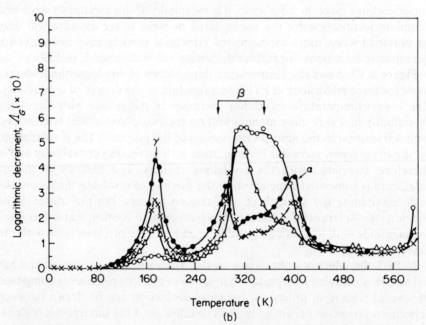

(b)

Figure 8.10. Temperature dependence of (a) shear modulus and (b) the logarithmic decrement Λ_G at ~1 Hz for PTFE samples of 92% (O), 76% (△), 64% (×) and 48% (●) crystallinity. [Redrawn with permission from McCrum, *J. Polymer Sci.*, **34**, 355 (1959.]

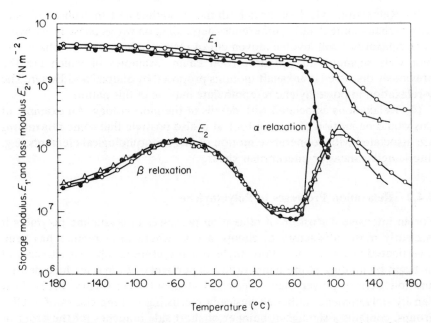

Figure 8.11. Storage modulus E_1 and loss modulus E_2 as a function of temperature at 138 Hz for PET samples of differing degrees of crystallinity (●, 5%; △, 34%; ○, 50%). [Redrawn with permission from Takayanagi, *Mem. Fac. Eng., Kyushu Univ.,* **23**, 1 (1963).]

not only that the α-relaxation occurs in the amorphous regions of the polymer but also that the presence of the crystallites now imposes a variety of considerable constraints on the amorphous regions, thus influencing the molecular movements associated with this relaxation process. These ideas have been confirmed by NMR measurements of molecular mobility in a series of corresponding samples[23].

The adequacy of the two-phase fringed micelle model was seriously called into question following the discovery of polyethylene single crystals, and must be severely modified to allow for chain-folding in the bulk, as illustrated in Figure 1.15. The implications of this new model are that we must identify the following types of relaxations:

(1) Relaxations which occur in the amorphous phase. These will include the glass transition, an example of which is the α-relaxation in polyethylene terephthalate. Another example of an amorphous transition is the γ-relaxation in PTFE.

(2) Relaxations which occur in the crystalline phase. These may be of two types: (a) those involving cooperative motions of the molecular chains along the length of the crystallite—these relaxations will be related to the lamellar thickness, and (b) those associated with defects such as end groups in the crystal—these relaxations could also involve chain movement.

(3) Relaxations which occur in both the crystalline and amorphous phases, with perhaps some detailed differences depending on which phase is involved. Such relaxations will involve movement of restricted length of the chain, e.g. four CH_2 segments in a hydrocarbon chain, examples of which are the previously discussed crankshaft motions proposed by Shatzki and Boyer. The γ-relaxation in polyethylene terephthalate may be of this nature.

(4) Relaxations associated with details of the morphology. An example of this would be motion of a chain fold. It is also possible that some relaxations are associated with cooperative motion of large morphological elements, e.g. interlamellar shear or interfibrillar shear.

8.4.2 Relaxation Processes in Polyethylene

For an intensive discussion of relaxation processes in crystalline polymers it is clearly most satisfactory to choose a case where the structure has been investigated in great detail. Polyethylene is therefore an obvious choice. Let us begin by looking at the temperature dependence of tan δ for high density and low density polyethylene measured at about 1 Hz (Figure 8.12). Low density polyethylene, although basically consisting of long chains of $-CH_2$ groups, contains a significant number of short side branches (of the order of three per 100 carbon atoms in a typical commercial polymer) together with a few long branches (about one per molecule). High density polyethylene, on

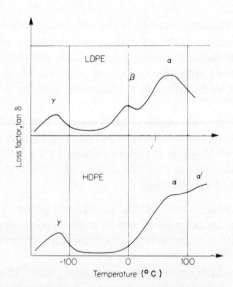

Figure 8.12. Schematic diagram showing α, α', β and γ relaxation processes in low density polyethylene (LDPE) and high density polyethylene (HDPE).

the other hand, is very much closer to being the pure $(CH_2)_n$ polymer, and the number of branches is often less than five per 1000 carbon atoms.

The low density polymer shows three clearly distinguishable loss peaks; these are conventionally labelled α, β and γ. In the high density polymer the γ-relaxation is very similar to that in the low density polymer; the β-relaxation is almost absent and the α-relaxation has been considerably modified. There is some controversy regarding the changes in the α-relaxation with structure. This arises partly because the analysis depends on whether $\tan \delta$ or G_2 is plotted as a function of temperature, and partly because the α-process appears to be a composite process consisting of at least two relaxation processes (usually called α and α') with different activation energies. These facts together mean that the observed shape of the loss peaks in the α-relaxation region can vary greatly according to the different workers.

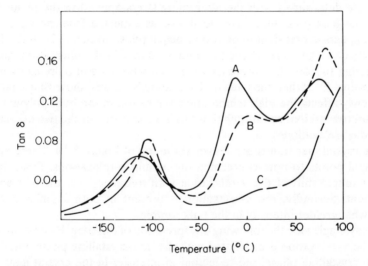

Figure 8.13. Loss factor $\tan \delta$ of polyethylene as a function of branching. Type A, 3.2 CH_3 groups per 100 CH_2 groups; type B, 1.6 per 100; type C, <0.1 per 100. [Redrawn with permission from Kline, Sauer and Woodward, *J. Polymer Sci.*, **22**, 455 (1956).]

The assignment of these three relaxations was first attempted in a general sense, by comparing the intensities of the loss peaks in different polyethylenes. The comparison shown in Figure 8.13 suggests that the β-relaxation is associated with the relaxation of side groups or short branch points. This was demonstrated quantitatively by Kline, Sauer and Woodward[24] who examined a series of polyethylenes of widely varying densities due to varying side group content.

The assignment of the α-relaxation to a crystalline relaxation was shown by the reduction in the intensity of the α-loss peak on chlorination of polyethylene[11]. This loss peak disappears at the degree of chlorination required to remove any X-ray diffraction evidence of crystallinity in the polymer.

The γ-relaxation peak was found to decrease in intensity as the crystallinity increased. It was therefore assigned to an amorphous relaxation.

These 'assignments' of the relaxation processes do not extend to a specific understanding of the molecular processes involved. The latter can only come from more detailed knowledge of the structure of polyethylene.

In this respect, our understanding has been much advanced by the comparison of the relaxations in the bulk high density polymer with those observed in mats of solution-crystallized high density polyethylene. These mats consist of an aggregate of polyethylene crystals in which the molecular chain axes are approximately perpendicular to the plane of the mat. Figure 8.14 shows results obtained by Takayanagi[25], in which the dynamical mechanical properties were determined with the alternating stress applied in the plane of the film. The bulk-crystallized sample shows, as expected from previous work, what appears at first sight to be two principal relaxations, at +70 and −120 °C respectively. These are labelled α (and α') and γ in the illustration, and it is interesting to note the difference in shape of the α- and α'-relaxations due to plotting E_2 rather than tan δ. Takayanagi's results show that change of frequency affects the high temperature relaxation in the bulk polymer in an asymmetric fashion, confirming that this is a composite of the two overlapping α- and α'-relaxations.

The crystal mat results are shown at the top of Figure 8.14. There are two principal points to note concerning the α- and γ-relaxations. First, there is now a simple shift of the α-relaxation with frequency, i.e. the α'-transition is absent. Secondly, the γ-relaxation is present in both samples but is of somewhat greater intensity in the bulk sample.

Takayanagi made the following interpretation of this data. First he suggested that the γ-relaxation is associated with the non-crystalline phase and defects in the crystalline phase, the reduction in intensity in the crystal mats being attributed to their greater degree of perfection. It was tentatively proposed that the molecular motion was a 'local twisting of the molecular chains'.

Secondly, Takayanagi attributed the α-relaxation to a similar molecular motion within the crystalline phase. Finally, he noted that the α'-relaxation is absent in the viscoelastic pattern of the crystal mats. He attributed this relaxation to the crystalline phase, and suggested that its absence in the crystal mats was due to the orientation of the applied stress within the crystal mat being unfavourable for the particular molecular process which he proposed. This is a translational motion of chain segments along the chain axis within the crystal lattice (akin to the Fisher and Schmidt process for the annealing of polyethylene single crystals[26]).

Sinnot[27] has also examined the behaviour of bulk and solution-crystallized high density polyethylene, and his results are summarized in Figures 8.15 and 8.16. Figure 8.15 shows a comparison between the behaviour of the bulk

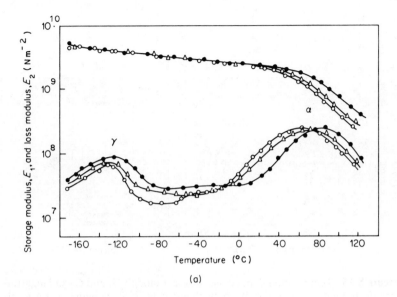

(a)

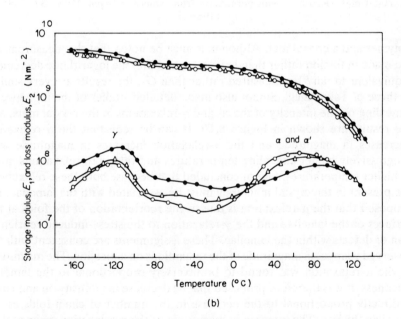

(b)

Figure 8.14. Temperature dependence of storage modulus E_1 and loss modulus E_2 at 3.5 Hz ($\bigcirc$), 11 Hz ($\triangle$) and 110 Hz ($\bullet$), (a) measured along the direction perpendicular to the chain axis (c-axis) for a single crystal mat of high density polyethylene; (b) for a bulk-crystallized specimen of high density polyethylene. [Redrawn with permission from Takayanagi, in *Proceedings of the Fourth International Congress of Rheology*, Interscience, New York, 1965, p. 161.]

184

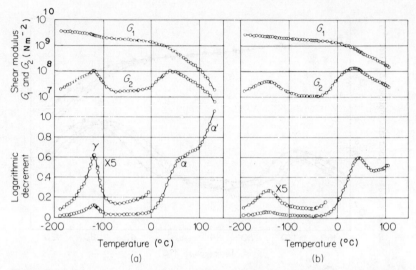

Figure 8.15. Temperature dependence of shear moduli G_1 and G_2 and logarithmic decrement for high density polyethylene: (a) bulk polymer; (b) a single crystal mat. [Redrawn with permission from Sinnot, *J. Appl. Phys.*, **37**, 3385 (1966).]

polymer and a crystal mat. Although it must be noted that the measurements are made in torsion rather than extension, and that the logarithmic decrement (equivalent to tan δ) is measured rather than G_2, the results are very similar to those of Takayanagi. Sinnot also made detailed studies of the influence of annealing on the intensity of the α- and γ-relaxations in the crystal mats, and the results are shown in Figure 8.16. It can be seen that the α-relaxation decreases in magnitude and the γ-relaxation increases in magnitude with progressively higher annealing temperatures and that both relaxations move to higher temperatures. Sinnot concluded that because both these relaxations are present in the crystal mats they must be associated with the lamellae. He proposed that the α-relaxation is due to the reorientation of the folds at the surfaces of the lamellae and the γ-relaxation to the stress-induced reorientation of defects within the lamellae. These assignments are consistent with the observed changes in magnitude of the relaxations on annealing. The magnitude of the α-relaxation was found to be inversely proportional to the lamellar thickness; it was therefore proposed that its decrease in intensity on annealing is directly proportional to the decrease in the number of chain folds as the lamellae thicken. The increase in magnitude of the γ-relaxation on annealing is attributed to the generation of defects within the lamellae, perhaps due to the motion of dislocations and point defects.

Further information of value comes from studies of the α- and α'-relaxations by McCrum and Morris[28]. Figure 8.17 compares the logarithmic decrement observed in a torsional pendulum experiment for two slowly cooled bulk-crystallized polyethylenes, one of which has had its lateral surface removed

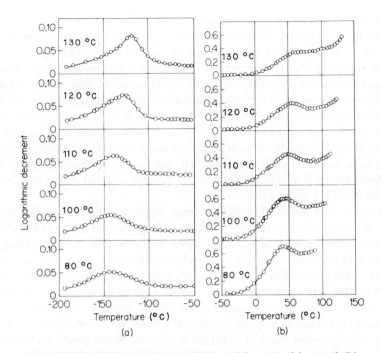

Figure 8.16. Influence of annealing on (a) γ-transition and (b) α-transition of solution-crystallized polyethylene mat. [Redrawn with permission from Sinnot, *J. Appl. Phys.*, **37**, 3385 (1966).]

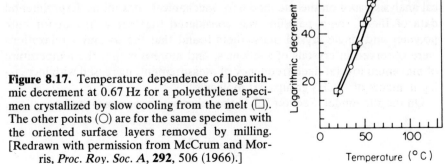

Figure 8.17. Temperature dependence of logarithmic decrement at 0.67 Hz for a polyethylene specimen crystallized by slow cooling from the melt ($\square$). The other points ($\bigcirc$) are for the same specimen with the oriented surface layers removed by milling. [Redrawn with permission from McCrum and Morris, *Proc. Roy. Soc. A*, **292**, 506 (1966).]

by milling. The consequent reduction in the α'-relaxation was explained by proposing that this relaxation is extremely sensitive to the orientation of the lamellae with respect to the direction of applied stress. Milling removes an oriented layer with the preferred orientation for maximum energy loss. It is a result consistent with that obtained by Takayanagi and Sinnot. McCrum and Morris proposed that the α'-relaxation can be interpreted as slip at the lamellar boundary. They envisage that the lamellae bend under the applied stress like an elastic beam in a viscous liquid. To explain the fact that their observed creep is totally recoverable, they assume that the lamellae are pinned along their length. This is equivalent to assuming a mechanical model where an elastic spring is now placed in parallel with the original spring and dashpot, i.e. the standard linear solid.

McCrum and Morris found that large variations in spherulitic structure have no influence on the α-relaxation. They agreed with Takayanagi in associating this relaxation with molecular motion within the crystallites. Both Sinnot and McCrum and Morris observed the effect of electron irradiation on the relaxations. The α- and α'-peaks are much affected by irradiation, whereas the γ-peak is unaffected. Sinnot concludes that this supports his assignment of the α-relaxation to motion of chain folds, since these are preferentially cross-linked. McCrum and Morris also claim support for their hypothesis, because the α'-relaxation is affected in the manner expected for an increase in the internal viscosity of bulk polymer due to cross-linking, i.e. for measurements at constant frequency it appears to move to higher temperatures.

8.4.3 Application of the Site Model

The above investigations leave us with a largely empirical knowledge of the relaxations in polyethylene. The α- and α'-relaxations are associated with the crystalline regions (which have a lamellar texture); the β-relaxation is associated with branch points in the amorphous regions; the γ-relaxation is associated with the non-crystalline regions and defects in the crystalline regions (i.e. the lamellae). A review article by Hoffman, Williams and Passaglia[29] has considered most of the available experimental data, and attempted a theoretical analysis based on the site model for mechanical relaxations. Experimental data of Illers[30] on n-paraffins was considered together with that for bulk polymer and single crystal mats. Illers found that the α- and γ-relaxations were observed in crystals of n-alkanes, and moreover that the temperature of maximum loss (at fixed frequency of measurement) increased with increasing numbers of $-CH_2$ groups per molecule.

On the site model (Section 7.3.1) the relaxation time τ is given by

$$\tau = \frac{1}{A'} e^{\Delta G_2/RT}$$

where ΔG_2 is the free energy of activation. This can be written in terms of an enthalpy ΔH_2 and entropy ΔS_2 of activation so that

$$\tau = \frac{1}{A'} e^{-\Delta S_2/R} e^{\Delta H_2/RT}$$

We will consider that the chain relaxation process in a linear paraffin involves the rotation of the chain through 180° together with its simultaneous *translation* along its axis to reach a new equilibrium position. If we assume that the chain rotation is identical to the rotation of a rigid rod,

$$\Delta H_2(n) = 2 \Delta H_{end} + (n-2) \Delta H_{CH_2},$$

where $\Delta H_2(n)$ now refers to a chain of n segments and ΔH_{end} and ΔH_{CH_2} refer to the enthalpy of each end group and that of each segment respectively. Similarly,

$$\Delta S_2(n) = 2 \Delta S_{end} + (n-2) \Delta S_{CH_2},$$

where $\Delta S_2(n)$, ΔS_{end} and ΔS_{CH_2} are the corresponding entropies. This gives

$$\Delta G_2(n) = 2(\Delta H_{end} - T \Delta S_{end}) + (n-2)(\Delta H_{CH_2} - T \Delta S_{CH_2}).$$

T_{max}, the temperature at which a loss peak occurs in a plot of loss versus T at a fixed measuring frequency f_m, where $2\pi f_m \tau_m = 1$, is given by

$$\ln \tau_m = \ln \frac{1}{A'} + \frac{(2 \Delta H_{end} + (n-2) \Delta H_{CH_2})}{RT_{max}} - \frac{(2 \Delta S_{end} + (n-2) \Delta S_{CH_2})}{R}$$

or

$$T_{max} = T_0 \frac{(a+n)}{(b+n)},$$

where

$$T_0 = \frac{\Delta H_{CH_2}}{\Delta S_{CH_2}}$$

is the value of T_{max} for very long chains ($n \to \infty$) and

$$a = \frac{2(\Delta H_{end} - \Delta H_{CH_2})}{\Delta H_{CH_2}},$$

$$b = \frac{[R \ln (A' \tau_m) + 2(\Delta S_{end} - \Delta S_{CH_2})]}{\Delta S_{CH_2}}.$$

This gives a value of T_{max} which rises rapidly for small n and then levels off to a value T_0 as n becomes large.

There is one necessary modification to this simple theory. The rigid rod approximation cannot hold for long chains, because here the rotation of one end would be expected to leave the other end unaffected. Thus $\ln \tau_m$ and

T_{max} will fall below the rigid rod curves at large values of n, and approach an asymptote.

Hoffman, Williams and Passaglia distinguish three different models for the relaxation:

(1) $\alpha_c - A$ *model*: chain-folded crystal with folds and interior chain coupled. The theoretical treatment here is similar to that given above for the n-paraffins, with

$$\Delta H_2(n) = \Delta H_{fold} + n \, \Delta H_{CH_2}$$

and

$$\Delta S_2(n) = \Delta S_{fold} + n \, \Delta S_{CH_2}$$

replacing the values previously given for $\Delta H(n)$ and $\Delta S(n)$.

(2) $\alpha_c - B$ *model*: n-paraffins and extended chain crystals. This is the model for which the theoretical treatment has been given.

(3) $\alpha_c - C$ *model*: chain-folded crystals with independent chain fold and interior chain relaxations. The interior chain relaxation is similar to the $\alpha_c - A$ and $\alpha_c - B$ models, with suitable modification of the theory. The independent chain fold relaxation will now be different in character, only involving the chain fold and hence having a much lower activation energy.

There are therefore three relaxation processes $\alpha_c - A$, $\alpha_c - B$ and the interior chain relaxation $\alpha_c - C$ which is of a different character. On a plot of T_{max} against n, using data from a wide variety of sources (Figure 8.18) it appears that all the processes are indistinguishable, and that a single characteristic law

$$T_{max} = T_0 \frac{(a+n)}{(b+n)}$$

can be obtained for suitably chosen a and b.

The $\alpha_c - C$ process corresponding to the fold motion is distinguishable in other ways, from its different activation energy (Takayanagi) and its variation in intensity with lamellar thickness in single crystals (Sinnot).

A similar correlation was shown to hold for the γ-relaxation in the n-paraffins and single crystals (Figure 8.19). Hoffman *et al.* emphasize, however, that it is difficult to accept that the molecular processes are identical in all these materials. The γ-relaxation in bulk polyethylene and in single crystal mats has been attributed to a relaxation of defects. If this relaxation is coupled to the main-chain motion, a theoretical model similar to that described here will be applicable, and n will relate to the number of participating segments, which will in turn relate to the lamellar thickness. It seems unlikely that there is a sufficient concentration of such defects in the n-paraffin crystals for the γ-relaxation observed in that case to be of a similar nature.

It therefore seems likely that the γ-relaxation is a composite relaxation, and that it involves not only defects in the crystals, but also molecules in the amorphous regions.

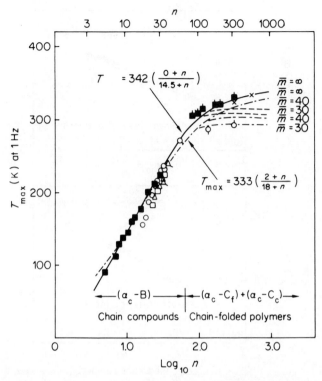

Figure 8.18. The temperature of maximum loss at 1 Hz plotted against log n (α_c processes). *Mechanical:* ■, single crystal mats of polyethylene; ×, bulk linear polyethylene; ■, n-paraffins. *Dielectric:* □, Esters; △, ethers; ○ and △, solid solutions of ketones and ethers, respectively, in n-paraffins; ○, solid solutions of ethers in low density polyethylene; ○, and ○, oxidized bulk polyethylene and single crystal mat polyethylene respectively. The upper calculated curve (——) corresponds to the relation $T_{max} = 342(0+n)/(14.5+n)$. The dashed lines ($\bar{m} = 40$ and $\bar{m} = 30$) falling off this curve correspond to the modification of this rigid rod approximation to include chain twisting. Similarly, the lower calculated curve (—·—) corresponds to $T_{max} = 333(2+n)/(18+n)$ and the lines $\bar{m} = 40$ and $\bar{m} = 30$ take into account twisting of the chains. [Redrawn with permission from Hoffman, Williams and Passaglia, *J. Polymer Sci. C*, **14**, 173 (1966).]

8.4.4   Use of Mechanical Anisotropy

The work of Takayanagi and McCrum–Morris indicated that the mechanical relaxations in an oriented sample can be dependent on the direction of the applied stress. Stachurski and Ward[31–33] have obtained experimental data on oriented sheets of low density and high density polyethylene which confirm

190

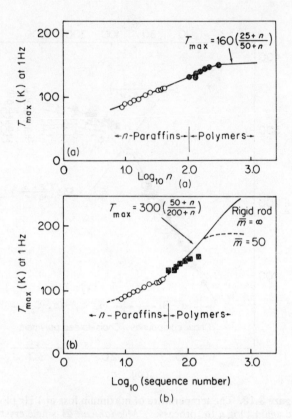

$$T_{max} = 160\left(\frac{25+n}{50+n}\right)$$

(a)

$$T_{max} = 300\left(\frac{50+n}{200+n}\right)$$

(b)

Figure 8.19. The temperature T_{max} of maximum mechanical loss for the γ_c process in single crystal mats of polyethylene. Data for low temperature process observed for n-paraffins in polystyrene matrix shown for comparison: (a) T_{max} versus $\log_{10} n$ where n is taken to be the number of C atoms for the n-paraffins, and the number of C atoms in a fold period for the single crystals; (b) T_{max} versus $\log_{10}$ (sequence number). The sequence number is taken to be n for the n-paraffins and $(n/2)$ for the single crystals. Dashed line on right-hand side shows effects of chain twisting for case $\bar{m} = 50$. [Redrawn with permission from Hoffman, Williams and Passaglia, *J. Polymer Sci. C*, **14**, 173 (1966).]

this expectation, and give further information concerning the origins of the α- and β-relaxations.

The key results are summarized in Figures 8.20 and 8.21. The dynamic mechanical measurements were taken on strips cut at chosen angles to the original draw direction. In the cold-drawn low density polyethylene sheet (Figure 8.20(a)) the relaxation process at about 0 °C shows striking anisotropy

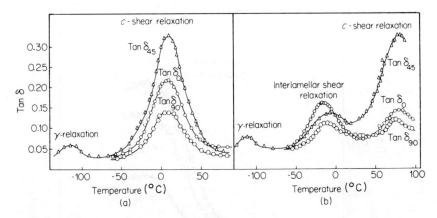

Figure 8.20. Temperature dependence of tan δ in three directions in (a) cold-drawn, and (b) cold-drawn and annealed low density polyethylene sheets at approximately 500 Hz. [Redrawn with permission from Stachurski and Ward, *J. Polymer Sci. A-2*, **6**, 1817 (1968).]

with maximum loss for the 45° sheet. This is the anisotropy to be expected for a relaxation which involves shear parallel to the draw direction in a plane containing the draw direction. Comparison with the cold-drawn and annealed low density polyethylene sheet (Figure 8.20(b)) where this relaxation has moved to about 70 °C, and with other specially oriented sheets, suggests that the c-axis orientation of the crystalline regions is the significant factor. For this reason the relaxation has been termed the 'c-shear relaxation', i.e. shear in the c-axis direction in planes containing the c-axis. The cold-drawn and annealed low density sheet also shows a second relaxation at about 0 °C with a different anisotropy. The losses are now greatest when the stress is applied along the 0° direction, and a similar anisotropy has been observed for high density sheet. (Figure 8.21). It was proposed that this relaxation process, which corresponds to the β-relaxation in low density polyethylene and to the α-relaxation in high density polyethylene, is an interlamellar shear process. This interpretation, which owes much to the examination of specially oriented orthorhombic sheets of low density polyethylene, is based on proposals that the lamellae in such sheets are arranged so that the lamellar planes make an acute angle of about 40° with the initial draw direction[34,35]. Applying the stress along the initial draw direction then gives the maximum resolved shear stress parallel to the lamellar planes.

These results are consistent with McCrum and Morris' interpretation of the α-relaxation in high density polyethylene and produce the rather surprising conclusion that this relaxation is similar mechanically to the β-relaxation in low density polymer. This does not mean that both relaxations are associated with identical *molecular* processes. Indeed it is likely that the α-relaxation in the high density polymer requires mobility of the fold surfaces (as proposed by Hoffman, Williams and Passaglia) whereas the β-relaxation in the low

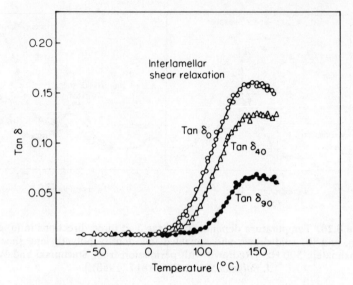

Figure 8.21. Temperature dependence of tan δ in cold-drawn and annealed high density polyethylene sheet in different directions at 50 Hz. [Redrawn with permission from Stachurski and Ward, *J. Macromol. Sci. B*, **3**, 445 (1969).]

density polymer requires mobility of chains close to branch points. Because the anisotropy of the α-relaxation in the low density polymer is related to the orientation of the crystalline regions it must be concluded that the chains taking part in the relaxation process thread these regions. It seems probable that the chains involved form interlamellar ties so that the stresses are transmitted throughout the bulk of the polymer.

These results emphasize that it is difficult to assign the mechanical relaxations to specific mechanisms on the basis of analogous behaviour in different polymers. In particular, the labelling of the relaxations α, β, γ, etc., in order of decreasing temperature may be misleading when it comes to comparisons between polymers. We have seen that in polyethylene the low density polymer shows α, β and γ relaxations in the isotropic state. Cold-drawn low density polymer shows only two relaxations in the same temperature range because the α-relaxation swamps the β-relaxation so that the latter is not discernable. Cold-drawn and annealed low density polyethylene does show α, β and γ relaxations and from the mechanical anisotropy the α-relaxation is identified as the c-shear relaxation and the β-relaxation as an interlamellar shear process.

Isotropic high density polyethylene shows α, α' and γ relaxations. From the orientation studies the α and α' relaxations show anisotropy which can be explained on the basis of interlamellar shear. It is therefore proposed that the nature of the α-relaxation in low density and high density polyethylene are quite different.

8.4.5 Conclusion

We can conclude that appreciable progress is being made in unravelling the complex relaxation processes in polyethylene and it is hoped that this account will serve as a guide-line for the interpretation of the relaxations in other crystalline polymers. Certainly there are no comparable data for other polymers, and this can be attributed to the comparatively smaller amount of structural information.

REFERENCES

1. K. Deutsch, E. A. Hoff and W. Reddish, *J. Polymer Sci.*, **13**, 365 (1954).
2. J. G. Powles, B. I. Hunt and D. J. H. Sandiford, *Polymer*, **5**, 505 (1964).
3. K. M. Sinnot, *J. Polymer Sci.*, **42**, 3 (1960).
4. J. Heijboer, *Physics of Non-Crystalline Solids*, North-Holland, Amsterdam, 1965, p. 231.
5. N. G. McCrum, B. E. Read and G. Williams, *Anelastic and Dielectric Effects in Polymer Solids*, Wiley, London, 1967.
6. A. B. Thompson and D. W. Woods, *Trans. Faraday Soc.*, **52**, 1383 (1956).
7. W. Reddish, *Trans. Faraday Soc.* **46**, 459 (1950).
8. G. Farrow, J. McIntosh and I. M. Ward, *Makromolek. Chem.*, **38**, 147 (1960).
9. K. H. Illers and H. Breuer, *J. Colloid Sci.*, **18**, 1 (1963).
10. P. I. Vincent, *Physics of Plastics*, Iliffe Books, London, 1965.
11. K. Schmieder and K. Wolf, *Kolloidzeitschrift*, **134**, 149 (1953).
12. T. G. Fox and P. J. Flory, *J. Appl. Phys.*, **21**, 581 (1950).
13. T. G. Fox and P. J. Flory, *J. Polymer Sci.*, **14**, 315 (1954).
14. L. E. Nielsen, *Mechanical Properties of Polymers*, Van Nostrand–Reinhold, New York, 1962.
15. M. Gordon and J. S. Taylor, *J. Appl. Chem.*, **2**, 493 (1952).
16. L. Mandelkern, G. M. Martin and F. A. Quinn, *J. Res. Natl Bur. Stand.*, **58**, 137 (1959).
17. T. G. Fox and S. Loshaek, *J. Polymer Sci.*, **15**, 371 (1955).
18. K. Wolf, *Kunststoffe*, **41**, 89 (1951).
19. A. H. Willbourn, *Trans. Faraday Soc.*, **54**, 717 (1958).
20. T. F. Shatzki, *J. Polymer Sci.*, **57**, 496 (1962).
21. R. F. Boyer, *Rubber Rev.*, **34**, 1303 (1963).
22. N. G. McCrum, *J. Polymer Sci.*, **34**, 355 (1959).
23. I. M. Ward, *Trans. Faraday Soc.*, **56**, 648 (1960).
24. D. E. Kline, J. A. Sauer and A. E. Woodward, *J. Polymer Sci.*, **22**, 455 (1956).
25. M. Takayanagi, in *Proceedings of the Fourth International Congress of Rheology*, Part 1, Interscience Publishers, New York, 1965, p. 161.
26. E. W. Fischer and G. F. Schmidt, *Zeit. Angew. Chem.*, **74**, 551 (1962).
27. K. M. Sinnot, *J. Appl. Phys.*, **37**, 3385 (1966).
28. N. G. McCrum and E. L. Morris, *Proc. Roy. Soc. A*, **292**, 506 (1966).
29. J. D. Hoffman, G. Williams and E. A. Passaglia, *J. Polymer Sci., C*, **14**, 173 (1966).
30. K. H. Illers, *Rheol. Acta.*, **3**, 194 (1964).
31. Z. H. Stachurski and I. M. Ward, *J. Polymer Sci. A-2*, **6**, 1083 (1968).
32. Z. H. Stachurski and I. M. Ward, *J. Polymer Sci. A-2*, **6**, 1817 (1968).
33. Z. H. Stachurski and I. M. Ward, *J. Macromol. Sci. B*, **3**, 445 (1969).
34. I. L. Hay and A. Keller, IUPAC Symposium, Prague, 1965, Preprint No. P.325; *J. Materials Sci.*, **2**, 538 (1967).
35. T. Seto and T. Hara, *Rept. Prog. Polymer Phys. (Japan)*, **7**, 63 (1967).

9

Non-linear Viscoelastic Behaviour

9.1 GENERAL INTRODUCTION

In many practical applications of plastics, although the ulitimate strains produced are recoverable, the viscoelastic behaviour does not satisfy the tests of linearity required by the Boltzmann superposition principle. This can occur for several reasons. In the first instance, there is the restriction of this representation to small strains† because of the changed definition of strain at large strains, and the consequence that superposition of strains cannot be expected to apply generally. This limitation applies in particular to studies on synthetic textile fibres where one may be interested in strains of at least 10% or in elastomers where the strains may be as high as 100%.

Secondly, although the experiments are restricted to small strains, it may still be that linear viscoelastic behaviour is not obtained. In this connection it is quite usual to observe linear viscoelastic behaviour at short times at given stress levels, but for the behaviour to be markedly non-linear for long times at the same stress levels.

There is not at present a representation of non-linear viscoelasticity which gives an adequate description of the behaviour and provides some physical insight into the origins of this behaviour. This is a subject where the divergence of the experimentalist and the theoretician is most marked. Faced with non-linear viscoelastic behaviour the experimentalist makes a number of measurements, necessarily finite, and then reduces his data empirically to a series of equations relating stress, strain and time. Although these equations can be extremely valuable in reducing the experimental data to manageable proportions they often do not reveal anything of the essential *nature* of the non-linearity, and may even be misleading in this respect.

The theoretician, on the other hand, will attempt to form a constitutive relation of a most general nature and examine how the form of this relation is determined by such features as 'short term' memory, material symmetry and invariance under rigid body rotation. The disadvantage of this approach is that in many cases it is too general. The experimentalist may well conclude that it is of no relevance to his particular problem, particularly if it does not appear to provide any physical insight into the situation.

---

† By 'small strains' we mean that the quadratic terms in the displacement gradients can be neglected (Chapter 3, Section 3.1).

As the subject of non-linear viscoelastic behaviour cannot be provided with an approach which satisfies all these requirements the various attempts to deal with the situation will be considered under three headings:

(1) *The engineering approach.* The design engineer requires the ability to predict behaviour exactly for a proposed situation in terms of as few initial experiments as possible. Empirical relations which describe the performance are adequate, and these need not have any physical significance.

(2) *The rheological approach.* Attempts are made to extend the formal descriptions of linear viscoelastic behaviour to non-linear behaviour. One line of development has been to preserve the separability of stress, strain and time in the functional representations. In its most sophisticated form this leads to single integral representations consistent with rigorous continuum mechanics. A second line of development has been based on the multiple integral representation which leads to considerable mathematical complexity.

(3) *The molecular approach.* The starting point in this case is the incorporation of a thermally activated rate process as the viscous element in a model representation. This approach has the attraction of possible identification of molecular mechanisms and hence links with structural understanding. Although there may be disadvantages in respect of the formal mathematical development, these are to some extent balanced by the advantages of built-in non-linearity and temperature dependence.

It must be admitted that at the present stage none of these approaches is entirely satisfactory. With some reservations therefore they will be considered in turn, noting their shortcomings where appropriate.

9.2 THE DESIGN ENGINEER'S APPROACH TO NON-LINEAR VISCOELASTICITY

9.2.1 Use of the Isochronous Stress–Strain Curves

For a linear viscoelastic solid the creep behaviour is completely specified at a given temperature by a measurement of the response to a constant stress over the required period of time. For a non-linear viscoelastic solid the behaviour over the range of stress required must be mapped out in detail over the required period of time. We will also see that because the Boltzmann superposition principle does not hold it is necessary to carry out systematic programmes of loading and unloading. The behaviour for any loading programme is not defined by the data obtained from a single step loading which gives a creep curve, or even from a two-step loading and unloading which gives creep and recovery.

In spite of these pitfalls, the design engineer starts with the stress–strain–time relationship obtained for creep under a constant stress. This produces the three-dimensional surface shown in Figure 9.1. It has been proposed by

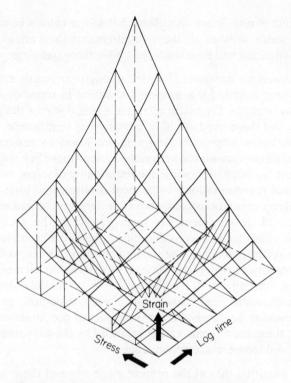

Figure 9.1. The stress–strain–time relationship obtained from creep.◺◹, Constant time section: isochronous stress–strain curve; ◿◺, constant stress section: creep curve. [Redrawn with permission from Turner, *Polymer Eng. Sci.*, **6**, 306 (1966).]

Turner[1] that this surface is in many practical cases defined to a sufficient degree of accuracy by a combination of two types of measurements:

(1) The relationship between stress and strain for a fixed time of measurement. This is the section in Figure 9.1 normal to the log time axis and is called the isochronous stress–strain curve. It is obtained by making a series of single step loading tests at different levels of stress and measuring the creep after a fixed time in each case.

(2) At least two creep curves at different stress levels, over a suitable time range, at the same temperature as the stress–strain curve.

Turner has presented results for creep of polypropylene which show how one isochronous stress–strain curve and two creep curves can be combined to allow a complete mapping of the creep behaviour (Figure 9.2). Although this procedure is very economical of experimental effort, data obtained from single step loading tests have severe limitations when we attempt to use them

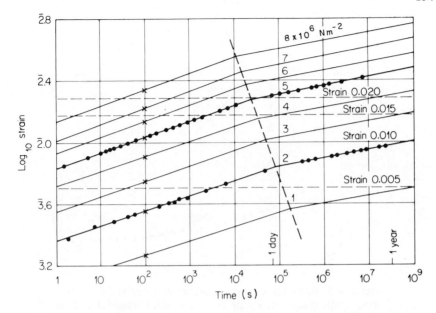

Figure 9.2. Tensile creep of polypropylene at 60 °C. The stress and time dependence are approximately separable and therefore creep curves at intermediate stresses can be interpolated from a knowledge of two creep curves ($\bullet$) and the isochronous stress–strain relationship ($\times$). [Redrawn with permission from Turner, *Polymer Eng. Sci.*, **6**, 306 (1966).]

to predict behaviour in more complex tests. The simplest of these is the recovery following the removal of load.

Consider the creep and recovery loading programme shown in Figure 5.7(c). The recovery at time t, $e_r(t - t_1)$ is defined as the *difference* between (i) the strain at time t under continuous application of the initial stress and (ii) the strain at time t due to the application of the initial stress at zero time followed by its removal at time t_1.

For a *linear* viscoelastic solid the arguments presented in Section 5.2.1 above show that

$$e_r(t - t_1) = e_c(t) - [e_c(t) - e_c(t - t_1)] = e_c(t - t_1), \qquad (9.1)$$

where $e_c(t)$ is the creep under the applied stress for a time t and $e_c(t - t_1)$ is the creep under the applied stress for a time $(t - t_1)$.

The recovery behaviour of polypropylene under typical conditions is shown in Figure 9.3, together with the predicted behaviour on the basis of equation (9.1). There is a very appreciable divergence. A rigorous treatment, to be given later, describes this divergence in terms of more complex memory functions than those provided by the Boltzmann superposition principle. At this stage in the discussion we shall develop an empirical description of the behaviour.

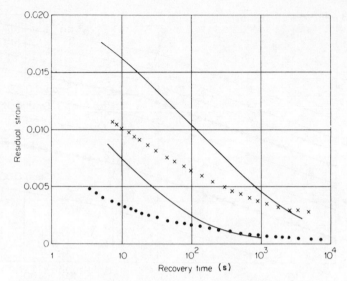

Figure 9.3. Recovery of polypropylene polymer at 20 °C after creep under a stress of $2 \times 10^7 \, \text{N m}^{-2}$. ×, after creep for 1000 s; ●, after creep for 100 s; ——, recovery predicted for linear viscoelastic behaviour. [Redrawn with permission from Turner, *Polymer Eng. Sci.*, **6**, 306 (1966).]

Turner[1] has proposed that two new quantities be introduced. The first one is called the 'fractional recovery' (FR) and is defined as

$$FR = \frac{\text{strain recovered}}{\text{maximum creep strain}} = \frac{e_c(t_1) - e'_r(t)}{e_c(t_1)},$$

where $e_c(t_1)$ is the creep strain at $t = t_1$, i.e. the time at which the load is removed, and $e'_r(t)$ is a new quantity called the residual strain. $e'_r(t)$ is the strain at a time t in a creep and recovery programme such as that shown in Figure 5.7(c), i.e. for a linear viscoelastic material

$$e'_r(t) = e_c(t) - e_c(t - t_1).$$

When FR is used as the parameter the family of diverse recovery curves obtained for fixed t_1 under different stresses can be brought into approximate coincidence. This is shown in Figure 9.4(a) and (b).

A second generalization is achieved when a second quantity, 'reduced time', is used. If we plot the FR versus log recovery time for one value of initial load, but for different times of application of this initial load (i.e. different t_1), a family of curves is obtained, as shown in Figure 9.5(a). Defining a reduced time t_R as

$$t_R = \frac{\text{recovery time}}{\text{creep time}} = \frac{t - t_1}{t_1},$$

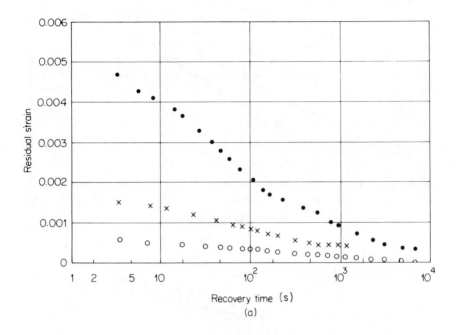

Recovery time (s)

(a)

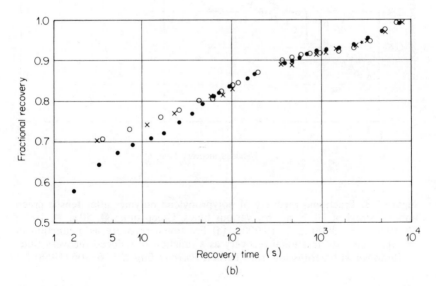

Recovery time (s)

(b)

Figure 9.4. Recovery of polypropylene polymer after tensile creep for 1000 s at three stress levels: ●, $10^7\,\text{N m}^{-2}$; ×, $5 \times 10^6\,\text{N m}^{-2}$; ○, $2 \times 10^6\,\text{N m}^{-2}$. The ordinate is either residual strain (a) or fractional recovery (b). [Redrawn with permission from Turner, *Polymer Eng. Sci.*, **6**, 306 (1966).]

200

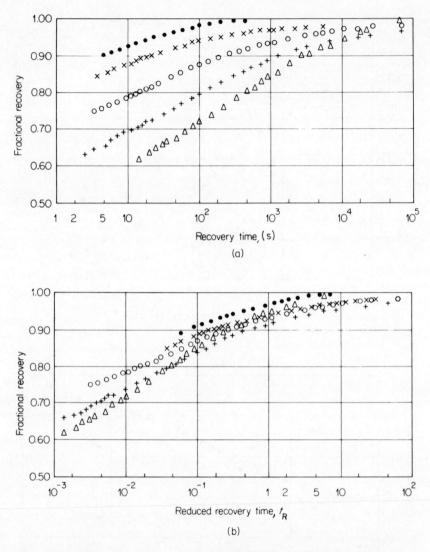

Figure 9.5. Fractional recovery of polypropylene polymer after tensile creep under a stress of $10^7 \, \text{N m}^{-2}$ for various times. Creep time: ●, 50 s; ×, 100 s; ○, 1000 s; +, 3000 s; △, 10 000 s. (a) Fractional recovery as a function of recovery time; (b) fractional recovery as a function of reduced recovery time. [Redrawn with permission from Turner, *Polymer Eng. Sci.*, **6**, 306 (1966).]

and replotting the recovery now as a function of t_R, gives approximate coincidence (Figure 9.5(b)).

Turner points out that this recovery behaviour, in terms of FR and reduced time, is to be expected if linear superposition can be applied to a material which obeys a power law relationship. The assumption is that although the creep under a variety of loads is non-linear, superposition still applies. We will see some further experimental justification for this hypothesis in later work (Section 9.3.2 below).

It is assumed that creep is determined by a power law of the form $e_c(t) = At^n$, where A is a (non-linear) function of stress.

The creep under load for a time $t = t_1$ is then $e_c(t_1) = At_1^n$ and the residual strain is given by

$$e_r'(t) = At^n - A(t - t_1)^n.$$

The FR is therefore

$$\frac{e_c(t_1) - e_r'(t)}{e_c(t_1)} = \frac{At_1^n - [At^n - A(t - t_1)^n]}{At_1^n}$$

$$= 1 - \left(\frac{t}{t_1}\right)^n + \left(\frac{t}{t_1} - 1\right)^n.$$

Since

$$t_R = \frac{t}{t_1} - 1$$

we have

$$FR = 1 + t_R^n - (t_R + 1)^n,$$

i.e. FR is a unique function of the reduced time t_R.

These two simplifications proposed by Turner mark the present limit of what has been achieved without recourse to a more sophisticated representation. The essential idea behind these simplifications is that the non-linearity in stress and time can be separated.

We will discuss shortly how more general representations of non-linear viscoelasticity require a reformulation of the Boltzmann superposition principle. This leads to a considerable increase in complexity and it can well be argued that for practical purposes this is unprofitable unless it leads to a greater physical understanding.

9.2.2 Power Laws for Non-linear Viscoelasticity

To describe the creep behaviour of glassy or tough polymers where the creep strains involved are small ($\sim 5\%$ say), as distinct from elastomers where the deformations are large ($\sim 100\%$ say), separable stress and time functions have been proposed.

Pao and Marin[2,3] followed the approach originally suggested by Marin and others for metals[4], where the total creep strain e is considered to consist of three independent components, an elastic strain e_1, a transient recoverable viscoelastic strain e_2, and a permanent non-recoverable plastic strain e_3. At constant stress σ, the elastic strain is given by $e_1 = \sigma/E$, where E is Young's modulus. The viscoelastic strain is defined by integrating the condition that the transient creep rate is a function of the stress σ and the transient creep strain, i.e. $de_2/dt = f(e_2, \sigma)$. The plastic strain is found by integrating the condition that the plastic strain rate is a function of stress only. For simplicity, the functions of stress for both the viscoelastic and the plastic strains are assumed to be simple power laws of stress, and most usually the same power law is adopted.

The total creep strain for loading under a constant stress σ is then

$$e = \sigma/E + K\sigma^n (1 - e^{-qt}) + B\sigma^n t, \tag{9.2}$$

where K, n, q and B are constants for the material.

Marin and his co-workers were primarily interested in generalizing this relationship to deal with the three-dimensional stress situation and then describing creep in bending, torsion, etc. Although a good fit to experimental data was not obtained, this extension of their representation will be discussed later in further detail.

Findley and his collaborators[5] have attempted to fit the creep of many plastics and plastic laminates to analytical relationships similar to those suggested for metals[6].

It was found that the creep strain e_c and time t could be related by an equation of the form

$$e_c(t) = e_0 + mt^n,$$

where e_0 and m are functions of stress for a given material and n is a material constant.

Further work suggested that the results could be represented by

$$e_c(\sigma, t) = e_0' \sinh \frac{\sigma}{\sigma_e} + m't^n \sinh \frac{\sigma}{\sigma_m},$$

where m', σ_e and σ_m are constants for the material.

This equation was a good fit to creep data obtained from single step loading tests only, i.e. it is exactly equivalent to Turner's stress–strain–time surface.

A similar relationship was found for the creep of nitrocellulose by Van Holde[7]. He proposed that

$$e_c(t) = e_0 + m't^{1/3} \sin \alpha\sigma,$$

where α is a constant. Since m' is a constant and $\sin \alpha\sigma$ is constant for a constant stress, at constant stress this relationship reduces to the Andrade creep law for metals[8]:

$$e_c(t) = e_0 + \beta' t^{1/3},$$

where β' is a constant.

If e_0 and β' are proportional to stress, the latter equation is consistent with *linear* viscoelastic behaviour. Plazek and his collaborators[9] have suggested that the Andrade creep law holds for several polymers and gels, although there is a divergence from linear behaviour at long times.

Findley's empirical equations are very useful to the design engineer for constant stress loading conditions as he can predict the creep for a given material if he is given the constants e_0, m and n or e_0', m', n, σ_e and σ_m.

The empirical approaches suggested so far have two principal limitations:

(1) They do not provide a general representation for creep, recovery and behaviour under complicated loading programmes.

(2) Creep data in these formulations cannot be simply related to stress relaxation and dynamic mechanical data.

9.3 THE RHEOLOGIST'S APPROACH TO NON-LINEAR VISCOELASTICITY

9.3.1 Large-strain Behaviour of Elastomers

A semi-empirical extension of linear viscoelastic theory has been used with considerable success by T. L. Smith[10] to explain the large-strain behaviour of elastomers[11].

Consider a Maxwell element, where the stress σ and the strain e are related by the equation

$$\frac{de}{dt} = \frac{\sigma}{\eta} + \frac{1}{E}\frac{d\sigma}{dt}$$

For a constant rate of increase of strain $de/dt = R$, it may be readily shown (see Section 5.2.5 above) that

$$\sigma = \eta R(1 - e^{-t/\tau}),$$

where $\tau = \eta/E$, i.e.

$$\sigma = R\tau E(1 - e^{-t/\tau}).$$

It would thus follow that for a continuous distribution of relaxation times $H(\tau)$ this generalizes to

$$\sigma = R \int_{-\infty}^{\infty} \tau H(\tau)(1 - e^{-t/\tau}) \, d \ln \tau.$$

For the constant strain rate R, write $R = e/t$. Then

$$\frac{\sigma}{e} = \frac{1}{t} \int_{-\infty}^{\infty} \tau H(\tau)(1 - e^{-t/\tau}) \, d \ln \tau + E_e$$

where the term E_e is added to denote the equilibrium modulus.

The quantity

$$\frac{\sigma}{e} = \frac{\sigma(e, t)}{e},$$

which is a function of time only, is called the 'constant strain-rate modulus' $F(t)$. Smith assumes that for large strain $F(t)$ can be written as

$$F(t) = \frac{g(e)\sigma(e, t)}{e},$$

i.e.

$$\log F(t) = \log \left\{ \frac{g(e)}{e} \right\} + \log \sigma(e, t), \tag{9.3}$$

where $g(e)$ is some function of strain which approaches unity as the strain approaches zero.

This approach is very successful empirically, as shown in Figure 9.6(a) and (b). The stress–strain curves at each strain rate are analysed by selecting the results for each value of strain and constructing plots of $\log \sigma$ (σ = nominal stress, i.e. load) versus $\log t$. The resulting parallel plots (Figure 9.6(b)) show that the results can be represented by equation (9.3), the displacements being the factor $\log g(e)/e$. This implies that $\log g(e)/e$ is independent of time. From what has been discussed of time–temperature equivalence it might well be anticipated that $g(e)/e$ will also be independent of temperature. Smith showed that this was true over a wide temperature range, only breaking down at the lowest temperatures.

Up to extensions of 100% the function $g(e)/e$ has a simple form. It was found that

$$\lambda\sigma = Ee,$$

where λ is the extension ratio (ratio of extended length to initial length). To the approximation that elastomers are incompressible, $\lambda\sigma$ is the true stress. The result that true stress is proportional to strain is physically plausible.

Above 100% extension Smith used an empirical formula

$$\sigma = E\frac{e}{\lambda^2} \exp A \left(\lambda - \frac{1}{\lambda} \right)$$

proposed by Martin, Roth and Stiehler[12].

These results are interpreted as implying that the non-linearity can be attributed to the large strains, and that the time and strain dependence are separable. It is not clear how applicable the treatment is to the general viscoelastic behaviour, i.e. to recovery or superposition. However, similar

---

Figure 9.6. (a) Tensile stress–strain curves of SBR vulcanized rubber at $-34.4\,°C$ and at strain rates between 8.89×10^{-3} and $8.89\ \text{min}^{-1}$. The stress ordinates are displaced by an amount A to enable distinction between curves for different strain rates. (b) Variation of log stress with log time at different strain values for SBR vulcanized rubber. The data were obtained from analysis of the curves shown in (a) and the strain values are indicated for each case. [Redrawn with permission from Smith, *Trans. Soc. Rheol.*, **6**, 61 (1962).]

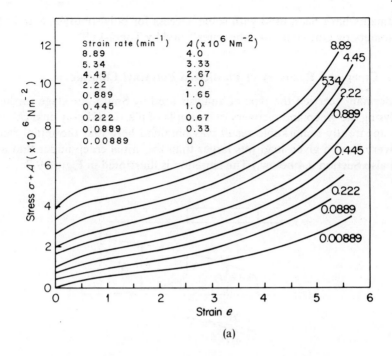

(a)

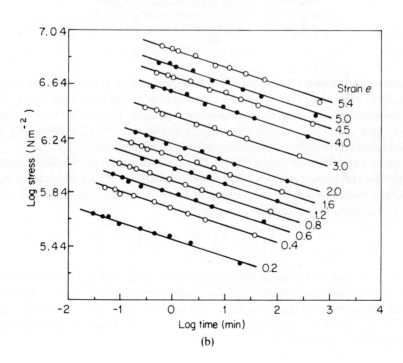

(b)

treatments have been used with some success for polyisobutylene and other elastomers by Guth and his colleagues[13] and by Tobolsky[14].

9.3.2 Creep and Recovery of Plasticized Polyvinyl Chloride

Leaderman[15] carried the type of analysis used by Smith one stage further in analysing the creep and recovery of a sample of plasticized polyvinyl chloride. The apparently remarkable result was obtained here that the initial rate of recovery from a given load was larger than the initial creep under that load. (See also Section 9.4 below.) The situation is illustrated in Figure 9.7.

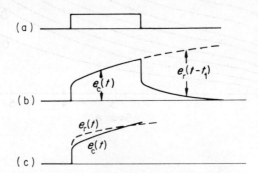

Figure 9.7. (a) Loading programme, (b) deformation and (c) direct comparison of creep $e_c(t)$ and recovery $e_r(t)$ for a non-linear viscoelastic solid.

Leaderman showed that if $\frac{1}{3}(\lambda - 1/\lambda^2)$ is used as a measure of the deformation, both creep and recovery, and creep curves at different load levels, can be described by a single time-dependent function. This is shown in Figure 9.8(a) and (b). The quantity $\frac{1}{3}(\lambda - 1/\lambda^2)$ is the equivalent quantity to the Lagrangian strain measure in the theory of finite elasticity.

Let us consider why using $\frac{1}{3}(\lambda - 1/\lambda^2)$ as a measure of the deformation brings the creep and recovery curves into coincidence. As Leaderman defines recovery (this is *not* how we have defined recovery previously in this textbook), recovery measures the quantity

$$\frac{1}{3}\left(\lambda_1 - \frac{1}{\lambda_1^2}\right) - \frac{1}{3}\left(\lambda_2 - \frac{1}{\lambda_2^2}\right),$$

where λ_1 is the extension at the time of unloading, and λ_2 is the extension at a chosen time after unloading. If e_1 is the conventional strain at the time of unloading and e_2 is the conventional strain at a chosen time after unloading, $\lambda_1 = 1 + e_1, \lambda_2 = 1 + e_2$ and the recovery as defined by Leaderman in terms of conventional strain is $e_1 - e_2$.

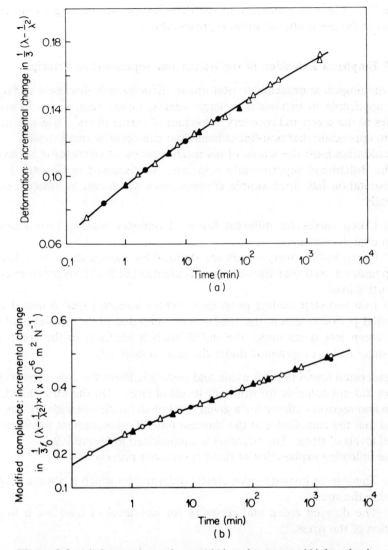

Figure 9.8. (a) Comparison of creep ($\triangle$) and recovery ($\blacktriangle$) for plasticized polyvinyl chloride under a constant nominal stress of $3.554 \times 10^5 \, \text{N m}^{-2}$. (b) Creep of plasticized polyvinyl chloride under constant nominal stress: $\bigcirc$, $f_0 = 4.443 \times 10^5 \, \text{N m}^{-2}$; $\bullet$, $f_0 = 3.554 \times 10^5 \, \text{N m}^{-2}$; $\triangle$, $f_0 = 2.667 \times 10^5 \, \text{N m}^{-2}$; $\blacktriangle$, $f_0 = 1.778 \times 10^5 \, \text{N m}^{-2}$. [Redrawn with permission from Leaderman, *Trans. Soc. Rheol.*, **6**, 361 (1962).]

Now a given change in the quantity $\frac{1}{3}(\lambda - 1/\lambda^2)$ at large λ (e.g. in the recovery situation where we change from λ_1 to λ_2), will involve a greater change in conventional strain $e_1 - e_2$ than it will at small λ (e.g. from $\lambda = 0$ in the creep situation). Thus recovery curves which coincide with creep curves using $\frac{1}{3}(\lambda - 1/\lambda^2)$ as a measure of the deformation will be larger than creep curves in the conventional strain representation.

9.3.3 Empirical Extension of the Boltzmann Superposition Principle

The rheological approaches to non-linear viscoelasticity discussed so far are only applicable to behaviour at large strains. Leaderman, in his extensive studies of the creep and recovery behaviour of textile fibres[16], was one of the first to appreciate that non-linear behaviour can occur at small strains.

Leaderman built the whole of his interpretation of viscoelastic behaviour on the Boltzmann superposition principle. As discussed in Chapter 5, this representation has three simple consequences which can be tested experimentally:

(1) Creep curves for different levels of one-step loading give a unique creep-compliance curve.

(2) Creep and recovery curves are identical for a given stress level, i.e. the creep under a particular constant load is identical to the recovery from creep under this load.

(3) In a two-step loading programme where a second load is added after an initial period of creep, the 'additional' creep due to the second load (i.e. total creep less creep under the initial load) is identical to the creep in a one-step loading programme under the second load only.

Leaderman found that for nylon and cellulosic fibres the creep compliance curves did not coincide for different levels of stress. On the other hand, the creep and recovery curves for a given level of stress did coincide. It was also found that the compliance at the shortest time of measurement was identical for all levels of stress. The situation is summarized in Figure 9.9.

The following explanation of these results was proposed:

(1) There is an instantaneous elastic deformation which is always proportional to the stress.

(2) The delayed creep and recovery for any level of load are a unique function of the stress.

This leads to a modified superposition principle,

$$e(t) = \frac{\sigma}{E} + \int_{-\infty}^{t} \frac{df(\sigma)}{d\tau} (t - \tau) \, d\tau, \tag{9.4}$$

where $f(\sigma)$ is an empirical function of stress.

For different fibres apparently arbitrary functions of stress were obtained.

Leaderman's extension of the Boltzmann superposition principle provided a satisfactory representation of the behaviour in creep and recovery for the fibres which he examined. We sill shortly see that it does not apply to the creep and recovery of all textile fibres. It will also appear that it does not

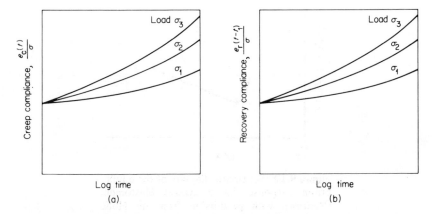

Figure 9.9. Comparison of creep compliance (a) and recovery compliance (b) at three load levels σ_1, σ_2, σ_3 for a non-linear viscoelastic material obeying Leaderman's modified Boltzmann superposition principle. Note that the creep and recovery curves for a given load level are identical.

describe the behaviour in more complicated loading programmes than creep and recovery.

9.3.4 More Complicated Single Integral Representations

The successful separation of the strain and time functions achieved by Smith and by Leaderman suggests that the non-linear behaviour may be dealt with by elaborating the measure of strain to embrace situations where finite strains occur. If we follow this hypothesis, another condition which it is reasonable to satisfy is that the stress–strain relationship, at very long times when equilibrium is reached, should be similar to that for the elastic behaviour of rubbers.

The elastic response of many rubbers has the form shown in Figure 9.10. A plot of $\sigma/(\lambda^2 - \lambda^{-1})$ against $1/\lambda$ shows a region with a linear relationship at low strains (high $1/\lambda$) and a minimum at large strains. The stress relaxation behaviour of certain elastomers, and plasticized polyvinyl chloride, presented in terms of isochronal stress–strain behaviour, takes a very similar form. This is shown in Figure 9.11, where data obtained by Bernstein et al.[17] and by Lianis and coworkers[18] have been selected. The similarity of the data serves also to emphasize that although these two sets of workers have adopted somewhat different starting points for their theoretical representations, the two theories must eventually converge. In fact, it will be seen that a series of time-dependent functions, each similar to the stress relaxation modulus of linear viscoelasticity, will be required.

Bernstein, Kearsley and Zapas[17] developed their theory (usually abbreviated to BKZ theory) in two forms, the first suitable for a solid and the second for a fluid. The first form is appropriate for elastomers at short times (and

210

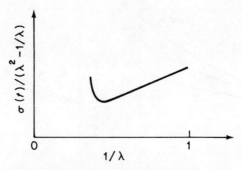

Figure 9.10. Schematic diagram of the equilibrium response for a typical elastomer. [Redrawn with permission from de Hoff, Lianis and Goldberg, *Trans. Soc. Rheol.*, **10**, 385 (1966).]

low temperatures) and the second at long times (and high temperatures). In both cases the material is considered to be incompressible.

The first form of BKZ theory derives the stress relaxation behaviour with two simplifying assumptions:

(1) There exist only *single* integral terms of the general form

$$\int_{-\infty}^{t} A(t-\tau)f[E(\tau)]\,d\tau,$$

where $E(\tau)$ is a generalized measure of strain.

(2) In equilibrium at long times the stress–strain relationship is consistent with a strain-energy function U such that

$$U = MJ_1 + \tfrac{1}{2}A_1J_1^2 + A_2J_2,$$

where A_1 and A_2 are the limiting values of the two time-dependent terms and

$$J_1 = \tfrac{1}{2}(I_1-3), \qquad J_2 = \tfrac{1}{4}(I_1-3)^2 + (I_1-3) - \tfrac{1}{2}(I_2-3).$$

I_1 and I_2 are the conventional strain invariants defined in Section 3.4.3. This form of the strain-energy function was first proposed by Signiorini[19] for rubbers.

The stress relaxation in simple extension for extension ratio λ is given by

$$\frac{\sigma(t)}{\lambda^2-\lambda^{-1}} = (\lambda^2-1)[\tfrac{1}{2}A_1(t)+A_2(t)] + \frac{1}{\lambda}[A_1(t)+A_2(t)] + A_3 - A_1(t). \quad (9.5)$$

The behaviour is therefore specified by two single integral independent stress relaxation functions $A_1(t)$ and $A_2(t)$ and a constant A_3.

The second form of BKZ theory describes an incompressible elastic fluid of the type proposed by Coleman and Noll[20] which has an elastic potential.

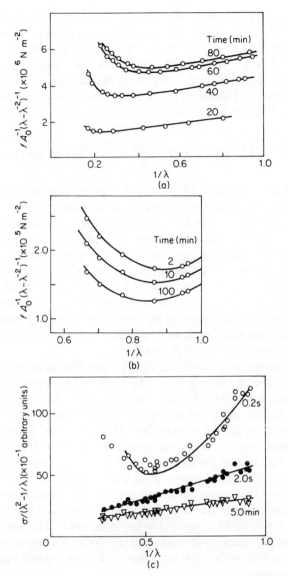

Figure 9.11. (a) Long-time equilibrium simple extension data on peroxide vulcanizates of natural rubber cured for different times as indicated. [Redrawn with permission from Bernstein, Kearsley and Zapas, *Trans. Soc. Rheol.*, **7**, 391 (1963).] (b) Isochronous data calculated from stress-relaxation curves on plasticized polyvinylchloride. [Redrawn with permission from Bernstein, Kearsley and Zapas, *Trans. Soc. Rheol.*, **7**, 391 (1963).] (c) Isochronous data for Estane polyurethane calculated from uniaxial stress-relaxation data. [Redrawn with permission from de Hoff, Lianis and Goldberg, *Trans. Soc. Rheol.*, **10**, 385 (1966).]

The stress at time t depends on the history of the relative deformation between the configuration at time t and all configurations at times prior to t. The effect of the configuration at a time $\tau < t$ depends on the time $(t - \tau)$, and the configuration at time τ is regarded as the preferred configuration, so that the elastic potential depends on E_τ, the strain measure at time τ, only through the strain invariants. This forms the link with finite elasticity theory.

The stress relaxation behaviour in simple extension is given by

$$\frac{\sigma(t)}{\lambda^2 - \lambda^{-1}} = (\lambda^2 - 1)[\tfrac{1}{2}A_1(t) + A_2(t)] + \frac{1}{\lambda}[A_1(t) + A_2(t)] + A_3(t) - A_1(t). \quad (9.6)$$

This is identical to equation (9.5), except that the constant term A_3 is now replaced by a single integral stress relaxation function $A_3(t)$.

Bernstein, Kearsley and Zapas[17] determined the stress relaxation behaviour in simple extension for polyisobutylene, vulcanized butyl rubber and plasticized polyvinyl chloride. Although the authors were modest in their assessment of the work, it can be firmly concluded that the second form of BKZ theory with a time-dependent A_3 term fitted the data very well. Zapas and Craft[21] later showed that double step and triple step stress relaxation results for polyisobutylene were very close to the behaviour predicted from simple step relaxation experiments. Creep and recovery, and constant strain rate data were also consistent with this representation.

In an even more ambitious study, Zapas[22] was able to correlate the results from biaxial strain and simple shear experiments by means of the potential function in the BKZ theory. This function is a time-dependent function analogous to the strain-energy function of finite elasticity and is assumed in the BKZ development of Coleman–Noll theory but has not been explicitly specified. Zapas' work is therefore analogous to the determination of the strain-energy function U for a rubber in terms of the strain invariants I_1 and I_2 (Section 3.5). The BKZ theory gives similar expressions for $\partial U / \partial I_1$ and $\partial U / \partial I_2$ in terms of the strain invariants, but U now depends on time as well as strain. Zapas showed that it was possible to represent and correlate data for vulcanized rubber to a good degree of accuracy. He was also able to correlate dynamic measurements on polyisobutylene solutions carried out in a torsion pendulum with constant shear rate experiments performed in a capillary rheometer.

In a parallel development, de Hoff, Lianis and Goldberg[18] took as their starting point the Coleman and Noll viscoelastic material in which stress relaxation can be described by 12 stress relaxation functions of the single integral type and three steady state coefficients. This representation has been called finite linear viscoelasticity. Lianis and his coworkers considered that the short term 'elastic' response will be similar in form to the equilibrium response of a rubber, which is shown schematically in Figure 9.10. It was shown that the three steady state coefficients could be obtained from the equilibrium response, and that there were only four independent relaxation functions. The stress relaxation behaviour of a polyurethane polymer and an ethylene–propylene rubber was examined. The polyurethane satisfied this simple representation where only the first strain invariant I_1 is involved. The behaviour of the ethylene–propylene rubber was more complicated but could be shown to fit a representation based on a linear combination of the two strain invariants I_1 and I_2. This representation involved five relaxation functions and additional tests were required to define these.

Lianis and his colleagues subsequently extended this work in a series of papers. In a study of stress relaxation for a styrene–butadiene rubber in combined tension–torsion, Goldberg and Lianis[23] first showed that the Lianis theory in its simplest form was identical to BKZ theory in the Signiorini form, and then proceeded to a treatment of the tension–torsion problem following that of Rivlin[24] for the finite deformation of an incompressible solid right circular cylinder. Step functions of twist and extension were applied and the subsequent stress relaxation behaviour determined. The results were very similar to those obtained for the equilibrium response of rubbers by Rivlin and Saunders[25], in that deviations from the simplest treatment were found at small strains, which would require a more complete expansion of strain invariants.

Goldberg and Lianis[26] used the same representation to describe the case of sinusoidal oscillations imposed on a finite strain, and McGuirt and Lianis[27] examined the extension of this representation first to isothermal single step relaxation over a range of temperatures and then to non-isothermal stress relaxation. In the latter case good agreement was obtained between experimental results for non-isothermal stress relaxation and those predicted from isothermal data.

Finally, Valanis and Landel[28] developed a very similar representation to that of Lianis, and hence to BKZ theory in the Signiorini form. The stress relaxation behaviour was given in terms of four relaxation functions, $G_1(t)$, $G_2(t)$, $G_3(t)$, $G_4(t)$, and current finite strain. Following the work of Smith[10], Valanis and Landel noted that time and strain are usually separable in rubbers, so that $G_1(t) = k_1 G(t)$, $G_2(t) = k_2 G(t)$, etc., where k_1, k_2, etc., are constants. Satisfactory fits were obtained to this simplified representation for a dimethyl-siloxane rubber, and in fact to a good approximation $k_4 = 0$, so that only three relaxation functions were required.

9.3.5 The Schapery Representation of Temperature-dependent Non-linear Viscoelasticity

Leaderman's results suggested that time–temperature equivalence held for the viscoelastic behaviour of the fibres he studied, and this was represented in terms of an activation energy. Eyring and coworkers[29] further showed that the data could be represented to a reasonable approximation by a standard linear solid (Section 5.2.7) with an activated dashpot. The activated dashpot gives the strain rate in the viscous element proportional to $\exp \Delta H / RT$, where

ΔH is the activation energy. This approach provides another starting point for consideration of non-linear viscoelastic behaviour and will be discussed in detail later. Here we wish to outline an attempt by Schapery to incorporate Leaderman's generalization of the Boltzmann single integral representation and temperature dependence into a comprehensive representation.

Consider first creep behaviour. Schapery[30] followed Leaderman in dividing the response into an immediate ·elastic compliance J_u ($J_{\text{unrelaxed}}$) and the delayed response $\Delta J(t) = J_1(t) - J_u$. The creep $e(t)$ is given by

$$e(t) = g_0 J_u \sigma(t) + g_1 \int_{-\infty}^{t} \Delta J(\psi - \psi') \frac{\mathrm{d}}{\mathrm{d}\tau} g_2[\sigma(\tau)] \, \mathrm{d}\tau, \qquad (9.7)$$

where ψ is the so-called reduced time defined by

$$\psi = \psi(t) = \int_0^t \mathrm{d}t'/a_\sigma[\sigma(t')], \qquad a_\sigma > 0,$$

and

$$\psi' = \psi(\tau) = \int_0^\tau \mathrm{d}t'/a_\sigma[\sigma(t')].$$

g_0, g_1, g_2 and a_σ are material properties which are functions of stress.

Schapery's representation contains several elements which are well known to apply in a general sense. For example, the compliance changes with increasing applied stress in the sense that the major relaxation moves to shorter times. Schapery attempts to provide a thermodynamic basis for his representation, proposing that g_0, g_1 and g_2 reflect third and higher order terms of the dependence of the Gibbs free energy on applied stress, and a_σ arises from similar high-order effects in both entropy production and free energy.

We have an analogous equation for stress relaxation:

$$\sigma(t) = h_e G_r e(t) + h_1 \int_{-\infty}^{t} \Delta G(\rho - \rho') \frac{\mathrm{d}}{\mathrm{d}\tau} [h_2 e(\tau)] \, \mathrm{d}\tau, \qquad (9.8)$$

where $\Delta G(t) = G(t) - G_r$ ($G_r = G_{\text{relaxed}}$) and the reduced time ρ is defined as

$$\rho = \rho(t) = \int_0^t \mathrm{d}t'/a_e[e(t')], \qquad a_e > 0,$$

and

$$\rho' = \rho(\tau) = \int_0^\tau \mathrm{d}t'/a_e[e(t')].$$

h_e, h_1, h_2 and a_e are material properties analogous to g_0, g_1, g_2 and a_σ but are now functions of strain, and are related by Schapery to thermodynamic functions.

For creep after single step loading and stress relaxation from single step strain, the non-linear creep compliance $J_n(t)$ and the non-linear stress relaxation modulus $G_n(t)$ are given by

$$J_n(t) = \frac{e(t)}{\sigma} = g_0 J_n + g_1 g_2 \, \Delta J(t/a_\sigma), \qquad (9.9)$$

$$G_n(t) = \frac{\sigma(t)}{e} = h_e G_r + h_1 h_2 \, \Delta G(t/a_e). \qquad (9.10)$$

A number of the representations already discussed can now be seen as special cases. For example, Leaderman's representation is the special case where $g_0 = a_\sigma = 1$. Findley's equation for creep is given by

$$g_0 = \frac{\sinh \sigma/\sigma_e}{\sigma/\sigma_e}, \qquad \frac{g_1 g_2}{a_\sigma^n} = \frac{\sinh \sigma/\sigma_m}{\sigma/\sigma_m}$$

and $\Delta J(\psi) = J_1 \psi^n$, where σ_e, σ_m, J_1 and n are material constants. Smith's representation for elastomers is given by $h_e = h_1 h_2$, $a_e = 1$, so that $G_n(t) = h_1 h_2 G(t)$, i.e. the separable product form for the strain and time variables.

Finally it is interesting to note that the effect of stress on the time dependence of stress relaxation for textile fibres has been shown by Meredith to approximate to the simple form

$$G_n(t) = G_r + \Delta G(t/a_e) = G(t/a_e).$$

This assumes that the effect of stress is to introduce a scaling factor in the time scale of the response only.

The Schapery theory incorporates time–temperature equivalence, in that it has the 'horizontal' shifts in time scale and the 'vertical' shifts. As discussed in Section 7.2, even in a linear viscoelastic material h_e and h_1 will in general vary with temperature. Moisture is another variable which in polymers such as nylons and polysulphones changes the relaxation behaviour in a similar manner to temperature and can be dealt with by the Schapery representation.

Equations (9.9) and (9.10) give g_0, a_σ and the product $g_1 g_2$ and h_e, a_e and the product $h_1 h_2$ as functions of stress and strain respectively. To separate the terms in the products $g_1 g_2$ and $h_1 h_2$ we require multiple loading programmes. It is interesting to note that the 'initial elastic recovery' $\Delta e_0 = g_2(g_1^{-1})\Delta J(t/a_\sigma)\sigma$ is greater than the initial response. This is a result often observed in polymers, but not predicted by Leaderman's simpler representation (equation (9.4)) (see also Section 9.3.6 below).

Schapery showed that the representation provided a good description of creep and recovery data for nitrocellulose (the Van Holde data discussed in Section 9.2.2) and for a fibre-reinforced phenolic resin, and of the Zapas and Craft data for the multistep tension straining of polyisobutylene; the latter providing a link with BKZ theory. Schapery also considered the case of multiaxial loading, taking the combined tension and torsion data of Onaran and Findley[28]. He concluded that the time dependence was carried through all the tests by multipliers f_{11} and f_{12} corresponding to g_1, g_2, etc., which related to the normal stresses (f_{11}) and shear stresses (f_{12}) and were functions of the octahedral shear stress only. Moreover, the dependences of the tensile and shear compliances were such that f_{12}/f_{11} was a constant ($\sim$1.35), implying a constant Poisson's ratio.

9.4 THE MULTIPLE INTEGRAL REPRESENTATION

The representations of non-linear viscoelastic behaviour discussed have all incorporated the assumption that for stress relaxation the strain and time dependence of the response are separable and for creep the stress and time dependence. We have seen how this assumption is retained even in the sophisticated BKZ theories of finite linear elasticity, and in the Schapery representation where non-linearity is introduced by modifying the single integral creep compliance or stress relaxation functions by an appropriate shift in time scale.

The inadequacy of such representations for describing the viscoelastic behaviour of polymers in other than the rubber-like response regime was brought out by a study of oriented polypropylene fibres. The results will be considered in detail, as they illustrate very well the formal inadequacies of previous representations.

Figure 9.12 shows the creep and recovery curves against time for a polypropylene fibre at a number of different stress levels; this illustrates two features of the behaviour very clearly.

(1) Creep and recovery only coincide at the lowest stress level. This suggests that there is a linear viscoelastic region at low stresses, but that the behaviour is strikingly non-linear at higher stresses.

(2) The 'instantaneous' or short term recovery is always greater than the instantaneous or short term creep.

The second feature eliminates the possibility of using Leaderman's original formulation. He considered that the creep and recovery curves could be separated into two parts; an instantaneous or elastic response and a delayed deformation. Clearly for this polypropylene fibre there is no definitive measure of the elastic response.

The following illustration (Figure 9.13(a) and (b)) shows the creep and recovery compliances for different levels of stress. The creep compliance curves are approximately independent of stress at lowest stress levels, as would be anticipated from Figure 9.12, but the recovery curves never coincide. Figure 9.14 shows the comparison of creep and recovery with additional creep. The additional creep curves are always in excess of the initial creep and show the same increase in 'instantaneous' strain as the recovery curves.

It can be seen that the simple tests of the Boltzmann superposition principle are not fulfilled for any case. Furthermore Leaderman's modification is not applicable for two reasons:

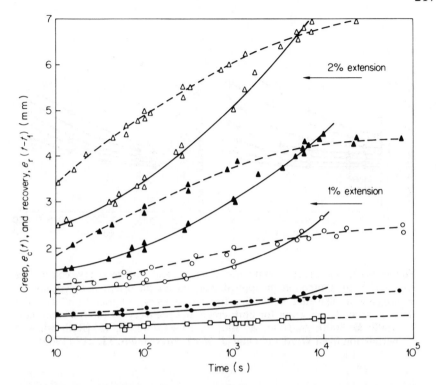

Figure 9.12. Successive creep (———) and recovery (– – –) for an oriented monofilament of polypropylene of total length 302 mm. The load levels are 587 g (△), 401.8 g (▲), 281 g (●) and 67.6 g (□). [Redrawn with permission from Ward and Onat, *J. Mech. Phys. Solids*, **11**, 217 (1963).]

(1) Creep and recovery curves are not identical.
(2) There is not a linear 'instantaneous response'.

These results led Ward and Onat[32] to examine the applicability of the multiple integral representations for viscoelasticity which had been proposed several years earlier by Green and Rivlin[33–35].

Consider the loading programme of Figure 9.15, in which incremental stresses $\Delta\sigma_1(\tau_1)$, $\Delta\sigma_2(\tau_2)$, $\Delta\sigma_3(\tau_3)$, ..., etc., are added at the times τ_1, τ_2, τ_3, etc. For a linear system the deformation $e(t)$ at time t is given by

$$e(t) = \Delta\sigma_1 J_1(t - \tau_1) + \Delta\sigma_2 J_1(t - \tau_2) \ldots,$$

where $J_1(t)$ is the creep compliance function. Let us now admit terms which arise from the *joint* contributions of the loading steps $\Delta\sigma_1(\tau_1)$, $\Delta\sigma_2(\tau_2)$, etc., to the final deformation.

These terms are taken to be of the form

$$+\Delta\sigma_1 \Delta\sigma_2 J_2(t - \tau_1, t - \tau_2) + \Delta\sigma_1 \Delta\sigma_2 \Delta\sigma_3 J_3(t - \tau_1, t - \tau_2, t - \tau_3) + \text{etc.},$$

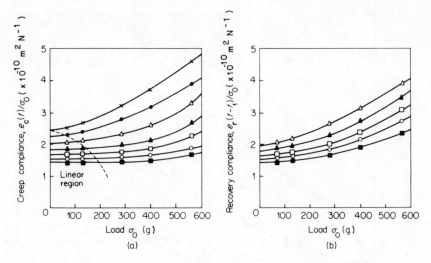

Figure 9.13. (a) Creep compliance $e_c(t)/\sigma_0$ and (b) recovery compliance $e_r(t-t_1)/\sigma_0$ as a function of applied load σ_0 for an oriented polypropylene monofilament. The time of loading for the recovery test was 9.3×10^3 s. The times in seconds are given for both creep and recovery by the following key: ×, 9300; ●, 3000; △, 1000; ▲, 300; □, 100; ○, 40; ■, 15. [Redrawn with permission from Ward and Onat, *J. Mech. Phys. Solids*, **11**, 217 (1963).]

where the 'memory functions' J_2, J_3, etc., are functions of *the differences in time* $t - \tau_1$, $t - \tau_2$, $t - \tau_3$, etc., between the instant in time at which the deformation is measured and the instant at which a given increment of stress $\Delta\sigma$ is applied. This condition was also used in setting up the Boltzmann principle.

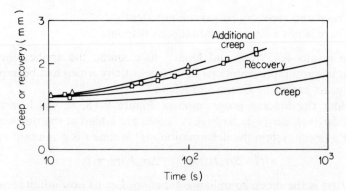

Figure 9.14. Comparison of creep and recovery curves for a load of 281 g with additional creep due to addition of a further 281 g after 3000 s (□) and 1000 s (△) respectively. [Redrawn with permission from Ward and Onat, *J. Mech. Phys. Solids*, **11**, 317 (1963).]

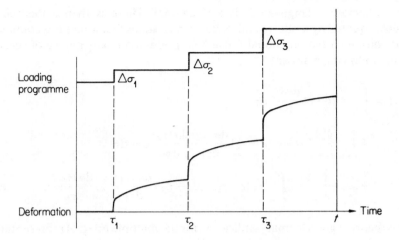

Figure 9.15. A multiple step loading programme.

For a continuous load history, the deformation $e(t)$ can then be written as an integral:

$$e(t) = \int_{-\infty}^{t} J_1(t-\tau) \frac{d\sigma(\tau)}{d\tau} d\tau$$

$$+ \int_{-\infty}^{t} \int_{-\infty}^{t} J_2(t-\tau_1, t-\tau_2) \frac{d\sigma(\tau_1)}{d\tau_1} \frac{d\sigma(\tau_2)}{d\tau_2} d\tau_1 d\tau_2$$

$$+ \ldots + \int_{-\infty}^{t} \ldots \int_{-\infty}^{t} J_N(t-\tau_1 \ldots t-\tau_N) \frac{d\sigma(\tau_1)}{d\tau_1} \ldots \frac{d\sigma(\tau_N)}{d\tau_N} d\tau_1 \ldots d\tau_N.$$

The first term is the Boltzmann superposition principle term, i.e. the linear term.

This representation is at first sight extremely complicated and the discussion will now be directed towards two features: (1) the mathematical rigour of the representation; (2) the practical application of the representation.

9.4.1 Mathematical Rigour

There are two theoretical aspects concerning this representation.

(1) It is being assumed that the elongation of the specimen at time t depends on all the previous values of the rate of loading to which the specimen has been subjected, i.e. the elongation is assumed to be a function of the history of the rate of loading:

$$e(t) = F\left[\frac{d\sigma(\tau)}{d\tau}\right]_{\tau=-\infty}^{t}$$

In mathematical language F is a functional. There is then a theorem by Fréchèt quoted by Volterra and Pérès[36] which states that where F is continuous and non-linear the functional F can be represented to any degree of accuracy in the following manner:

$$e(t) = \int_{-\infty}^{t} J_1(t-\tau) \frac{d\sigma(t)}{d\tau} d\tau$$

$$+ \int_{-\infty}^{t} \int_{-\infty}^{t} J_2(t-\tau_1, t-\tau_2) \frac{d\sigma(\tau_1)}{d\tau_1} \frac{d\sigma(\tau_2)}{d\tau_2} d\tau_1 d\tau_2$$

$$+ \ldots + \int_{-\infty}^{t} \ldots \int_{-\infty}^{t} J_N(t-\tau_1, \ldots t-\tau_N) \frac{d\sigma(\tau_1)}{d\tau_1} \ldots \frac{d\sigma(\tau_N)}{d\tau_N} d\tau_1 \ldots d\tau_N.$$

This theorem gives formal justification for the multiple integral representation.

(2) The representation as discussed so far is for a one-dimensional situation. Green and Rivlin[33] have given a more complete development which does not suffer from this limitation.

Green and Rivlin[33] consider stress relaxation rather than creep. It is assumed that the stress at time t depends on the displacement gradients at time t and at N previous instants of time in the interval 0 to t. After considering the restrictions imposed by invariance under a rigid rotation, Green and Rivlin allow N to approach infinity, and obtain a multiple integral representation for general non-linear viscoelastic behaviour. Their representation describes stress relaxation. For the most general type of deformation it is not possible simply to invert the relationship to describe creep. This is because the stress functional involves the displacement gradients. Thus, during the deformation the stress components, in a fixed coordinate system, will change depending on the rotation of the body.

These objections do not apply to a one-dimensional situation, to which the remainder of this chapter is devoted.

9.4.2 The Practical Application of the Multiple Integral Representation

The immediate objection to the multiple integral representation is that it is in principle so general that the data *must* be fitted, because there is an infinite number of curve-fitting constants.

This representation is therefore only useful if the following apply:

(1) The data can be fitted to a small number of multiple integral terms.
(2) The behaviour under complex loading programmes can be predicted from a few simple loading programmes.
(3) The representation should at least assist an empirical understanding of the relationships between viscoelastic behaviour and the structure of the polymer.

With these qualifications in mind, the results for a polypropylene fibre[32,37,38] will be examined in detail.

For creep at a constant level of stress σ_0 the multiple integral representation gives

$$e_c(t) = J_1(t)\sigma_0 + J_2(t, t)\sigma_0^2 \ldots J_N(t, \ldots t)\sigma_0^N \ldots . \tag{9.11}$$

This immediately suggests that it will be of greater interest to examine the creep data in a different fashion, by plotting compliance $e(t)/\sigma_0$ against stress σ_0 for various fixed values of t, i.e. isochronous compliance curves.

The results are shown in Figure 9.13(a). For a linear viscoelastic material, there would be a series of straight lines parallel to the load axis. In fact a small region approximates to this and has been so designated. In general the curves are parabolic, which suggests that

$$\frac{e_c(t)}{\sigma_0} = A + B\sigma_0^2,$$

where $A = J_1(t)$ and $B = J_3(t, t, t)$ are functions of time only.

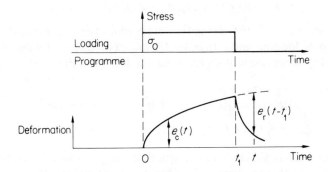

Figure 9.16. Creep and recovery programme.

It appears therefore that the non-linear behaviour of this particular specimen may be represented by retaining the first- and third-order terms only. If this is correct the recovery response (Figure 9.16) in a creep and recovery test where $\sigma = 0$, $\tau < 0$; $\sigma = \sigma_0$, $0 < \tau < t_1$; $\sigma = 0$, $\tau > t_1$, is given by

$$e_r(t - t_1) = J_1(t - t_1)\sigma_0 + J_3(t - t_1, t - t_1, t - t_1)\sigma_0^3$$

$$+ 3[J_3(t, t, t - t_1) - J_3(t, t - t_1, t - t_1)]\sigma_0^3 \ldots . \tag{9.12}$$

This shows that if our representation is adequate, recovery will take the form

$$\frac{e_r(t - t_1)}{\sigma_0} = A' + B'\sigma_0^2,$$

where $A' = J_1(t - t_1)$ and

$$B' = J_3(t - t_1, t - t_1, t - t_1) + 3[J_3(t, t, t - t_1) - J_3(t, t - t_1, t - t_1)]$$

are functions of time only.

The results are shown in Figure 9.13(b), and demonstrate that this is correct. Equation (9.12) also shows that the recovery $e_r(t - t_1)$ in this test would be expected to be greater than the creep $e_c(t - t_1)$ in a creep test under constant stress σ_0 because we would expect $J_3(t, t, t - t_1) > J_3(t, t - t_1, t - t_1)$. The latter inequality is self-evident for linear creep behaviour, where it only states that $J(t) > J(t - t_1)$. This follows from the experimental fact that $J(t)$ is an increasing function of time t. The difference between 15 s recovery and 15 s creep was measured, and, as would be expected by comparing equations (9.11) and (9.12), this (additional recovery compliance), $(e_r - e_c)/\sigma$ is also a parabolic function of the stress.

The third type of test employed is a two-step loading test with the loading programme

$$\sigma = 0, \quad \tau < 0; \qquad \sigma = \sigma_0, \quad 0 < \tau < t_1; \qquad \sigma = 2\sigma_0, \quad \tau > t_1.$$

The additional creep response is defined as

$$e'_c(t - t_1) = e'(t, \sigma_0, t_1, \sigma_0) - e_c(t, \sigma_0), \qquad t > t_1,$$

where $e'(t, \sigma_0, t_1, \sigma_0)$ is the elongation measured after the application of the second step of loading and $e_c(t, \sigma_0) = J_1(t)\sigma_0 + J_3(t, t, t)\sigma_0^3$. Evaluation of the multiple integral representation retaining the first- and third-order terms only gives

$$e'_c(t - t_1) = J_1(t - t_1)\sigma_0 + J_3(t - t_1, t - t_1, t - t_1)\sigma_0^3$$
$$+ 3[J_3(t, t, t - t_1) + J_3(t, t - t_1, t - t_1)]\sigma_0^3 \ldots. \tag{9.13}$$

The results shown in Figure 9.17 confirm that only first- and third-order terms are required for this specimen. Comparison of equations (9.12) and (9.13) implies that the additional creep should be greater than the recovery for a given stress level, and this is confirmed by the experimental results shown in Figure 9.14.

This analysis affords a consistent if somewhat more complicated interpretation of the complexity of the viscoelastic behaviour, and offers some insight into the failure of simple superposition laws as discussed by Turner and reviewed in Section 9.2.1 above.

Attempts have also been made[38] to predict the behaviour in complex loading programmes from creep, recovery and superposition of identical loads. The predictions for these same polypropylene fibres for loading programmes involving superposition of a constant stress σ_0 and stresses of $2\sigma_0$ and $3\sigma_0$ respectively, after a fixed interval of time, are shown in Table 9.1. The results are somewhat surprising. Although the creep and recovery are markedly non-linear and very significantly different (Table 9.2) the corrections obtained

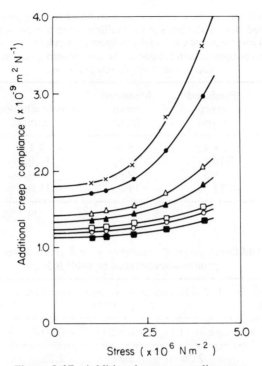

Figure 9.17. Additional creep compliance as a function of applied stress for an oriented polypropylene monofilament. [Redrawn with permission from Ward and Wolfe, *J. Mech. Phys. Solids*, **14**, 131 (1966).]

by this sophisticated representation are small. It is only just convincing that this representation is better than a simple addition superposition law (compare second and fourth columns in Table 9.1). These results show, as the multiple integral representation emphasizes, that creep and superposition measurements can be very misleading with regard to recovery behaviour. In fact, a modified superposition rule proposed by Pipkin and Rogers[39], on the basis of work by Findley and Lai[40], can bring the predicted and measured creep under complex loading programmes to very close agreement in this case. Their proposal is to add the *incremental* difference between the creep under stress σ_1 to that under stress σ_2 (see Figure 9.18).

This gives for a loading programme σ_1 at zero time followed by σ_2 at time $t = t_1$:

$$e(t) = e_c(t, \sigma_1) + e_c(t - t_1, \sigma_1 + \sigma_2) - e_c(t - t_1, \sigma_1).$$

Such simplifications are extremely interesting, but they cannot be applied to an unknown polymer without first establishing the relative magnitude of the different multiple integral terms.

Table 9.1. Comparisons of measured creep in complex loading programmes with the creep response predicted on the basis of the multiple integral representation (predicted creep, second column) and with the creep response calculated by simple addition of individual creep responses (fourth column). The last column gives the correction factor due to the multiple integral term.

Loading programme	Predicted creep (mm)	Measured creep (mm)	Simple addition (mm)	Correction factor (mm)
1 $\sigma_0 + 2\sigma_0$	8.4_6	8.7_2	8.1_1	0.35
2 $2\sigma_0 + \sigma_0$	10.6_7	10.6_8	10.0_5	0.62
3 $\sigma_0 + 3\sigma_0$	8.3_4	8.6_8	7.9_8	0.36
4 $3\sigma_0 + \sigma_0$	11.5_9	11.5_6	11.0_6	0.53

Table 9.2. Additional creep e'_c, recovery e_r and creep e_c for the loading programmes detailed in Table 9.1.

Loading programme (from Table 9.1)	Individual terms e'_c, e_r, and e_c (mm) for $t - t_1 = 100$ s			Correction factor (mm)
	$e'_c(t - t_1)$	$e_r(t - t_1)$	$e_c(t - t_1)$	
1	2.36	2.27	2.23	0.35
2	6.10	5.09	4.63	0.62
3	1.73	1.68	1.66	0.36
4	7.37	6.08	5.37	0.53

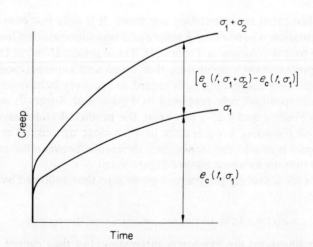

Figure 9.18. Comparison of creep under applied load σ_1 and $(\sigma_1 + \sigma_2)$ for a non-linear viscoelastic solid.

9.4.3 Further Applications of the Multiple Integral Representation

Several workers have examined further the applications of the multiple integral representation. The formal theoretical development is discussed fully in a text-book by Lockett[41] and Findley and coworkers have been prominent in carrying out experimental studies. In two related publications, Lai and Findley[42,43] described a method for predicting uniaxial stress relaxation from non-linear creep data, and then examined the behaviour of a polyurethane polymer in the light of their predictions. Only terms from the first three orders were retained, and moreover a product form was assumed for the creep kernels, with

$$G_1(t - \tau_1) = a_1 - b_1(t - \tau_1)^m,$$

$$G_2(t - \tau_1, t - \tau_2) = a_2 - b_2(t - \tau_1)^{m/2}(t - \tau_2)^{m/2},$$

$$G_3(t - \tau_1, t - \tau_2, t - \tau_3) = a_3 - b_3(t - \tau_1)^{m/3}(t - \tau_2)^{m/3}(t - \tau_3)^{m/3}.$$

To calculate the stress relaxation from creep, a term by term inversion was carried out by substitution of the stress relaxation $\sigma(t)$ into the integral equation for creep. A reasonably good fit was obtained. Multiple step loading was also examined, comparing two approximate procedures, assuming (a) the product forms for the kernel functions, (b) the modified incremental superposition principle discussed above. It was concluded that the latter fitted the data better, although both approximations gave good results.

Although the temperature dependence of viscoelastic behaviour is an extremely important guide to any structural or molecular interpretation, there have been comparatively few studies of this nature. Morgan and Ward[44] carried out creep, recovery and load superposition experiments in the temperature range 28–60 °C on an oriented polypropylene monofilament similar to that studied by Ward and Onat. It was found that the creep curves for a given stress at different temperatures could be superposed by a simple horizontal shift. Moreover the shifts were found to be identical for all stress levels. No vertical shift was required, which is reasonable in view of the large changes in creep compliance with temperature. Figure 9.19 shows that the same shift factors also produced excellent superposition of the recovery data. Furthermore, the multiple integral representation would imply, in the simplest formulation, that superposition experiments conducted at different temperatures would require the same shift factors to be applied to the loading time t as to the response times t and $t - t_1$. Figure 9.20 confirms that the shift factor required to superpose creep data e_c at two temperatures (curve A) can also superpose additional creep data e_c' following superposition of an identical load at two temperatures (curve B).

Similar results have also been obtained for isotropic polyethylene terephthalate, where again temperature shift factors were found to be independent of stress. It would not be wise to assume, however, that such simplicity will hold in all cases. In both these investigations the temperatures were in the glass

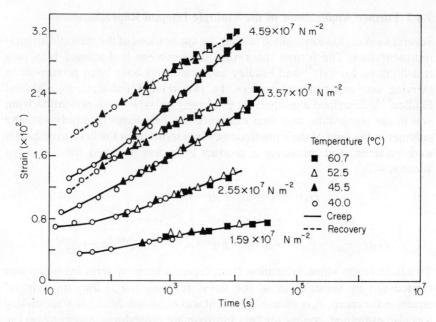

Figure 9.19. Master creep and recovery curves for an oriented polypropylene monofilament reduced to 40 °C. [Redrawn with permission from Morgan and Ward, *J. Mech. Phys. Solids*, **19**, 165 (1971).]

transition temperature range, and it may well be that more complicated representations are required elsewhere.

Finally, it is worth noting that there have been very few attempts to use the multiple integral representation in systematic studies of the influence of structure on the viscoelastic behaviour of polymers. Some preliminary work along these lines by Hadley and Ward[37] showed that the nature of the non-linear behaviour in polypropylene fibres was very much affected by the degree of molecular orientation and possibly by the morphological structure. Results for different levels of molecular weight were similar, although the absolute magnitude of the compliance level was significantly reduced with increasing molecular weight.

9.4.4 Implicit Equation Approach

Some unsatisfactory implications of arbitrarily truncating the multiple integral representation were examined in a publication by Brereton, Croll, Duckett and Ward[45], which proposed an *implicit* equation approach so that creep, stress relaxation and constant strain rate data could be readily related. As will be discussed more fully in Chapter 11, constant strain rate data are of particular importance in studying the yield behaviour of polymers.

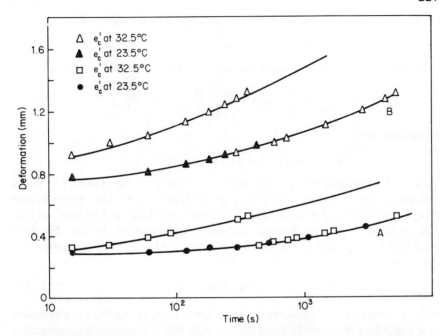

Figure 9.20. Superposition data for an oriented polypropylene monofilament. Curve A, initial creep e_c; curve B, additional creep e'_c. [Redrawn with permission from Morgan and Ward, *J. Mech. Phys. Solids*, **19**, 165 (1971).]

The multiple integral representation for constant strain rate behaviour takes the form

$$\sigma(t) = \int_{-\infty}^{t} G_1(t-\tau_1) \frac{de}{d\tau_1} \, d\tau_1 + \int_{-\infty}^{t} \int_{-\infty}^{t} G_2(t-\tau_1, t-\tau_2) \frac{de_1}{d\tau_1} \frac{de_2}{d\tau_2} + \dots \quad (9.14)$$

If $de/d\tau = \dot{e}_0$, the constant strain rate, then

$$\sigma(t) = \dot{e}_0 I_1(t) + \dot{e}_0^2 I_2(t, t),$$

where

$$I_1(t) = \int_{-\infty}^{t} G_1(t-\tau_1)\tau_1 \, d\tau_1,$$

etc. Equation (9.14) can therefore be regarded as a polynomial in $\dot{e}_0$, for which the coefficients of the various terms may be evaluated at the various instants of time. An oriented polyethylene terephthalate sheet was selected for detailed study, and the coefficients obtained from a curve-fitting procedure based on Forsythe's method[46] using orthogonal Chebyshev polynomials. The remarkable result was obtained that over an appreciable period of time there existed a factorization of up to nine kernel terms such that

$$I_n(t_1 t_1 \dots t) = (-p)^{n-1} I_1^n(t),$$

i.e. the stress can be represented by a convergent series such that

$$\sigma(t) = \dot{e}_0 I_1(t) - p\dot{e}_0^2 I_1^2(t) + p^2 \dot{e}_0^3 I_1^3(t) + \ldots + p^8 \dot{e}_0^9 I_1^9(t).$$

If it is assumed that this series extends to infinity, then its sum gives

$$\sigma(t) = \frac{\dot{e}_0 I_1(t)}{1 + p\dot{e}_0 I_1(t)}$$

or equivalently

$$\sigma(t) = I_1(t)\dot{e}_0 - pI_1(t)\dot{e}_0\sigma(t).$$

This is a simple integral equation with a constant p of dimensions I^{-1} which generates the multiple integral representation. For the polyethylene terephthalate sheet p was not quite a constant but depended on time, so that a slightly more general integral equation was required. It was therefore proposed that the behaviour should be described by an implicit equation of the form

$$a\sigma + be + c\sigma e = 0 \qquad (9.15)$$

with two material response functions, a linear term $j(t)$ related symbolically to a and b by $j = a/b$ and a term $j_2(t) = -c/b$, which expresses the interaction between stress and strain, and admits that this will depend on the history of the treatment of the sample. It was found adequate to make the further approximation of assuming that this interaction term $\int j_2(t - \tau_1, t - \tau_2)\sigma e$ has its overwhelming contribution from products containing the diagonal term $j_2(t - \tau, t - \tau)$, i.e.

$$j_2(t_1, t_2) \simeq j_2(t_1)\delta(t_1 - t_2).$$

This implies that the polymer has a strongly fading memory for non-linear effects, these depending primarily on *current* stress and strain.

It was found that the creep rate for the polyethylene terephthalate sheet increased markedly with increasing stress. This result is consistent with the creep behaviour predicted by this implicit representation where the creep strain $e(t)$ for creep under a constant stress σ_0 is given by

$$e(t) = \frac{j(t)\sigma_0}{1 - k\sigma_0}.$$

This equation predicts that creep becomes infinite for $\sigma_0 > 1/k$, which was identified with a critical stress for yielding or flow.

To exemplify the predictions of this implicit representation it was assumed that the response functions $j(t)$ and $j_2(t)$ took a simple form analogous to the standard linear solid representation for linear viscoelastic behaviour. This gives

$$j(t) = A\delta(t) + B \exp(-t/\tau_1), \qquad (9.16)$$

$$j_2(t) = C\delta(t) + D \exp(-t/\tau_2), \qquad (9.17)$$

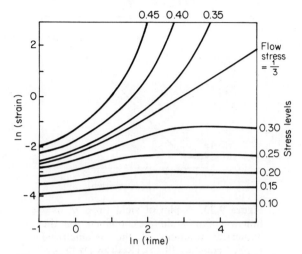

Figure 9.21. A plot of ln (strain) against ln (time) as predicted by the implicit representation (9.15) parametrized using equations (9.16) and (9.17) for various applied stress levels. [Redrawn with permission from Brereton, Croll, Duckett and Ward, *J. Mech. Phys. Solids*, **22**, 97 (1974).]

where τ_1, τ_2 are retardation times for the linear and the non-linear responses respectively.

The principal predictions of this representation are then summarized in Figures 9.21–9.23.

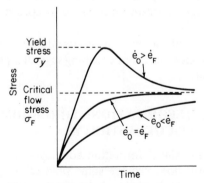

Figure 9.22. Stress–strain curves at various applied strain rates for a material described by the implicit representation (9.15), parametrized using equations (9.16) and (9.17). [Redrawn with permission from Brereton, Croll, Duckett and Ward, *J. Mech. Phys. Solids*, **22**, 97 (1974).]

230

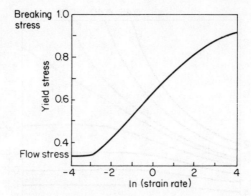

Figure 9.23. A plot of yield stress σ_y as a function of ln (strain rate) obtained from the numerical solutions of the parametrized implicit equation (9.15). [Redrawn with permission from Brereton, Croll, Duckett and Ward, *J. Mech. Phys. Solids*, **22**, 97 (1974).]

First, Figure 9.21 shows that there is a critical stress level above which creep to failure can occur. Below this level there is a finite equilibrium response at long times. This stress level has been termed the *flow* stress σ_F. Secondly, Figure 9.22 shows that the stress–strain curve in constant strain rate tests changes with strain rate. The flow stress δ_F is now seen as akin to a drawing stress (see Section 11.5), and at sufficiently high strain rates a peak in true stress is observed corresponding to the yield stress σ_y. Thirdly, Figure 9.23 shows that the yield stress σ_Y increases approximately linearly with ln (strain rate) over a wide range of strain rates, which is consistent with observation (Section 11.6.1).

These results show that the implicit representation can provide a comprehensive phenomenological base for describing several aspects of the non-linear viscoelastic behaviour.

9.5 MULTIAXIAL DEFORMATION: THREE-DIMENSIONAL NON-LINEAR VISCOELASTICITY

Consideration of multiaxial deformation adds complexity to the discussion, but we shall see that it does provide some physical weight for the origin of non-linear behaviour. Furthermore, it is of relevance to the technological applications of polymers to consider complex strain and stress situations.

9.5.1 Octahedral Shear Stress Theories

The first approach to three-dimensional creep behaviour stems from the intuitive hypothesis that creep is caused by the shearing of molecules past one another. This leads naturally to an octahedral shear stress theory, and an early formulation was by Pao and Marin[3] who extended their uniaxial creep law (equation (9.2) above) to the three-dimensional situation. The principal strains e_1, e_2, e_3 are related to the principal components of stress σ_1, σ_2, σ_3 by expressions such as

$$e_1 = \frac{1}{E}[\sigma_1 - \nu(\sigma_2 - \sigma_3)] + \frac{1}{2}(2\sigma_1 - \sigma_2 - \sigma_3)(J_2)^{(n-1)/2}[K(1 - e^{-qt}) + Bt]. \quad (9.18)$$

There are three major assumptions in this representation:

(1) Because only shear stresses cause creep, σ^n in the one-dimensional representation is replaced by $(\sqrt{J_2})^n$, where

$$J_2 = \frac{1}{2}[(\sigma_1 - \sigma_2)^2 + (\sigma_2 - \sigma_3)^2 + (\sigma_3 - \sigma_1)^2]$$

and is therefore proportional to the octahedral shear stress.

(2) The time-dependent part of the deformation is accompanied by zero volume change. The Poisson ratio ν refers, of course, only to the 'immediate elastic deformation'.

(3) The response is identical in tension and compression.

The most comprehensive experimental studies which enable this representation to be tested have been undertaken by Benham and his coworkers[47-49]. Their experiments included uniaxial tension and compression, shear and hydrostatic compression, on isotropic polyvinyl chloride, polymethyl methacrylate and polypropylene. Typical results for the creep of polymethyl methacrylate[47] are shown in Figure 9.24. It can be seen that the behaviour in tension and compression is not identical, showing that assumption (3) above is not justified. The other two assumptions of Pao and Marin do, however, make a useful starting point. Benham suggested that the shear and uniaxial creep behaviour could be related to a good approximation using the concept of an equivalent strain proportional to the octahedral shear strain and an equivalent stress proportional to the octahedral shear stress. To take into account the difference between behaviour in tension and compression the mean value of the octahedral shear strains in tension and compression was equated to the octahedral strain in pure shear. The results for polymethyl methacrylate are shown in Figure 9.25, and confirm that this procedure is quite satisfactory.

Benham and McCammond also showed that for polymethyl methacrylate and polypropylene, the volume did not change with time, which is consistent with assumption (2) above. For polyvinyl chloride a change in volume with time was observed, particularly at the highest stresses, but this was attributed to the formation of voids.

Another important result obtained by Benham and his colleagues is shown in Figure 9.26. This shows the volumetric isochronous stress–strain curves for polymethyl methacrylate obtained from the volume change under hydrostatic

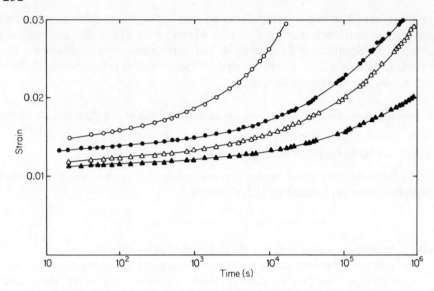

Figure 9.24. Creep curves for polymethyl methacrylate. Stress, $4.2 \times 10^7 \, \text{N m}^{-2}$: $\bigcirc$, tension, axial strain; $\bullet$, compression, axial strain. Stress $3.5 \times 10^7 \, \text{N m}^{-2}$: $\triangle$, tension, axial strain; $\blacktriangle$, compression, axial strain. [Redrawn with permission fron Benham and McCammond, *Plastics Polymers*, **39**, 130 (1971).]

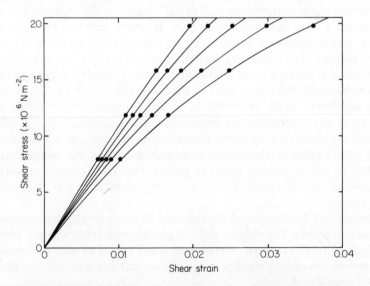

Figure 9.25. Prediction of shear strain from uniaxial data for polymethyl methacrylate. ———, From shear–creep tests; $\bullet$, from uniaxial data. [Redrawn with permission from Benham and Mallon, *J. Strain Analysis*, **8**, 277 (1973).]

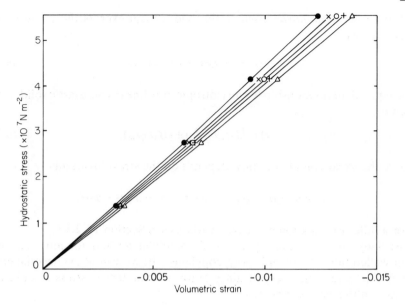

Figure 9.26. Volumetric isochronous stress–strain curves for polymethyl methacrylate: ●, 10^2 s; ×, 10^3 s; ○, 10^4 s; +, 10^5 s; △, 10^6 s. [Redrawn with permission from Mallon and Benham, *Plastics Polymers*, **40**, 77 (1972).]

pressure. It can be seen that the viscoelastic behaviour is linear to a very good approximation. It is also important to note that the magnitude of the response is much lower than that for shear deformation, i.e. the magnitude of the bulk compliance is comparatively low.

9.5.2 Single Integral Representation of Three-dimensional Non-linear Viscoelasticity

In studies contemporaneous with those of Benham and coworkers, Sternstein and Ho[50] identified the effect of hydrostatic strain on the stress relaxation behaviour of polymethyl methacrylate in combined tension–torsion tests where the ratio of the octahedral shear strain to the mean normal strain was varied. It was concluded that the magnitude of the mean normal strain was the governing factor in determining the stress relaxation behaviour.

Further developments by Buckley and McCrum[51] and by Buckley and Green[52] have provided a more satisfactory theoretical framework which can embrace the results of Benham and coworkers and Sternstein and Ho.

Benham's results suggest that it is reasonable to consider that the total creep response is the sum of two contributions, a linear viscoelastic term from the hydrostatic component of the applied stress and a non-linear viscoelastic term, which also depends to some extent on the hydrostatic component, but relates primarily to the deviatoric component of the applied stress.

For an isotropic *linear* viscoelastic solid we can express the strain $e(t)$ at time $t > 0$ as

$$e(t) = \tfrac{1}{2}J(t)\boldsymbol{\sigma}' + \tfrac{1}{3}B(t)\sigma_m\mathbf{I}. \tag{9.19}$$

The equivalent relationship for an isotropic non-linear viscoelastic solid would then be

$$e(t) = \tfrac{1}{2}J(t, J_1, J_2) + \tfrac{1}{3}B(t)\sigma_m\mathbf{I}, \tag{9.20}$$

where the shear compliance now depends on the stress invariants

$$J_1 = \sigma_1 + \sigma_2 + \sigma_3, \qquad J_2 = \sigma_1\sigma_2 + \sigma_2\sigma_3 + \sigma_3\sigma_1.$$

(For a fuller explanation of stress invariants see Section 11.2.1.)

Buckley and McCrum[51] performed combined tension–torsion tests on a thin-walled tube of isotropic polypropylene. The measured extensional creep strain $e_\tau(t)$ and shear creep strain $\gamma(t)$ for an axial tensile stress σ and shear stress τ in the wall of tube are

$$e_\tau(t) = [\tfrac{1}{3}J(t, J_1, J_2) + \tfrac{1}{9}B(t)]\sigma,$$
$$\gamma(t) = J(t, J_1, J_2)\tau. \tag{9.21}$$

$B(t)$ was determined by extrapolating the ratios $e_\tau(t)/\sigma$ and $\gamma(t)/\tau$ obtained in pure tension and pure torsion tests respectively to the limits $\sigma \to 0$ and $\tau \to 0$. It was then possible to obtain $J(t, J_1, J_2)$ independently from either the tensile response $e_\tau(t)$ or the shear response $\gamma(t)$. These were termed J_e and J_γ respectively. Their values at $100\,\text{s}$ are plotted in Figure 9.27 for various proportional loading ratios. It can be seen that $J_e = J_\gamma$ to a very good approximation.

The attraction of Buckley and McCrum's representation of non-linear viscoelastic behaviour is threefold:

(1) It separates the effects of the hydrostatic and deviatoric components of stress, which is useful in that these may well be expected to relate to different mechanisms of deformation.

(2) It implies that the non-linearity of the viscoelasticity enters through the shear deformation behaviour. In a further paper Buckley and Green[52] showed that Buckley and McCrum's representation for isotropic polypropylene was consistent with a rigorous theoretical treatment for situations of proportional loading. The basic implication is that the non-linearity can be adequately described by allowing the shear creep compliance to change with stress history.

(3) There is a further influence of the hydrostatic component of stress in that it affects the shear deformation behaviour. This has kinship with the effect of the hydrostatic component of stress on the yield behaviour of polymers, and affords a useful link between non-linear viscoelasticity and yield.

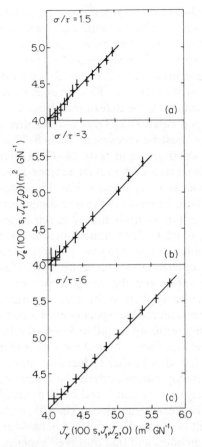

Figure 9.27. Comparison of measured values of $J(100 \text{ s}, J_1, J_2, 0)$ obtained from $e_\tau(100 \text{ s})$, J_e, and $\gamma(100 \text{ s})$, J_γ, using equation (9.21). Estimated error limits are indicated. [Redrawn with permission from Buckley and McCrum, *J. Mater. Sci.*, **9**, 2064 (1974).]

9.5.3 The Three-dimensional Multiple Integral Representation

A basic attraction of the Buckley and McCrum representation discussed in Section 9.5.2 is that it is a single integral representation. In certain cases, however, remarkable synergistic effects have been observed due to combinations of applied stresses. A description of such effects can be made in terms of the multiple integral representation. Lockett[41,53] has given the general multiple integral representation for the creep and stress relaxation of isotropic material up to and including third-order terms. He showed that the representation contains (1) two first-order kernels, which are functions of $t - \tau_1$ (these are the shear compliance/stress relaxation modulus and bulk compliance/bulk stress relaxation modulus of linear viscoelasticity); (2) four second-order kernels, which are functions of $t - \tau_1$, $t - \tau_2$; (3) six third-order kernels, which are functions of $t - \tau_1$, $t - \tau_2$, $t - \tau_3$. This gives a total of 12 terms, kernel functions ψ_1 to ψ_{12}, which can be determined from 17 tests. For example, we shall consider the creep behaviour. Tests 1–7 are then single stress tests, but more than one strain must be measured. Tests 8–12 are biaxial stress tests, test 13 is a triaxial stress test and tests 14–17 involve a combination of a normal stress and a shear stress. The actual number of tests performed depends on the number of values of t_1 chosen. For 10 values of t_1, 463 tests are required, compared with 78 tests for the one-dimensional situation.

It would seem from this analysis that so much experimentation would be required that this approach to three-dimensional viscoelastic behaviour would only be of an academic interest. However, Findley and his coworkers[31,54,55] have shown that this representation can be used, provided that certain simplifying assumptions are made and the creep responses parameterized so that fitting procedures can be applied to the data. Onaran and Findley[31] carried out tension–torsion tests on tubular specimens of a polyvinyl chloride copolymer. By only admitting terms up to and including the third order, the number of independent kernel functions required for a three-dimensional representation is reduced to 12, and is further reduced to nine for single step loading only. Onaran and Findley parameterized the tensile and torsional creep by expressions of the form $e = e_0 + mt^n$, and carried out 20 tests, nine of which were used to find the kernel functions. They then showed that there was reasonable agreement between the predicted behaviour for all 20 tests, based on the nine kernel functions, and that observed experimentally. Furthermore, the theory did predict the remarkable synergistic effects which were observed. These were an abrupt increase in tensile strain when torque is added to a specimen under tension, and an abrupt increase in torsional strain where a specimen under torque is subjected to tensile load. Further studies of the same polymer suggested that a product form for the kernels can be assumed to a reasonable approximation where

$$J_1 = w_1(t - \tau_1)^n,$$

$$J_2 = w_2(t - \tau_1)^{n/2}(t - \tau_2)^{n/2},$$

$$J_3 = w_3(t - \tau_1)^{n/3}(t - \tau_2)^{n/3}(t - \tau_3)^{n/3}.$$

In a further study, Nolte and Findley[55] examined the behaviour of a polyurethane, in tension, torsion, combined tension–torsion and compression. It was found adequate to assume that the volume change is *linearly* dependent on the stress history, which reduces the number of kernel functions from 12 to seven (neglecting kernels of order higher than three). This assumption is in line with the results of Mallon and Benham discussed above.

9.6 CREEP AND STRESS RELAXATION AS THERMALLY ACTIVATED PROCESSES: THE EYRING EQUATION

A simple starting point for the molecular approach to non-linear viscoelastic behaviour is closely allied to the attempts made to gain a molecular understanding of solution viscosities on the basis of the theory of thermally activated rate processes. This approach is due to Eyring and his coworkers[29].

It is assumed that deformation of the polymer involves the motion of chain molecules or parts of a chain molecule over potential energy barriers.

The basic molecular process could be either intermolecular (e.g. chain-sliding) or intramolecular (e.g. a change in the conformation of the chain). The situation is illustrated schematically in Figure 9.28. With no stress acting, a dynamic equilibrium exists, chain segments moving with a frequency ν over the potential barrier in each direction where $\nu = \nu_0 \, e^{-\Delta H/RT}$.

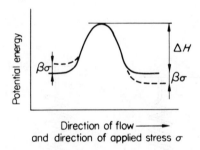

Figure 9.28. The Eyring model for creep.

This equation is identical to equation (7.2) above, describing the frequency of a molecular event. ΔH is an activation energy and ν_0 involves the fundamental vibration frequency and the entropy contribution to the free energy.

It is assumed that the applied stress σ produces linear shifts $\beta\sigma$ of the energy barriers in a symmetrical fashion.

We then have a flow

$$\nu_1 = \nu_0 \exp\left(-\frac{(\Delta H - \beta\sigma)}{RT}\right)$$

in the forward direction (i.e. the direction of application of the stress) and

$$\nu_2 = \nu_0 \exp\left(-\frac{(\Delta H + \beta\sigma)}{RT}\right)$$

in the backward direction. This gives a net flow

$$\nu^1 = \nu_1 - \nu_2 = \nu_0 \, e^{-\Delta H/RT}\{e^{\beta\sigma/RT} - e^{-\beta\sigma/RT}\}$$

in the forward direction.

If we assume that the net flow in the forward direction is directly related to the rate of change of strain, we have

$$\frac{\mathrm{d}e}{\mathrm{d}t} = \dot{e} = \dot{e}_0 \, e^{-\Delta H/RT} \sinh \frac{v\sigma}{RT}, \tag{9.22}$$

where $\dot{e}_0$ is a constant pre-exponential factor and the symbol β is replaced by v, which is termed the activation volume for the molecular event.

This equation defines an 'activated' non-Newtonian viscosity. Eyring developed his ideas around 1940. It was therefore natural that he should have attempted to incorporate them into the spring and dashpot models of viscoelasticity which were in vogue at that time.

Consider the standard linear solid (Figure 9.29(a)) and replace the dashpot with viscosity n_m by the activated dashpot defined by constants A, α (Figure 9.29(b)). This leads to a more complicated relationship between stress and strain than that for the standard linear solid, giving non-linear viscoelastic behaviour.

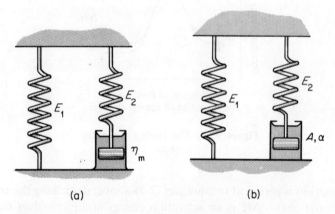

(a)   (b)

Figure 9.29. (a) The standard linear solid; (b) Eyring's modification of the standard linear solid with the activated dashpot.

Eyring and his collaborators took Leaderman's data for the creep of silk and other fibres[16] (these results will be discussed further here). They showed that this model gave a good fit at a given level of stress by suitable choice of the four parameters E_1, E_2, A and α, over the four decades of time observed. An attempt to fit the data to the three-parameter standard linear solid was only successful over about one and a half decades.

The particular attraction of this model with the 'activated' dashpot arises from the following. Although creep curves are sigmoidal over a very long time interval when plotted on a logarithmic time scale, over the middle time region they are to a good approximation a straight line. Now it so happens that this model gives just this algebraic form for the creep (i.e. $e = a' + b' \log t$).

Since Eyring undertook this work, the limitation of model representations of *linear* viscoelasticity have been better appreciated, and it is accepted that exact fitting of data requires a relaxation time spectrum. Nevertheless, there have been several recent publications which have revived interest in the Eyring approach, for two reasons. First, the parameters of the Eyring equations, particularly the activation energy and activation volume, may give some indication of the underlying molecular mechanisms. Secondly, the activated rate process may provide a common basis for the discussion of creep and yield behaviour.

Sherby and Dorn[56] examined the creep behaviour of polymethyl methacrylate under constant stress in the temperature range 263–320 K. The creep rate was determined at different temperatures by applying step temperature changes. In this way the so-called Sherby–Dorn plots of creep rate versus total creep strain could be constructed for various temperatures at a given stress level (Figure 9.30). These data could then be superposed by assuming that temperature dependence at each stress level followed a simple activated process, so that there is a master curve for each stress level giving the relationship between strain rate and strain (Figure 9.31). Sherby and Dorn

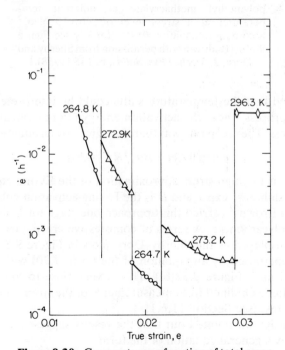

Figure 9.30. Creep rates as a function of total creep strain for polymethyl methacrylate at indicated temperatures for a stress level of $5.6 \times 10^7 \, \text{N m}^{-2}$. [Redrawn with permission from Sherby and Dorn, *J. Mech. Phys. Solids*, **6**, 145 (1958).]

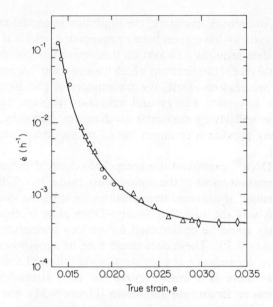

Figure 9.31. Superposition of creep data for polymethyl methacrylate at different temperatures at a stress level of $5.6 \times 10^7 \, \mathrm{N \, m}^{-2}$ according to equation (9.19) (for key see Figure 9.30). [Redrawn with permission from Sherby and Dorn, *J. Mech. Phys. Solids*, **6**, 145 (1958).]

further showed that the temperature shifts could be interpreted in terms of an activated process where the activation energy fell in a linear fashion with increasing stress. The creep rate was therefore given by an equation of the form

$$\dot{e} \exp\left[(\Delta H - B\sigma)/RT\right] = F(e). \tag{9.23}$$

This equation is the high stress approximation of the Eyring equation (9.19) above where $\sinh x \doteq \frac{1}{2} \exp x$ and B is the Eyring activation volume.

Mindel and Brown[57] carried this approach one stage further in their study of creep in polycarbonate. A series of compressive creep curves at different stress levels displayed as the Sherby–Dorn plots in Figure 9.32(a), could be excellently superposed by an equation of the form (9.20) with an activation volume of $5.7 \, \mathrm{nm}^3$ (Figure 9.32(b)). This is very close to the values of the activation volume obtained from measurements of the strain rate dependence of the yield stress (see Section 11.6.1).

Mindel and Brown pointed out that the results suggest that creep can be represented by a general equation of the form

$$\dot{e} = f_1(T) f_2(\sigma/T) f_3(e), \tag{9.24}$$

where $f_1(T)$, $f_2(\sigma/T)$ and $f_3(e)$ are separate functions of the variables T, σ and e.

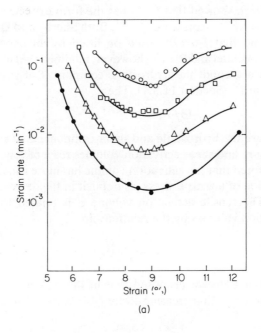

(a)

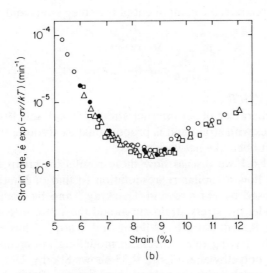

(b)

Figure 9.32. (a) Creep rates for polymethyl methacrylate in compression at various stress levels: $\bigcirc$, $7.1 \times 10^7 \, \text{N m}^{-2}$; $\square$, $7.0 \times 10^7 \, \text{N m}^{-2}$; $\triangle$, $6.9 \times 10^7 \, \text{N m}^{-2}$; $\bullet$, $6.8 \times 10^7 \, \text{N m}^{-2}$. (b) Superposition of creep data shown in (a) using an activation volume of 5.7 nm^3. [Redrawn with permission from Mindel and Brown, *J. Mater. Sci.*, **8**, 863 (1973).]

Sherby and Dorn showed that $f_1(T)$ has the form expected for a thermally activated process $f_1(T) = \text{const.} \times e^{-\Delta H/RT}$. Both Sherby and Dorn and Mindel and Brown found that $f_2(\sigma/T)$ could be fitted by an exponential so that $f_2(\sigma/T) = e^{\sigma/RT}$; Mindel and Brown, however, modified the simple exponential form to take into account the hydrostatic component of stress in a manner suggested by Ward[58] (Section 11.5.1). Then

$$f_2(\sigma/T) = e^{-p\Omega/RT} e^{\tau V/RT},$$

where p and τ are the hydrostatic and shear components of stress and Ω and V are the pressure and shear activation volumes respectively.

It should be noted that the activation volume has more usually been defined for the application of a tensile stress, as implicit in the derivation of equation (9.22) above. The tensile activation volume v is related to the shear and pressure activation volumes by the relationship

$$v = \frac{V}{2} + \frac{\Omega}{3}.$$

Finally, Mindel and Brown proposed that in the region in which the creep rate is falling rapidly with increasing strain $f_3(e)$ takes the form

$$f_3(e) = \text{const. } e^{-ce_R},$$

where e_R is the recoverable component of the creep strain and c is a constant. We then have

$$\dot{e} = \dot{e}_0 \, e^{-(\Delta H - \tau V + p\Omega)/RT} e^{-ce_R}$$

$$= \dot{e}_0 \, e^{-[\Delta H - (\tau - \tau_{int})V + p\Omega]/RT}$$

where $ce_R = \tau_{int} V/RT$.

τ_{int} has the character of an internal stress. Mindel and Brown point out that τ_{int} increases with strain and is proportional to absolute temperature, as expected for a rubber-like network stress.

In Section 11.6.1 we discuss how these results for creep relate to yield behaviour, and how a similar representation to that of Mindel and Brown has been proposed by Haward and Thackray[59] and by Fotheringham and Cherry[60] to model the stress–strain curve and the strain rate dependence of the yield stress. Recent work by Wilding and Ward[61,62] has also shown the usefulness of the Eyring rate process in modelling the creep behaviour of highly oriented polyethylene. Figure 9.33 shows Sherby–Dorn plots for the creep of an ultrahigh modulus polyethylene sample. It can be seen that at high strains (corresponding to long creep times) the creep rate reaches a constant value. It was shown that for low molecular weight polymers the stress and temperature dependence of this final creep rate could be modelled very well indeed by a single activated process with activation volume $\sim 0.08 \text{ nm}^3$. This is quite reasonable in molecular terms, as it could be just the volume swept out by a single molecular chain moving through the lattice by a discrete

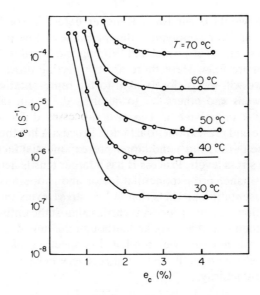

Figure 9.33. Sherby–Dorn plots of creep of ultra high modulus polyethylene at different temperatures. [Redrawn with permission from Wilding and Ward, *Plastics and Rubber Processing and Applications*, **1**, 167 (1981).]

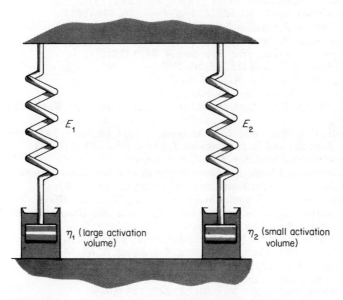

Figure 9.34. The two-process model for permanent flow creep. [Redrawn with permission from Wilding and Ward, *Plastics and Rubber Processing and Applications*, **1**, 167 (1981).]

distance, such as would occur for a Reneker defect[63]. For high molecular weight polymers and for copolymer, this permanent flow process was only activated at high stress levels, which led Wilding and Ward to the representation shown in Figure 9.34. Here there are two Eyring processes coupled in parallel, which we will see is closely akin to the representation proposed by Roetling[64], Bauwens and others[65,66] to describe the strain rate dependence of the yield stress in polymers. The two processes differ with regard to activation volume and pre-exponential factor. Process A has the smaller tensile activation volume (~ 0.05 nm^3) and larger pre-exponential factor, and is only activated at high stress levels. Process B has a larger tensile activation volume (~ 1 nm^3) and a smaller pre-exponential factor and is operative at low stress levels. This representation predicts that at low stress levels there will be very little permanent flow because process B carries almost the entire load. Wilding and Ward confirmed that this is so, and although the time dependence of this low strain creep was not very well modelled (a spectrum of relaxation times would be required) the overall creep and recovery behaviour could be represented very satisfactorily.

REFERENCES

1. S. Turner, *Polymer Eng. Sci.*, **6**, 306 (1966).
2. Y. H. Pao and J. Marin, *J. Appl. Mech.*, **19**, 478 (1952).
3. Y. H. Pao and J. Marin, *J. Appl. Mech.*, **20**, 245 (1953).
4. J. Marin, *J. Appl. Mech. Trans. ASME*, **59**, A21 (1937).
5. W. N. Findley and G. Khosla, *J. Appl. Phys.*, **26**, 821 (1955).
6. P. G. Nutting, *J. Franklin Inst.*, **235**, 513 (1943).
7. K. Van Holde, *J. Polymer Sci.*, **24**, 417 (1957).
8. E. N. da C. Andrade, *Proc. Roy. Soc. A*, **84**, 1 (1910).
9. D. J. Plazek, *J. Colloid Sci.*, **15**, 50 (1960).
10. T. L. Smith, *Trans. Soc. Rheol.* **6**, 61 (1962).
11. I. H. Hall, *J. Polymer Sci.*, **54**, 505 (1961).
12. G. M. Martin, F. L. Roth and R. D. Stiehler, *Trans. Inst. Rubber Ind.*, **32**, 189 (1956).
13. E. Guth, P. E. Wack and R. L. Anthony, *J. Appl. Phys.*, **17**, 347 (1946).
14. A. V. Tobolsky and R. D. Andrews, *J. Chem. Phys.*, **13**, 3 (1945).
15. H. Leaderman, *Trans. Soc. Rheol.*, **6**, 361 (1962).
16. H. Leaderman, *Elastic and Creep Properties of Filamentous Materials and Other High Polymers*, Textile Foundation, Washington, D.C., 1943.
17. B. Bernstein, E. A. Kearsley and L. P. Zapas, *Trans Soc. Rheol.*, **7**, 391 (1963).
18. P. H. de Hoff, G. Lianis and W. Goldberg, *Trans. Soc. Rheol.*, **10**, 385 (1966).
19. A. Signiorini, *Ann. Mat. Pura Applicata*, **39**, 147 (1955).
20. B. D. Coleman and W. Noll, *Rev. Mod. Phys.*, **33**, 239 (1961).
21. L. J. Zapas and T. Craft, *J. Res. Natl. Bur. Stand. A*, **69**, 541 (1965).
22. L. J. Zapas, *J. Res. Natl. Bur. Stand. A*, **70**, 525 (1966).
23. W. Goldberg and G. Lianis, *J. Appl. Mech.*, **37**, 53 (1970).
24. R. S. Rivlin, *Phil. Trans. Roy. Soc. A*, **242**, 173 (1949).
25. R. S. Rivlin and D. W. Saunders, *Phil. Trans. Roy. Soc. A*, **243**, 251 (1951).
26. W. Goldberg and G. Lianis, *J. Appl. Mech.*, **35**, 433 (1968).
27. C. W. McGuirt and G. Lianis, *Int. J. Eng. Sci.*, **7**, 579 (1969).

28. K. C. Valanis and R. F. Landel, *Trans. Soc. Rheol.*, **11**, 243 (1967).
29. G. Halsey, H. J. White and H. Eyring, *Text. Res. J.*, **15**, 295 (1945).
30. R. A. Schapery, *Polym. Eng. Sci.*, **9**, 295 (1969).
31. K. Onaran and W. N. Findley, *Trans. Soc. Rheol.*, **9**, 299 (1965).
32. I. M. Ward and E. T. Onat, *J. Mech. Phys. Solids*, **11**, 217 (1963).
33. A. E. Green and R. S. Rivlin, *Arch. Rat. Mech. Analysis*, **1**, 1 (1957).
34. A. E. Green and R. S. Rivlin, *Arch. Rat. Mech. Analysis*, **4**, 387 (1960).
35. A. E. Green, R. S. Rivlin and A. J. M. Spencer, *Arch. Rat. Mech. Analysis*, **3**, 82 (1959).
36. V. Volterra and J. Pérès, *Théorie générale des fonctionelles*, Gauthier–Villars, Paris, 1936, p. 61.
37. D. W. Hadley and I. M. Ward, *J. Mech. Phys. Solids*, **13**, 397 (1965).
38. I. M. Ward and J. M. Wolfe, *J. Mech. Phys. Solids*, **14**, 131 (1966).
39. A. C. Pipkin and T. G. Rogers, *J. Mech. Phys. Solids*, **16**, 59 (1968).
40. W. N. Findley and J. S. Y. Lai, *Trans. Soc. Rheol.*, **11**, 361 (1967).
41. F. J. Lockett, *Non-Linear Viscoelastic Solids*, Academic Press, London, New York, 1972.
42. J. S. Y. Lai and W. N. Findley, *Trans. Soc. Rheol.*, **12**, 243 (1968).
43. J. S. Y. Lai and W. N. Findley, *Trans. Soc. Rheol.*, **12**, 259 (1968).
44. C. J. Morgan and I. M. Ward, *J. Mech. Phys. Solids*, **19**, 165 (1971).
45. M. G. Brereton, S. G. Croll, R. A. Duckett and I. M. Ward, *J. Mech. Phys. Solids*, **22**, 97 (1974).
46. G. E. Forsythe, *J. Soc. Ind. Appl. Maths*, **5**, 74 (1957).
47. P. P. Benham and D. McCammond, *Plastics Polymers*, **39**, 130 (1971).
48. P. J. Mallon and P. P. Benham, *Plastics Polymers*, **40**, 77 (1972).
49. P. P. Benham, *Polymer Eng. Sci.*, **13**, 398 (1973).
50. S. S. Sternstein and T. C. Ho, *J. Appl. Phys.*, **43**, 4370 (1972).
51. C. P. Buckley and N. G. McCrum, *J. Mater. Sci.*, **9**, 2064 (1974).
52. C. P. Buckley and A. E. Green, *Phil. Trans. Roy. Soc. A*, **281**, 543 (1976).
53. F. J. Lockett, *Int. J. Eng. Sci.*, **3**, 59 (1965).
54. W. N. Findley and K. Onaran, *Trans. Soc. Rheol.*, **12**, 217 (1968).
55. K. G. Nolte and W. N. Findley, *J. Appl. Mech.*, **37**, 441 (1970).
56. O. D. Sherby and J. E. Dorn, *J. Mech. Phys. Solids*, **6**, 145 (1958).
57. M. J. Mindel and N. Brown, *J. Mater. Sci.*, **8**, 863 (1973).
58. I. M. Ward, *J. Mater. Sci.*, **6**, 1397 (1972).
59. R. N. Haward and G. Thackray, *Proc. Roy. Soc. A*, **302**, 453 (1968).
60. D. G. Fotheringham and B. Cherry, *J. Mater. Sci.*, **13**, 951 (1978).
61. M. A. Wilding and I. M. Ward, *Polymer*, **19**, 969 (1978).
62. M. A. Wilding and I. M. Ward, *Polymer*, **22**, 870 (1981).
63. D. H. Reneker, *J. Polym. Sci.*, **59**, 539 (1962).
64. J. A. Roetling, *Polymer*, **6**, 311 (1965).
65. C. Bauwens–Crowet, J. C. Bauwens and G. Homès, *J. Polym. Sci. A2*, **7**, 735 (1969).
66. R. E. Robertson, *J. Appl. Polymer Sci.*, **7**, 443 (1963).

10

Anisotropic Mechanical Behaviour

10.1 THE DESCRIPTION OF ANISOTROPIC MECHANICAL BEHAVIOUR

An oriented polymer is in the strictest terms an anisotropic non-linearly viscoelastic material. A comprehensive understanding of anisotropic mechanical behaviour is therefore a very considerable task. In this chapter we will restrict the discussion to cases where the strains are small.

The mechanical properties of an anisotropic elastic solid for small strains are defined by the generalized Hooke's law:

$$\varepsilon_{ij} = s_{ijkl}\sigma_{kl}, \qquad \sigma_{ij} = c_{ijkl}\varepsilon_{kl},$$

where the s_{ijkl} are the compliance constants and the c_{ijkl} are the stiffness constants. This has been discussed in Chapter 2. The use of this representation does not necessarily restrict the discussion to time-independent behaviour. The compliance and stiffness constants could be time dependent, defining creep compliances and relaxation stiffnesses in step function loading experiments, or complex compliances and complex stiffnesses in dynamic mechanical measurements. For simplicity, the methods of measurement are usually carefully standardized, e.g. by measuring each creep compliance after the same loading programme and the same time interval. It will be assumed that for such measurements there is an exact equivalence between elastic and linear viscoelastic behaviour, as proposed by Biot[1].

In an elastic material, the presence of symmetry elements leads to a reduction in the number of independent elastic constants and corresponding reductions will be assumed for anisotropic linear viscoelastic behaviour, although there is not enough experimental evidence to confirm that exactly the same rules hold in every case[2].

There are two important points to emphasize regarding the description of data:

(1) In practice an abbreviated notation is often used in which

$$e_p = s_{pq}\sigma_q, \qquad \sigma_p = c_{pq}e_q.$$

As explained in Section 2.5, σ_p represents $\sigma_{xx}, \sigma_{yy}, \ldots, \sigma_{xy}$, etc., and e_q represents $e_{xx}, e_{yy}, \ldots, e_{xy}$, etc., and in the compliance and stiffness matrices s_{pq} and c_{pq} p, q take the values $1, 2, \ldots, 6$.

The conversion rules from the s_{ijkl} and c_{ijkl} notation to the abbreviated notation are given in Section 2.5 above.

It is important to remember that the engineering strains e_p are not the components of a tensor. Similarly, the 6×6 compliance matrix s_{pq} does not represent a tensor and therefore tensor manipulation rules do not apply. As will be demonstrated, for working out problems involving transformation of coordinates from one system of axes to another it is always desirable to use the original tensor notation in terms of ε_{ij}, σ_{kl} and s_{ijkl} or c_{ijkl}.

(2) It is usually more convenient to work in terms of compliance constants than stiffness constants. This is because in the experimental procedures it is easier to apply a simple stress of a given type, e.g. a tensile stress or a shear stress, and measure the corresponding strains, e.g.

$$e_1 = s_{11}\sigma_1 + s_{12}\sigma_2 + s_{13}\sigma_3 + s_{44}\sigma_4 + s_{55}\sigma_5 + s_{66}\sigma_6.$$

The compliance constants s_{11}, s_{12}, s_{13}, etc., can be found by applying stresses σ_1, σ_2, σ_3, etc., and measuring e_1 in each case. The procedures will become clearer as the various experimental methods are discussed.

10.2 MECHANICAL ANISOTROPY IN POLYMERS

10.2.1 The Elastic Constants for Specimens Possessing Fibre Symmetry

Studies of mechanical anistropy in polymers have for the most part been restricted to drawn fibres and uniaxially drawn films, both of which show isotropy in a plane perpendicular to the direction of drawing. The number of independent elastic constants is reduced to five (Reference 3, p. 138). Choosing the z direction as the axis of symmetry the compliance matrix s_{pq} reduces to

$$\begin{pmatrix} s_{11} & s_{12} & s_{13} & 0 & 0 & 0 \\ s_{12} & s_{11} & s_{13} & 0 & 0 & 0 \\ s_{13} & s_{13} & s_{33} & 0 & 0 & 0 \\ 0 & 0 & 0 & s_{44} & 0 & 0 \\ 0 & 0 & 0 & 0 & s_{44} & 0 \\ 0 & 0 & 0 & 0 & 0 & 2(s_{11}-s_{12}) \end{pmatrix}.$$

The various compliance constants are illustrated diagrammatically in Figure 10.1. The situation is most easily appreciated for a fibre specimen, but a uniaxially oriented sheet possesses identical symmetry.

The relationships of these compliance constants to the better known Young's moduli and Poisson's ratios are as follows:

(1) Consider application of a stress along the z direction, i.e. along the fibre axis, or the draw direction for the polymer film. Then $e_{zz} = s_{33}\sigma_{zz}$ and

$$\text{Young's modulus, } E_3 = \frac{\sigma_{zz}}{e_{zz}} = \frac{1}{s_{33}}, \quad \text{giving} \quad s_{33} = \frac{1}{E_3}.$$

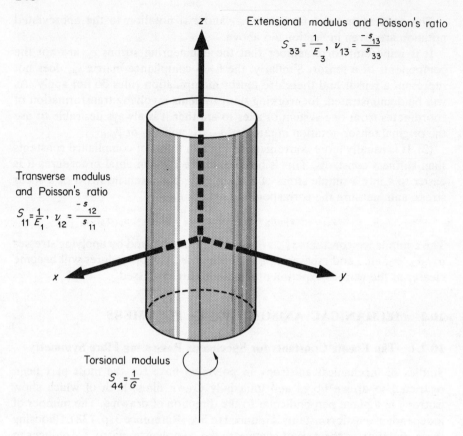

Extensional modulus and Poisson's ratio

$$S_{33} = \frac{1}{E_3}, \quad v_{13} = \frac{-s_{13}}{s_{33}}$$

Transverse modulus and Poisson's ratio

$$S_{11} = \frac{1}{E_1}, \quad v_{12} = \frac{-s_{12}}{s_{11}}$$

Torsional modulus

$$S_{44} = \frac{1}{G}$$

Figure 10.1 The fibre compliance constants.

(2) Similarly the strain in the plane transverse to the fibre axis for a stress σ_{zz} along the fibre axis is given by $e_{xx} = e_{yy} = s_{13}\sigma_{zz}$ and

$$\text{Poisson's ratio,} \quad v_{13} = -\frac{e_{xx}}{e_{zz}} = -\frac{s_{13}}{s_{33}}.$$

(The negative sign ensures that Poisson's ratio is the conventionally positive quantity since e_{xx} is negative, i.e. a contraction.)

(3) In a similar manner, s_{11}, s_{12} and s_{13} are related to the modulus E_1 (the transverse modulus) and the corresponding Poisson's ratios $v_{21} = v_{12}$ and $v_{31} = v_{13}$ for application of a stress in a plane perpendicular to the fibre axis, i.e.

$$s_{11} = \frac{1}{E_1} \quad \text{and} \quad v_{21} = -\frac{s_{21}}{s_{11}} = -\frac{s_{12}}{s_{11}}, \quad v_{31} = -\frac{s_{31}}{s_{11}} = -\frac{s_{13}}{s_{11}}$$

(4) The shear compliance is the reciprocal of the shear or torsional modulus G. There are two equivalent shear compliances $s_{44} = s_{55} = 1/G$. These relate to torsion about the symmetry axis z, i.e. shear in the yz or xz planes.

The shear compliance s_{66} relates to shear in the xy plane and is related to the compliance constants s_{11} and s_{12}, such that $s_{66} = 2(s_{11} - s_{12})$. This relationship expresses the fact that these specimens are isotropic in a plane perpendicular to the symmetry axis, i.e. that the elastic behaviour in this plane is specified by only two elastic constants as for an isotropic material. It will be seen that this property is very important in determining the elastic constants for fibres.

10.2.2 The Elastic Constants for Specimens Possessing Orthorhombic Symmetry

Oriented polymer films which are prepared by either rolling, rolling and annealing, or some commerical one-way draw processes, may possess orthorhombic rather than transversely isotropic symmetry. For such films the elastic behaviour is specified by nine independent elastic constants. Choose the initial drawing or rolling direction as the z axis for a system of rectangular Cartesian coordinates; the x axis to lie in the plane of the film and the y axis normal to the plane of the film (Figure 10.2). The compliance matrix is

$$\begin{pmatrix} s_{11} & s_{12} & s_{13} & 0 & 0 & 0 \\ s_{12} & s_{22} & s_{23} & 0 & 0 & 0 \\ s_{13} & s_{23} & s_{33} & 0 & 0 & 0 \\ 0 & 0 & 0 & s_{44} & 0 & 0 \\ 0 & 0 & 0 & 0 & s_{55} & 0 \\ 0 & 0 & 0 & 0 & 0 & s_{66} \end{pmatrix}.$$

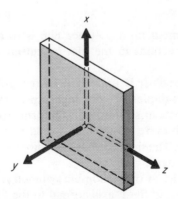

Figure 10.2. Choice of axes for a polymer sheet possessing orthorhombic symmetry.

There are three Young's moduli,

$$E_1 = \frac{1}{s_{11}}, \qquad E_2 = \frac{1}{s_{22}} \quad \text{and} \quad E_3 = \frac{1}{s_{33}},$$

and six Poisson's ratios,

$$\nu_{21} = -\frac{s_{21}}{s_{11}}, \qquad \nu_{31} = -\frac{s_{31}}{s_{11}}, \qquad \nu_{32} = -\frac{s_{32}}{s_{22}},$$

$$\nu_{12} = -\frac{s_{12}}{s_{22}}, \qquad \nu_{13} = -\frac{s_{13}}{s_{33}}, \qquad \nu_{23} = -\frac{s_{23}}{s_{33}},$$

corresponding to situations where a tensile stress is applied along the x, y and z directions.

There are three independent shear moduli $G_1 = 1/s_{44}$, $G_2 = 1/s_{55}$ and $G_3 = 1/s_{66}$ corresponding to shear in the yz, xz, and xy planes respectively. For a sheet of general dimensions, torsion experiments where the sheet is twisted about the x, y, or z axis will involve a combination of shear compliances. This will be discussed in greater detail later, when methods of obtaining the elastic constants are described.

10.3 MEASUREMENT OF ELASTIC CONSTANTS

The measurement of elastic constants is a very different undertaking for the two situations of a sheet and a fibre. The experimental methods employed for these two cases will therefore be discussed separately.

10.3.1 Measurements on Films or Sheets

Extensional Moduli

The simplest measurement on a polymer film is to determine the Young's modulus in various directions in the film by cutting long thin strips in the selected directions.

For anistropic materials it is important to recognize that it is generally necessary to measure samples of very high aspect ratio† to minimize 'end effects'. These end effects arise from non-uniform stress conditions near the clamps which are much more severe than would be anticipated on the basis of St Venant's principle. The situation has been discussed in detail by Horgan[4,5] and by Folkes and Arridge[6].

We will consider a film of orthorhombic symmetry, the x and z axes lying in the plane of the film and the y axis normal to the film as in Section 10.2.2 above.

---

† Aspect ratio is the ratio of length to width or thickness.

Consider a long strip cut in a direction making an angle θ with the z direction (Figure 10.3(a)).

The Young's modulus for this strip $E_0 = 1/s_\theta$ where s_θ is the compliance in a direction making an angle θ with the z direction.

To calculate s_θ in terms of the compliance constants we will use the full tensor notation.

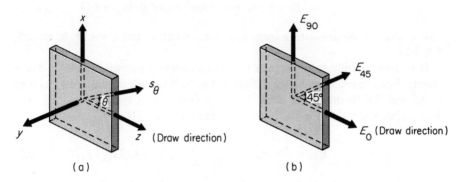

Figure 10.3. (a) The compliance s_θ is at an angle θ to the initial draw direction in the plane of the sheet. (b) The E_0, E_{45} and E_{90} moduli.

The compliance constants s_{ijkl} referred to one system of Cartesian axes are related to those s'_{pqmn} referred to a second system of Cartesian axes by the tensor transformation rule:

$$s'_{pqmn} = a_{pi}a_{qj}a_{mk}a_{nl}s_{ijkl},$$

where a_{pi}, a_{qj}, ... define the cosines of the angles between the p axis in the second system and the i axis in the first, the q axis in the second system and the j axis in the first, ... and p, q, m, n take the values 1, 2, 3 in the second system of axes and i, j, k, l take the values $1'$, $2'$, $3'$ in the first system of axes.

We will take the direction of the strip to be the z' direction in a second system of Cartesian axes. Then $s_\theta = s_{3'3'3'3'}$ is given by

$$s_{3'3'3'3'} = a_{3'1}a_{3'1}a_{3'1}a_{3'1}s_{1111} + a_{3'3}a_{3'3}a_{3'3}a_{3'3}s_{3333}$$

$$+ a_{3'1}a_{3'1}a_{3'3}a_{3'3}s_{1133} + a_{3'3}a_{3'3}a_{3'1}a_{3'1}s_{3311}$$

$$+ a_{3'1}a_{3'3}a_{3'3}a_{3'1}s_{1331} + a_{3'3}a_{3'1}a_{3'1}a_{3'3}s_{3113}$$

$$+ a_{3'3}a_{3'1}a_{3'3}a_{3'1}s_{3131} + a_{3'1}a_{3'3}a_{3'1}a_{3'3}s_{1313}.$$

Note that all compliance terms containing the suffix 2 will vanish, because $a_{3'2} = 0$.

The change in coordinate systems corresponds to a rotation of the coordinate axes through an angle θ about the y direction as axis.

We therefore put $a_{3'1} = \sin \theta$ and $a_{3'3} = \cos \theta$ and

$$s_{3'3'3'3'} = \sin^4 \theta s_{1111} + \cos^4 \theta s_{3333} + 2 \sin^2 \theta \cos^2 \theta s_{1133}$$
$$+ 4 \sin^2 \theta \cos^2 \theta s_{1313}.$$

In the abbreviated notation

$$s_0 = s_{3'3'} = \sin^4 \theta s_{11} + \cos^4 \theta s_{33} + \sin^2 \theta \cos^2 \theta (2s_{13} + s_{55}). \qquad (10.1)$$

(Note factor 4 in converting from s_{ijkl} to s_{pq} when p and $q = 4, 5, 6$, i.e. 23, 13, 12.)

It is thus possible to undertake three independent measurements on these sheets. For convenience choose these to be the Young's modulus on strips at 0, 45 and 90° to the initial draw direction and denote these by E_0, E_{45} and E_{90}, respectively (see Figure 10.3(b)). From equation (10.1)

$$E_0 = \frac{1}{s_{33}}, \quad E_{90} = \frac{1}{s_{11}} \quad \text{and} \quad \frac{1}{E_{45}} = \frac{1}{4}[s_{11} + s_{33} + (2s_{13} + s_{55})]. \qquad (10.2)$$

Such measurements yield immediately two of the nine independent elastic constants, s_{11} and s_{33}, and give the combination $(2s_{13} + s_{55})$ but do not involve s_{12}.

For a transversely isotropic sheet where z is the symmetry axis, there are only five independent elastic constants, and $s_{55} = s_{44}$.

The Transverse Stiffness

The stiffness normal to the plane of the sheet c_{22} has been determined by measuring the compressional strain of narrow strips under load in a compressional creep apparatus[7]. The load is applied to the compression cage A (Figure 10.4) via two level arms pivoted about a common fulcrum B. The load is placed on the weight pan at the end of the larger arm C, and supported by the rod D. This rod is held in position by an electromagnet E, and until released, prevents the load from being applied to the samples.

Because the apparatus was originally designed for compression of much thicker samples, an intermediate steel spacer was inserted in the compression cage. To improve the accuracy of the measurements, two identical samples were compressed in each experiment as indicated in the diagram.

Experiments comparing narrow strips cut in the x and z directions (Figure 10.3), together with a theoretical analysis of the frictional effects, indicated that for polyethylene terephthalate sheets the frictional constraints prevented any strains developing in either the 1 or the 3 directions. In this case $e_1 = e_3 = 0$ and $\sigma_2 = c_{22}e_2$ or

$$e_2 = \left[\begin{array}{c} s_{22} + s_{12}(s_{13}s_{23} - s_{12}s_{33}) \\ -s_{23}(s_{11}s_{23} - s_{12}s_{23})/(s_{11}s_{33} - s_{13}^2) \end{array} \right] \sigma_2$$

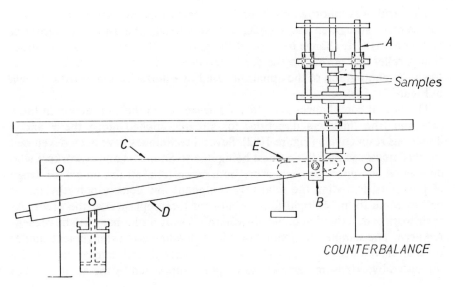

Figure 10.4. Schematic diagram of compression apparatus. [Redrawn with permission from Wilson, Cunningham, Duckett and Ward, *J. Mater. Sci.*, **11**, 2189 (1976).]

Lateral Compliances and Poisson's Ratios

For a polymer film possessing orthorhombic symmetry there are three lateral compliances s_{12}, s_{13} and s_{23} which relate to the six Poisson ratios defined in Section 10.2.2.

The lateral compliance s_{13} defines the contraction in the x direction for a stress applied along the z direction (Figure 10.2). This has been determined by measuring the change in shape of a grid of perpendicular lines printed on the surface of oriented polymer film, one set of lines being parallel to the draw direction[8]. The procedure was to align accurately an electron microscope grid on the surface of the sheet by viewing the sheet between crossed polars in a polarizing microscope. A thin coat of aluminium was then deposited by placing the sheet in a vacuum coating unit. The sample, in the form of a long narrow strip, was cut parallel to the set of lines of the grid containing the draw direction using a special cutting device. The sample was extended in a robust but friction-free extensometer, by holding it between two clamps, one of which is fixed to the frame and the other is attached to sliding rods which move through four linear bearings (Figure 10.5(a)).

The grid was photographed under a comparatively small 'zero load', sufficient to straighten the sample, and at a fixed time after application of further loads. The change in shape of the grid gave the extension and contraction parallel and normal to the draw direction and hence s_{33} and s_{13} respectively. A photograph of the apparatus used by Ladizesky and Ward is shown in Figure 10.5(b).

The lateral compliances s_{12} and s_{23} correspond to the contraction in the y direction (the thickness direction) for stresses applied along the x and z directions respectively (Figure 10.2). Several techniques have been developed for this measurement, the earliest being due to Saunders and coworkers who developed an apparatus to measure both extensional strains on a wide range of polymers, including the relatively compliant low density polyethylene at one extreme to highly rigid fibre-reinforced thermoplastics at the other. As their work is described in detail elsewhere[9,10], only a summary will be presented here. The samples are loaded by a lever loading arm arrangement similar to that developed by Turner[11]. The extensional and lateral strains are measured by extensometers, whose weight is supported by the main frame of

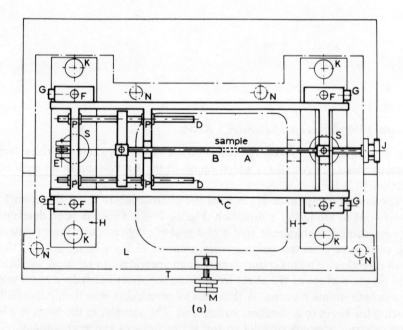

(a)

Figure 10.5. (a) Overhead projection of Poisson's ratio apparatus. A is a sample grip fixed to frame C. B is sample grip attached to sliding rods D, which move through linear bearing C. E is pulley for loading weights. G, J, M and S are positioning screws. [Redrawn with permission from Ladizesky and Ward, *J. Macromol. Sci.*, **B, 5,** 661 (1971).]

Figure 10.5. (b) Photograph of Poisson's ratio apparatus.

the machine so that they do not affect the loading of the sample to any significant degree. The mode of operation of the extensometers can be understood by reference to Figure 10.6. The tensile extensometer consists of two arms (1) rotating freely in a vertical plane. The arms are supported on five bearings at their mid-points, and are attached to the specimen at one end by screw pins (2). At the other end their displacement is monitored by a displacement capacitance transducer. The lateral extensometer (showed to measure change in thickness s_{12} or s_{23}) operates in a horizontal plane in an identical manner. The rotating arms (4) contact the specimen via brass domes (5). Thin glass cover slides are inserted between these domes and the specimen faces to prevent indentation of the specimen.

Movements of <0.25 μm can be readily detected which gives accurate tensile strain measurements in the range 0.1–7.5%. The lateral extensometers can be used to measure changes in width as well as thickness.

Figure 10.6. Photograph of the extensometry system of Clayton, Darlington and Hall: 1, upper arm of tensile extensometer; 2, specimen; 3, brass contact pieces on lateral extensometer arms; 4, lower arm of tensile extensometer; 5, glass plates with 'shoulders' resting on ends of lateral extensometer arms. [Redrawn with permission from Clayton, Darlington and Hall, *J. Phys. E.*, **6**, 218 (1973).]

A second method for determination of the lateral compliances s_{12} and s_{23} uses a Michelson interferometer[12], and is particularly suitable for thin samples, with the obvious proviso that the films must possess a fairly high degree of optical clarity. A schematic diagram of the apparatus is shown in Figure 10.7. Consider that a sample of thickness t is inserted into one arm of the interferometer, which operates in a vertical fringe mode, with air only in both parts,

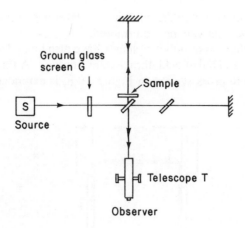

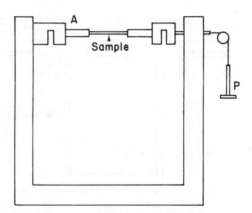

Figure 10.7. Arrangement of Michelson interferometer apparatus for measuring Poisson's ratio. [Redrawn with permission from Wilson, Cunningham and Ward, *J. Mater. Sci.*, **11**, 2181 (1976).]

to produce fringe shifts. When the sample extends under load the resultant fringe shift Δm is given by

$$\Delta m = \frac{2}{\lambda}\left[(n_i - 1)\,\Delta t + t\,\Delta n_i\right],$$

where Δt, Δn_i are the changes in thickness and refractive index respectively. For a stress σ applied along the 2 direction, $\Delta t_3 = s_{23}t\sigma_3$. Similarly $\Delta t_1 = s_{21}t\sigma_1$ for stress σ applied along the x direction. The change in refractive index $\Delta n_i = \pi'_{ij}\sigma_j$, where π'_{ij} is a photoelastic constant. Because the fringe shift depends on both Δt and Δn_i it is necessary to make measurements in both air and water, or in two liquids, so that the lateral compliances and the stress-optical coefficients can be determined.

The lateral compliances of polymer sheets have also been determined using a specially constructed Hall effect lateral extensometer[13]. A thin polymer strip of polymer S, seen in cross-section in Figure 10.8, is extended between two

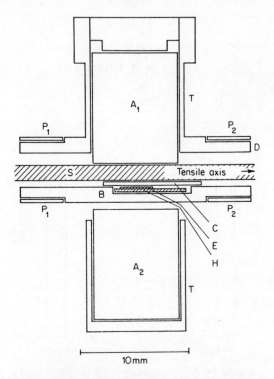

Figure 10.8. Scale diagram of the Hall effect lateral extensometer. [Redrawn with permission from Richardson and Ward, *J. Polymer Sci., Polymer Phys. Ed.*, **16**, 667 (1978).]

Alnico permanent magnets A_1 and A_2 mounted in a brass tube T. The like poles of the magnets are adjacent, so that the magnetic field between the magnets has a null point, and the field gradient is twice that of a single magnet. The polymer strip is held in contact with the magnet A_1 by a stainless steel plate B, containing the Hall effect device H, covered by a thin stainless steel plate C. The Hall plate is positioned so that the magnetic field sensing element E is on the axis of the magnets. Continuous contact between the plate containing the Hall effect device and the specimen is ensured by exerting a small lateral compressive force by means of two pairs of phosphor bronze springs P_1 and P_2. Lateral strains down to 10^{-3} could be measured in sheets ~ 0.5 mm thickness.

Torsion of Oriented Polymer Sheets

Torsion of oriented polymer sheets was undertaken by Raumann[14] to determine the shear compliances s_{44} and s_{66} for uniaxially oriented (transversely isotropic) low density polyethylene. Torsion of oriented sheets can also be used to determine the shear compliances s_{44}, s_{55} and s_{66} for sheets possessing orthorhombic symmetry. As this situation is more general than that of transverse isotropy it will be considered first.

For the orthorhombic sheets, a solution can only be found to the elastic torsion problem when the sheets are cut as rectangular prisms with their surfaces normal to the three axes of orthorhombic symmetry, and where the torsion axis coincides with one of these three axes.

A typical situation is illustrated in Figure 10.9. Torsion about the z axis involves the shear compliances in the yz and xz planes which are s_{44} and s_{55}, respectively.

The St Venant theory (see Reference 15, p. 201) gives the torque Q_z required to produce the twist T in a specimen of length l, thickness a and

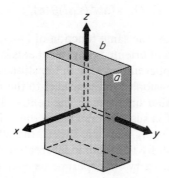

Figure 10.9. The orthorhombic sheet.

width b:

$$Q_z = \frac{ab^3 T}{s_{55}l}\beta(c_z) = \frac{ba^3 T}{s_{44}l}\beta(\overset{+}{c}_z),$$

i.e. $Q_z/T =$ torsional rigidity of the specimen, where

$$c_z = \frac{1}{\overset{+}{c}_z} = \frac{a}{b}\left(\frac{s_{55}}{s_{44}}\right)^{1/2}$$

and $\beta(c_z)$ is a rapidly converging function of c_z which for $c_z > 3$ can be approximated to

$$\beta(c_z) = \frac{1}{3}\left\{1 - \frac{0.630}{c_z}\right\}.$$

For a transversely isotropic sheet (with z direction as axis of symmetry) a similar expression describes torsion about an axis perpendicular to the symmetry axis. In this case $s_{44} = s_{55}$ and the torque Q_z is given by

$$Q_z = \frac{bc^3 T}{s_{66}l}\beta\,(\overset{*}{c}) = \frac{cb^3 T}{s_{44}l}\beta\,(\overset{+}{c}),$$

where

$$\overset{+}{c} = \frac{1}{\overset{*}{c}} = \frac{c}{b}\left(\frac{s_{44}}{s_{66}}\right)^{1/2},$$

with b the thickness, c the width and l the length of the specimen.

These formulae show that the relative contribution of the various shear compliances to the torque depend on their relative magnitude and the aspect ratios a/b or b/c. In principle, therefore, both compliances can be obtained from measurements on sheets of different aspect ratios, and this has been done for several polymers[16–18].

For transversely isotropic sheets a much simpler formula applies for torsion around the symmetry axis z where the torque Q_z is given by

$$Q_z = (ab^3/s_{44}l)\beta(c),$$

where $c = a/b$. $\beta(c)$ is now the same function of $c = a/b$ only.

In practice, the torsion of oriented polymer sheets is complicated by several effects which cause an apparent increase in the stiffness of the sample[18].

First, there is the extension of lines parallel to the twist axis at the sample edge with respect to similar lines nearer the centre. This is analogous to the bifilar or multifilar effect in suspensions. When the twist axis is parallel to the symmetry axis in transversely isotropic sheets, the effect of such small axial stresses can be dealt with by the theoretical treatment of Biot[19]. In general, however, it is necessary to carry out experiments over a range of axial stresses, and extrapolate to zero axial stress.

Secondly, planes normal to the twist axes warp into characteristic patterns, but the grips at each end prevent such warping locally. Although Timoshenko and Goodier[20] have given a theoretical treatment for the effective increase in stiffness, it is again satisfactory to adopt a more empirical procedure. Following Folkes and Arridge[6] it is considered that such end effects are confined to a block with sample compliance s' and length p at each end of the sample of total length l, the central region of homogeneous stress having the true sample compliance s^0. This gives a linear variation of measured overall sample compliance s with reciprocal length

$$s = s^0 + (2p/l)(s' - s^0).$$

s^0 can be found by taking measurements on samples of different length and extrapolating to zero reciprocal length[18].

Simple Shear of Oriented Polymer Sheets

In view of the complications arising in the torsion of oriented polymer sheets due to the effects of axial stress and end effects associated with the grips, there is considerable incentive to determine the shear compliances s_{44} and s_{66} by simple shear. Figure 10.10 shows a diagram of an apparatus designed

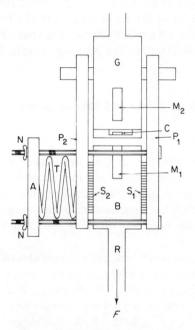

Figure 10.10. Diagram of the Hall effect simple shear apparatus. C is the Hall plate between magnets M_1 and M_2. Samples S_1 and S_2 are mounted between plates P_1 and P_2 and the moveable block B pulled by force F. [Redrawn with permission from Lewis, Richardson and Ward, *J. Phys. E.*, **12**, 189 (1979).]

by Lewis, Richardson and Ward[21]. Identical samples s_1 and s_2 are mounted between outer brass plates P_1 and P_2 and an inner movable block B. The samples are held in place by a calibrated spring T, adjusted by four wing nuts N acting on a plate A. The shear stress is applied to the samples by a downward force F acting through shaft R on block B. Conveniently the force F may be applied by means of weights, running the shaft through linear bearings to minimize friction. The shear displacement is sensed by a Hall plate C mounted between the like poles of two magnets M_1 and M_2 of approximately equal magnetic moment. The principle of use of the Hall plate is therefore identical to that described for the lateral extensometer described above. The Hall voltage is measured by an incremental gaussmeter, the apparatus being calibrated by placing non-ferrous spacers of known thickness between the block B and G.

The measured values for the shear compliance of polymers are found to depend on the magnitude of the lateral compressive stress. Results are therefore obtained for a range of lateral stresses, and the true shear compliance found by extrapolation to zero lateral stress. It has been shown that the values of shear compliance obtained in this way agree with those obtained from torsion of sheets[18].

10.3.2 Measurements on Fibres and Monofilaments

Extensional Modulus $E_3 = 1/s_{33}$

Dynamic mechanical measurements have been used to study the influence of molecular orientation on the extensional moduli of fibres drawn to different draw ratios† and also to compare the extensional moduli for a wide range of textile fibres produced by conventional manufacturing processes. The most extensive studies of this type are those of Wakelin and coworkers[22] and Meredith[23].

Detailed measurements of the extensional modulus of monofilaments have been made by longitudinal wave-propagation methods, where the relationship of the extensional modulus to molecular orientation and crystallinity has been examined. Early investigations using this technique were made by Kolsky and Hillier[24], Ballou and Smith[25], Nolle[26] and Hamburger[27]. The experimental method of Hillier and Kolsky, which was very similar to that of Ballou and Smith, is described in detail in Section 6.5.

More recently the measurement of the extensional modulus has been reexamined as a possible method for the measurement of molecular orientation in textile yarns by Charch and Moseley[28,29] and by Morgan[30]. Morgan has developed Hamburger's pulse-propagation method.

The Torsional Modulus $G = 1/s_{44}$

A convenient dynamic method for measuring the torsional modulus of synthetic fibre filaments was developed by Wakelin, Voong, Montgomery and Dusenbury[22].

A simpler method is that adopted by Meredith[23], where the fibre undertakes free torsional vibrations supporting known inertia bars at its free end.

---

† The draw ratio is the ratio of the length of a line parallel to the draw direction in the drawn material to its length before drawing. For synthetic fibres it is often determined by measuring the ratio of the initial diameter D_i to final diameter D_f, assuming that volume is conserved, i.e. draw ratio = D_i/D_f.

The Extensional Poisson's Ratio $v_{13} = -s_{13}/s_{33}$

Measurements of the extensional Poisson's ratio v_{13} have been attempted using optical diffraction and mercury-displacement techniques by Davis[31] and Frank and Ruoff[32], respectively. Satisfactory data were only obtained for nylon because this fibre is a particularly favourable case, showing no permanent deformation up to 5% extension.

More recently, measurements have been made by observing in a microscope the radial contraction, together with the corresponding lateral extension of a fibre monofilament[33]. The monofilament was extended between two moveable grips which were part of a specially constructed microscope stage. Two ink marks were placed on the monofilament to act as reference points for the measurement of length and changes in length. An immersion liquid was used to reduce diffraction effects at the edges of the monofilament. The method was of limited accuracy, and errors of at least 10% were reported for 95% confidence limits on the mean value.

The Transverse Modulus $E_1 = 1/s_{11}$

The two remaining elastic constants for fibres, the compliances s_{11} and s_{12}, have to be determined by more sophisticated methods. Both can be obtained from the compression of fibre monofilaments between parallel plates under conditions of plane strain. The transverse modulus is involved in the contact width $2b$[33,34] (Figure 10.11(a)).

The monofilament is a transversely isotropic solid; thus it is isotropic in a plane perpendicular to the fibre axis. This implies that under compressive loading normal to the fibre axis the stresses in the transverse plane will be

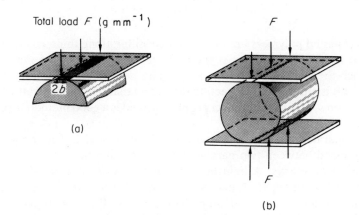

Figure 10.11. The contact zone in the compression of a fibre monofilament (a) and for consideration of deformation in the central zone of the compressed monofilament it is sufficient to assume line contacts (b).

identical in form to those for the compression of an isotropic cylinder. As the length of monofilament under compression is comparatively long, friction ensures that the compression occurs under plane strain conditions. There is, therefore, no change in dimension along the fibre axis ($e_{zz} = 0$) and only a normal stress acts along the fibre axis σ_{zz} which can be found in terms of the normal stresses σ_{xx} and σ_{yy} in the plane perpendicular to the fibre axis. We have

$$\sigma_{zz} = -\frac{s_{13}}{s_{33}}(\sigma_{xx} + \sigma_{yy}).$$

All the stresses can therefore be obtained from the solution to the problem of compression of an isotropic cylinder. The corresponding strains can then be obtained using the constitutive equations $e_p = s_{pq}\sigma_q$.

The contact zone is arranged to be small compared with the radius of the monofilament. It is therefore adequate to assume that we are dealing with the contact between two semi-infinite solids and follow Hertz's classic solution for the compression of an isotropic cylinder[35]. In this solution the displacement of the cylinder within the contact zone is assumed to be parabolic and the boundary conditions are satisfied along the boundary plane only. For purely algebraic reasons it is most convenient to use the complex variable method of McEwen[36] to obtain an analytical solution for b. It was shown by Ward et al.[33] that

$$b^2 = \frac{4FR}{\pi}\left(s_{11} - \frac{s_{13}^2}{s_{33}}\right),$$

where F is the load per unit length of monofilament in kilograms per metre, and R is the radius of the monofilament. This expression may be written as

$$b^2 = \frac{4FR}{\pi}(s_{11} - \nu_{13}^2 s_{33}).$$

Highly oriented polymers are usually much stiffer along their axis than transverse to it. The quantity s_{33} is therefore usually very small compared with s_{11}. Since the Poisson's ratio ν_{13} is typically near to 0.5, it follows that the term $\nu_{13}^2 s_{33}$ is only a small correction factor and that the contact width depends primarily on s_{11}. Thus the contact problem provides a good method in principle for determining s_{11}.

The apparatus is shown schematically in figure 10.12. The monofilament is compressed between two parallel glass plates on a microscope stage. This is arranged as follows: A light but rigid metal bar is attached at one end to a pivot capable of slight vertical adjustments. A small vertical hole is bored through this lever arm, and immediately below the hole is cemented an optically flat block of glass, of thickness sufficient for only negligible distortion during loading. By hanging weights from the free end of the lever arm the monofilament is compressed between this block and a lower transparent glass flat placed over a small hole in a rigid base plate.

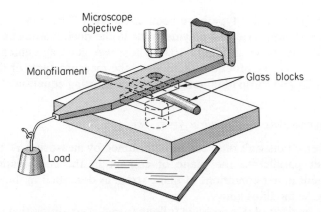

Figure 10.12. Schematic diagram of the compression experiment.

The loading device is fastened to the stage of a microscope equipped with a vertical illuminator, and the 45° plane mirror gives an enlarged image of the contact zone on a screen. At low loads, asperites and irregularities of the surface are very evident, and interference fringes are observed on each side of the contact zone. These fringes can be used to extrapolate to the true contact width, but this is a small correction which is well within the experimental error. A typical result for a polyethylene terephthalate monofilament is shown in Figure 10.13, showing the anticipated proportionality between the applied load and the square of the contact width.

It can be shown[38] that the total diametral compression in the axial direction u_1 (i.e. parallel to the direction of the applied load) is

$$u_1 = \frac{-4F}{\pi}\left(s_{11} - \frac{s_{13}^2}{s_{33}}\right)(0.19 + \sinh^{-1}(R/b). \tag{10.3}$$

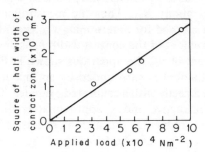

Figure 10.13. Compression of polyethylene terephthalate monofilament (diameter 2.82×10^{-4} m): measurement of contact zone width as a function of applied load.

Measurements of this type have been undertaken[38] by compressing large diameter (~ 10 mm) oriented polymer rods in the dead loading compression creep apparatus shown in Figure 10.4. There was good agreement between values of $(s_{11} - s_{13}^2/s_{33})$ found directly from the measurements of the contact zone width $2b$ and those obtained from u_1 and b, using equation (10.3).

The Transverse Poisson's Ratio $v_{12} = s_{12}/s_{11}$

The transverse Poisson's ratio can be determined by measuring u_2, the change in diameter parallel to the plane of contact in the compression of the monofilament under conditions of plane strain as described in the section on the transverse modulus above.

A simple analysis of this problem follows from the condition that the contact zone can be arranged to be small compared with the radius of the monofilament. To calculate the deformations in the diametral plane it is then adequate to consider the problem as the compression of a cylinder under concentrated loads (Figure 10.11(b)). For an isotropic cylinder this is a well known problem to be found in text-books on elasticity (see Reference 37, p. 107). It is necessary to satisfy the boundary conditions on the surface of the cylinder, and this is done by addition of an isotropic tension in the plane perpendicular to the fibre axis.

The stresses for the transversely isotropic monofilament correspond exactly to those for the isotropic case. It is therefore very straightforward to calculate the strains and hence evaluate the diametrical expansion u_2.

It is found that

$$u_2 = F\left\{ \left(\frac{4}{\pi} - 1 \right)\left(s_{11} - \frac{s_{13}^2}{s_{33}} \right) - \left(s_{12} - \frac{s_{13}^2}{s_{33}} \right) \right\}.$$

For most oriented monofilaments, s_{13}^2/s_{33} is small compared with s_{11}, as discussed previously. Hence u_2 will depend primarily on s_{12}, with a substantial term in s_{11}, which is about $\frac{1}{4}s_{11}$. Thus the measurement of the diametral expansion provides a method for determining s_{12}, provided that s_{11} is determined from a measurement of the contact width b, as described in (4) above.

For the measurement of u_2, the apparatus shown in Figure 10.12 is again used, but the monofilament is surrounded by an immersion liquid, and the diameter is measured directly with a calibrated eyepiece. The immersion liquid is chosen to have refractive index approximately equal to that of the monofilament, hence reducing diffraction effects without making the monofilament invisible. Very careful focusing of the microscope is necessary in these experiments. Inaccuracy in focusing can cause errors in the diameter measurements of the order of u_2 itself.

A typical set of results for polyethylene terephthalate is shown in Figure 10.14. It can be seen that the change in diameter is proportional to the applied load, as predicted theoretically.

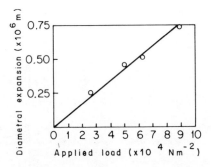

Figure 10.14. Compression of polyethylene terephthalate monofilament (diameter 2.82×10^{-4} m): measurement of diametral expansion as a function of applied load.

10.3.3 Ultrasonic Measurements

Measurements of the velocity and attenuation of elastic waves at ultrasonic frequencies offer a comprehensive method for the determination of mechanical anisotropy in oriented polymers. A brief outline will be given of the different methods which have been adopted.

Chan et al.[39] describe the preparation of sample discs from a uniaxially oriented polymer rod. To obtain all the elastic constants requires a fairly substantial initial sample with dimensions of at least ~10 mm edge cube. Three discs of diameter ~12 mm and thickness 4–8 mm can then be cut so that the symmetry axes of the discs are inclined at 0, 45 and 90°, respectively, to the symmetry axes of the uniaxial rod (Figure 10.15). By use of suitably cut quartz transducers bonded to the discs, three different linearly polarized elastic waves can be generated to propagate along the geometric axes of each disc, one longitudinal and two transverse waves. For each sample one can therefore in principle measure nine different velocities v_{ab}, where a, b refer to the directions of polarization and propagation respectively.

If we define $Q_{ab} = \rho v_{ab}^2$, where ρ is the density and v_{ab} the wave velocity, it follows that

$$c_{11} = Q_{xx}, \quad c_{33} = Q_{zz}, \quad c_{44} = Q_{xz} \quad \text{and} \quad c_{66} = \tfrac{1}{2}(c_{11} - c_{12}) = Q_{yx}. \quad (10.4)$$

(The directions x, y, z are defined in Figure 10.9 above.) Hence four of the five independent elastic constants for the uniaxially oriented rod can be found from the velocities of waves propagating along the principal axes of the sample. c_{13} can be obtained from measurements on the 45° disc, using a general formula[40] for the velocity of an elastic wave propagating at an angle to the principal axis,

$$c_{13} = \tfrac{1}{2}[4(Q_{vv} - Q_{uv})^2 - (c_{33} - c_{11})]^{1/2} - c_{44}. \quad (10.5)$$

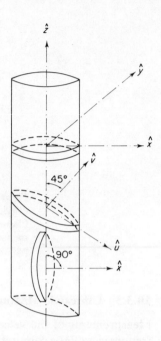

Figure 10.15. Schematic diagram illustrating the sample discs employed in the ultrasonic measurements. [Redrawn with permission from Chan, Chen, Choy and Ward, *J. Phys. D*, **11**, 481 (1975).]

It can be appreciated that the errors in c_{11}, c_{33}, c_{44} and c_{66} relate directly to the corresponding velocities squared, and hence have simply twice the fractional error of the latter. In practice this was $\pm 10\%$. But for c_{12} and c_{13} and the compliance constants s_{ij} the situation is clearly more complicated. It turns out that the errors in s_{11} and s_{44} are quite small ($\pm 10\%$) but fairly large for s_{33} and very large ($\geqslant 30\%$) for c_{12}, c_{13}, s_{12} and s_{13}. In spite of these limitations useful results have been obtained using this method[39].

The velocity measurements were made by the pulse echo-overlap technique[41–43]. It is also possible to determine $\tan \delta$ by making attenuation measurements, in this case by pulse-height comparison of successive echoes with standard pulses[41,44].

An alternative ultrasonic technique has been described by Markham[45], and applied by Rawson and Rider[46]. The sample is immersed in a tank of water containing an ultrasonic transmitter and receiver. For a sample in the form of a rectangular parallelopiped, and normal incidence, the wave velocity v in the polymer is given by[45]

$$\frac{1}{v} = \frac{1}{v_w} - \frac{\tau}{d},$$ (10.6)

where v_w is the wave velocity in water, d is the thickness of the sample and τ is the difference in transit time of the pulse from transmitter to receiver with and without the sample in the beam. For non-normal incidence, Snell's law gives

$$v = v_w \frac{\sin r}{\sin i} \tag{10.7}$$

and it may be shown[45] that

$$\tan r = \frac{\sin i}{\cos i + \tau v_w / d}, \tag{10.8}$$

where i, r are the angles of incidence and refraction respectively.

For a sufficiently large sample (~ 10 mm cube) the elastic constants can be obtained in a manner exactly analogous to that outlined for the method of Chan et al. above. For example, with the z axis of the sample vertical (i.e. normal to the propagation direction) the longitudinal and transverse velocities in the xy plane are measured and we have

$$c_{11} = Q_{xx} \quad \text{and} \quad c_{66} = \tfrac{1}{2}(c_{11} - c_{12}) = Q_{yx}$$

as in equation (10.4) above. With the x axis vertical, for wave normals in the z direction we have

$$c_{33} = Q_{zz}.$$

The samples used by Rawson and Rider were only ~ 4 mm thickness in the y direction so that difficulty was experienced in obtaining velocity measurements in the x and z directions, and rather more complicated procedures had to be adopted to obtain c_{33} and also c_{44}. c_{13} was obtained from measurements at $45°$, using equation (10.5) above. The experimental errors quoted by Rawson and Rider were considerably smaller than those of Chan et al., being $\pm 0.4\%$ for c_{11} and $\pm 3\%$ for the other constants.

Wright et al.[47] have also used an ultrasonic pulse method to determine the wave velocities and hence the elastic constants of oriented polymers. Their method was similar to that of Rawson and Rider in that thick polymer sheets were immersed in a liquid through which an ultrasonic beam was propagated, but they detected the part of the beam reflected from the surface of the sample and hence measured the critical angle of incidence. The wave velocity is then obtained from equation (10.7) with $r = 90°$.

10.4 INTERPRETATION OF MECHANICAL ANISOTROPY: GENERAL CONSIDERATIONS

There are three major factors which can determine the mechanical anisotropy of oriented polymers: (1) the structure of the molecular chain, and if the polymer crystallizes, the crystal structure; (2) the molecular orientation, and in a crystalline polymer, the morphology; (3) relaxation processes, i.e. thermally activated processes in both the crystalline and non-crystalline regions. These factors will now be discussed in turn.

10.4.1 Chain Structure and Crystal Structure

Linear synthetic polymers consist of long chain molecules which often take the form of either a fully extended linear chain or a helix. The molecular chain in the crystalline regions of polyethylene is a planar zig-zag, in isotactic polypropylene it is a threefold helix. In both cases there is a large difference between the forces involved when the structure is deformed parallel to the chain or helix axis and perpendicular to this axis.

The stiffness constants for the polyethylene crystal have been estimated by Odajima and Maeda[48], and more recently by Tadokoro[49], and lead to the following stiffness and compliance matrices:

$$c_{ij} = \begin{bmatrix} 7.99 & 3.28 & 1.13 & 0 & 0 & 0 \\ 3.28 & 9.92 & 2.14 & 0 & 0 & 0 \\ 1.13 & 2.14 & 315.92 & 0 & 0 & 0 \\ 0 & 0 & 0 & 3.19 & 0 & 0 \\ 0 & 0 & 0 & 0 & 1.62 & 0 \\ 0 & 0 & 0 & 0 & 0 & 3.62 \end{bmatrix} \text{GN m}^{-2},$$

$$s_{ij} = \begin{bmatrix} 14.5 & -4.78 & -0.019 & 0 & 0 & 0 \\ -4.78 & 11.7 & -0.062 & 0 & 0 & 0 \\ -0.019 & -0.062 & 0.317 & 0 & 0 & 0 \\ 0 & 0 & 0 & 31.4 & 0 & 0 \\ 0 & 0 & 0 & 0 & 61.7 & 0 \\ 0 & 0 & 0 & 0 & 0 & 27.6 \end{bmatrix} \times 100 \text{ m}^2 \text{ GN}^{-1}$$

It can be seen that the crystal modulus along the chain direction $E_3^c = 1/s_{33} \doteq c_{33}$ is $\sim 300 \text{ GN m}^{-2}$. The high value arises because the deformation involves primarily the bending and stretching of covalent bonds. The tensile modulus perpendicular to the chain direction, and the shear moduli, on the other hand are very much lower (~ 1–10 GN m^{-2}) and relate to the much weaker van der Waals or dispersion forces between the chains. In polypropylene, the tensile modulus in the helix axis direction is much lower than that along the chain axis in polyethylene, because it involves rotation around bonds as well as bending of bonds. However, at $\sim 50 \text{ GN m}^{-2}$, it is still much larger than the tensile and shear moduli perpendicular to the helix axis which are of similar magnitude and origin to those in polyethylene.

These considerations indicate that the intrinsic mechanical anistropy of oriented polymers can be very high, and in a very few cases, notably polyethylene, this has been achieved in practice. Table 10.1 shows collected data for ultrahigh draw linear polyethylene, which confirm the theoretical expectations.

The determination of the elastic constants for the crystalline regions of polymers is of considerable importance because the results clearly provide a base-line against which practical achievements can be judged. In addition to theoretical calculations[48-52], the moduli can be determined from the changes

Table 10.1. Elastic constants of ultrahigh draw polyethylene.

	20 °C	−196 °C	Theoretical (Tadokoro)
Axial modulus (GN m^{-2})	70	160	316
Transverse modulus (GN m^{-2})	1.3	—	8–10
Shear modulus (GN m^{-2})	1.3	1.95	1.6–3.6
Poisson's ratio	0.4	—	0.5

in the X-ray diffraction pattern on stressing an oriented sample[53], by Raman spectroscopy[54] and inelastic neutron scattering[55]. Comprehensive reviews of the subject have been presented elsewhere[56–58].

10.4.2 Orientation and Morphology

For most oriented polymer systems, however, and this includes commercially available synthetic fibres and films, the degree of mechanical anisotropy is much less than these considerations of the molecular chain would imply. In particular, the very high intrinsic modulus along the chain axis direction is not achieved. Crystalline polymers are essentially composite materials with alternating crystalline and non-crystalline regions. Although the crystalline regions can become very highly aligned in the fabrication processes, the non-crystalline regions are less oriented. Even if the overall orientation of the chain segments as determined by, say, infrared dichroism or birefringence is apparently quite high, there will still be very few chains where long lengths of the molecule are aligned parallel to a single direction. As we have seen, it is these molecules which will be critical in increasing the stiffness, because there is such a large difference between the stresses involved in bond stretching and bending, and other modes of deformation. Peterlin[59] has recognized this by his hypothesis that the Young's modulus of an oriented fibre is essentially determined by the proportion of taut tie molecules (i.e. extended chains) which produce links between the crystalline blocks in the fibre direction. An alternative proposal, which appears to be valuable in understanding drawn polyethylene, is that there are crystalline bridges[60] (narrow clumps of crystalline chains) linking the crystalline blocks. These provide a degree of overall crystal continuity so that an appreciable fraction of the crystal chain modulus can now be obtained.

In polymers such as polymethyl methacrylate, which do not crystallize, the degree of mechanical anisotropy correlates very well with molecular orientation determined by spectroscopic techniques and birefringence[61]. In this case there is so much disorder that it seems unlikely that a significant proportion of the chains achieve the high alignment of a crystalline polymer such as polyethylene or polypropylene. Some polymers, such as polyethylene terephthalate, which show comparatively low overall crystallinity, may, however, occupy an intermediate position. In PET the mechanical anisotropy produced by drawing correlates well with overall molecular orientation[62]. However, it could still be that as far as the stiffness along the draw direction is concerned, the proportion of fully extended and aligned molecular chains (i.e. tie molecules) plays a vital role and that this proportion systematically increases with overall molecular orientation.

10.4.3 Relaxation Processes

As we have seen in Chapter 8 relaxation processes can be associated with either main-chain segmental motions or side group motions, and in a crystalline polymer associated with either the crystalline or the amorphous phase. From the viewpoint of the mechanical anisotropy this means that the relaxation processes can reflect either the molecule orientation, e.g. the orientation of the chain axis, or in a crystalline polymer the morphological structure, e.g. the lamellar orientation where the non-crystalline interlamellar material softens preferentially.

10.5 INTERPRETATION OF MECHANICAL ANISOTROPY: QUANTITATIVE MODELS

Recognition of the importance of the two major factors, (1) molecular orientation and (2) the composite nature of a crystalline polymer, led to the two different starting points for the interpretation of mechanical anisotropy in oriented polymers.

10.5.1 Molecular Orientation and the Aggregate Model

The aggregate model[63] considers that the polymer consists of an aggregate of anisotropic units. The mechanical properties of the units are considered to be those of a highly oriented polymer, obtained from experimental results. In the case of a crystalline polymer, as already discussed these will not coincide with the crystal moduli, the difference being very large in the case of the chain direction. This model originated as a single phase model, and is clearly in principle most obviously applicable to amorphous polymers. It has, however, been particularly successful in polyethylene terephthalate[64] and in low density polyethylene[63,65], where it predicts the effect of the c-shear relaxation well (Section 8.4.4 above). The aggregate model has also been applied with success to graphite[66] and to Kevlar fibres[67].

Charch and Moseley[28,29] and Morgan[30] showed that a simplified version of the aggregate model provided a semi-theoretical basis for using sonic modulus as a measure of molecular orientation in fibres, similar to birefringence. Their treatment was extended by Samuels[68] to a two-phase model, where it is effectively assumed that the compliances of the crystalline and non-crystalline phases are additive in the draw direction. This model was shown to describe data for a wide range of oriented polypropylene fibres very well, and was also later applied to polyethylene terephthalate.

10.5.2 Composite Solid Models

The Takayanagi model[69] recognizes the two-phase nature of crystalline polymers but it is only concerned with the tensile moduli parallel and perpendicular to the draw direction in a highly oriented polymer. It is consistent with the Peterlin model of crystalline blocks alternating with disordered material within the microfibrils[70]. The crystalline and non-crystalline phases are therefore in series (additivity of compliances) parallel to the draw direction and in parallel (additivity of stiffnesses or moduli) for the direction perpendicular to the draw direction. The crystalline phase is considered to be fully oriented, and the effect of tie molecules or crystal bridges can be taken into account by permitting some proportion of crystalline material to be in parallel with the non-crystalline material.

A limitation of the Takayanagi model is that it does not explicitly take into account shear deformation. Hence it cannot deal with the orientation of the lamellae, and distinguish between the behaviour when the lamellar normals are parallel to the draw direction or at any appreciable angle to this direction. The lamellar orientation was considered by Ward and coworkers[71-73] and it was shown that interlamellar shear is a key deformation mechanism which must be treated formally if a satisfactory understanding of the mechanical anisotropy of these materials is to be achieved.

In the case of high modulus polyethylene, Barham and Arridge[74] have proposed an alternative composite model, analogous to a short fibre composite. The oriented polymer is considered to be composed of needle-like crystals of high aspect ratio (the fibre phase) embedded in a matrix composed of oriented non-crystalline and some crystalline material.

It is in the light of these attempts at interpretation, and especially the aggregate model and the Takayanagi model, that the experimental results for oriented polymers will be considered.

10.6 EXPERIMENTAL STUDIES OF MECHANICAL ANISOTROPY IN POLYMERS: DRAWN POLYMERS AND THE AGGREGATE MODEL

The first attempts to determine a complete set of independent elastic constants for oriented polymers were Raumann and Saunders' measurements on low density polyethylene sheets[4,75] and the combination of measurements on fibres and films of polyethylene terephthalate undertaken by Pinnock and Ward[76,77]. Later work included measurements on polyethylene terephthalate sheets by Raumann[78] and extensive measurements on various fibre monofilaments by Ward and coworkers[33,34,79]. In addition to these more complete surveys there

274

are a few measurements of the extensional and torsional moduli of filaments (e.g. extensional and torsional modulus measurements on nylon and polyethylene terephthalate) by Wakelin and coworkers[22], the sonic modulus measurements of Kolsky and Hillier[24], Ballou and Smith[25] and Morgan[30] and torsional measurements by Meredith[23]. All these measurements were confined to room temperature. More recently, the low density polyethylene data were extended to cover a range of temperatures[65]. In addition there are a number of measurements on the viscoelastic behaviour of polymers by Takayanagi and coworkers[69] and by Ward and coworkers[72,77,80,81]. There are also results of ultrasonic measurements on oriented amorphous polymers by Wright et al.[47] and by Rawson and Rider[46], and on crystalline polymers by Choy and his colleagues[39,41].

Most of the measurements of mechanical anistropy in polymers are for transversely isotropic systems, but there are a number of studies on sheets possessing orthorhombic symmetry[17,18,22,23,37,38,71,82,83].

These experimental studies will now be discussed in some detail.

10.6.1 Low Density Polyethylene Sheets

Raumann and Saunders[75] prepared a series of oriented low density polyethylene sheets by uniaxial stretching of isotropic sheets to varying final extensions. They measured the tensile modulus in directions making various angles with the initial draw direction, and presented their results in two types of diagram. Figure 10.16 shows the plot of Young's modulus $(1/s_\theta)$ for deformation in

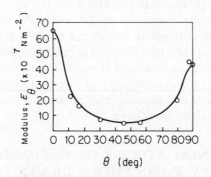

Figure 10.16. Comparison of the observed variation of modulus E_0 with angle θ to draw direction and the theoretical relation (i.e. full curve) calculated from E_0, E_{45} and E_{90} for low density polyethylene sheet drawn to a draw ratio of 4.65. [Redrawn with permission from Raumann and Saunders, *Proc. Phys. Soc.*, **77**, 1028 (1961).]

the plane of the sheet for a highly oriented sample. The unusual feature is that the sheet shows the lowest stiffness in a direction making an angle of about 45° to the initial draw direction, whereas intuitively one might expect the lowest stiffness direction to be at right angles to the draw direction, the latter being the direction of overall molecular orientation. Recalling the compliance equation (10.1),

$$s_\theta = s_{11} \sin^4 \theta + s_{33} \cos^4 \theta + (2s_{13} + s_{44}) \sin^2 \theta \cos^2 \theta,$$

this experimental result implies that $(2s_{13} + s_{44})$ is much greater than either s_{11} or s_{33}, since when $\theta = 45°$ these terms will be equally weighted; see equation (10.2) for E_{45}.

The second type of diagram is a plot of E_0, E_{45} and E_{90} as a function of draw ratio (Figure 10.17). Again the results are somewhat unexpected in that E_0 first falls with increasing draw ratio and $E_{90} > E_0$ at low draw ratios, E_{90}

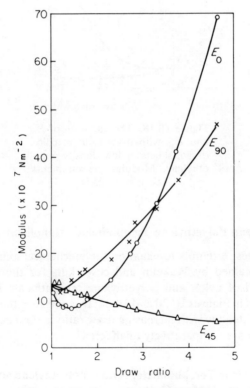

Figure 10.17. The variation of E_0, E_{45} and E_{90} with draw ratio in cold-drawn sheets of low density polyethylene. Modulus measurements taken at room temperature. [Redrawn with permission from Raumann and Saunders, *Proc. Phys. Soc.*, **77**, 1028 (1961).]

and E_0 crossing at a draw ratio of about 3, as E_0 rises to a larger final value than E_{90} at the highest draw ratio.

Recent work by Gupta and Ward[65] has shown that this unexpected behaviour is specific to the room temperature measurements on this polymer. Reducing temperature eventually produced a less complicated pattern for E_0, E_{45} and E_{90} as a function of draw ratio (Figure 10.18). We will shortly show that this is the more conventional situation. At the same time the polar diagram of the modulus in a highly oriented sheet changed markedly with decreasing temperature (Figure 10.19).

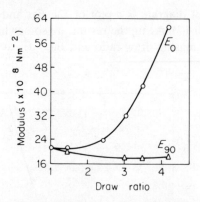

Figure 10.18. The variation of E_0 and E_{90} with draw ratio in cold-drawn sheets of low density polyethylene. Modulus measurements taken at $-125\,°C$.

10.6.2 Nylon and Polyethylene Terephthalate Monofilaments

One of the earliest attempts to examine the mechanical anisotropy of fibres were results obtained by Wakelin and coworkers for the extensional and torsional moduli of nylon and polyethylene terephthalate filaments[22]. The results are shown in Figures 10.20 and 10.21. It can be seen that the extensional moduli increase steadily with increasing draw ratio in both cases, whereas the torsional moduli are comparatively unaffected.

10.6.3 Polyethylene Terephthalate, Nylon, Polyethylene and Polypropylene Monofilaments

In a comprehensive study of several fibre monofilaments, Hadley, Pinnock and Ward[33,34,79] attempted to determine the five independent elastic constants for oriented polyethylene terephthalate, nylon, low and high density polyethylene and polypropylene. All measurements were confined to room tem-

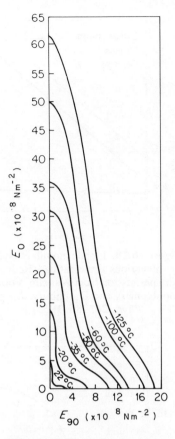

Figure 10.19. Polar representation of the mechanical anisotropy in a highly oriented low density polyethylene sheet at different temperatures.

perature, and the elastic constants were obtained as a function of molecular orientation as determined by the draw ratio and birefringence.

The results are summarized in Table 10.2 and Figures 10.22–10.26. (The calculated curves will be discussed in Section 10.6.5.)

The precise development of mechanical anisotropy in these fibres does depend on details of their chemical composition and the exact nature of the drawing process. However, the following general features can be distinguished.

The principal effect of increasing draw ratio (i.e. increasing molecular orientation) is to increase the Young's modulus E_3 measured in the direction of the fibre axis. Thus E_3 for the highly oriented fibres is much greater than the isotropic extensional modulus for unoriented polymers. In nylon and polyethylene terephthalate there is a corresponding but small decrease in the

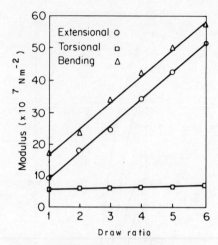

Figure 10.20. Elastic moduli of Nylon 66 of various draw ratios. [Redrawn with permission from Wakelin, Voong, Montgomery and Dusenbury, *J. Appl. Phys.*, **26**, 786 (1955).]

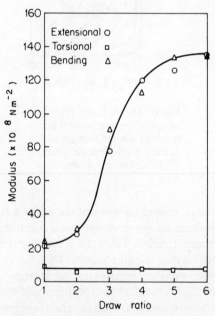

Figure 10.21. Elastic moduli of polyethylene terephthalate of various draw ratios. [Redrawn with permission from Wakelin, Voong, Montgomery and Dusenbury, *J. Appl. Phys.*, **26**, 786 (1955).]

Table 10.2. Elastic compliances of oriented fibres (units of compliance are 10^{-10} m^2 N^{-1}; errors quoted are 95% confidence limits) (see Reference 79).

Material	Birefringence (Δn)	s_{11}	s_{12}	s_{33}	s_{13}	s_{44}	$\nu_{13} = -\dfrac{s_{13}}{s_{33}}$	$\nu_{12} = -\dfrac{s_{12}}{s_{11}}$
Low density polyethylene film 14	—	22	−15	14	−7	680	0.50	0.68
Low density polyethylene 1	0.0361	40±4	−25±4	20±2	−11±2	878±56	0.55±0.08	0.61±0.20
Low density polyethylene 2	0.0438	30±3	−22±3	12±1	−7±1	917±150	0.58±0.08	0.73±0.20
High density polyethylene 1	0.0464	24±2	−12±1	11±1	−5.1±0.7	34±1	0.46±0.15	0.52±0.08
High density polyethylene 2	0.0594	15±1	−16±2	2.3±0.3	−0.77±0.3	17±2	0.33±0.12	1.1±0.14
Polypropylene 1	0.0220	19±1	−13±2	6.7±0.3	−2.8±1.0	18±1.5	0.42±0.16	0.68±0.18
Polypropylene 2	0.0352	12±2	−17±2	1.6±0.04	−0.73±0.3	10±2	0.47±0.17	1.5±0.3
Polyethylene terephthalate 1	0.153	8.9±0.8	−3.9±0.7	1.1±0.1	−0.47±0.05	14±0.5	0.43±0.06	0.44±0.09
Polyethylene terephthalate 2	0.187	16±2	−5.8±0.7	0.71±0.04	−0.31±0.03	14±0.2	0.44±0.07	0.37±0.06
Nylon 66	0.057	7.3±0.7	−1.9±0.4	2.4±0.3	−1.1±0.15	15±1	0.48±0.05	0.26±0.08

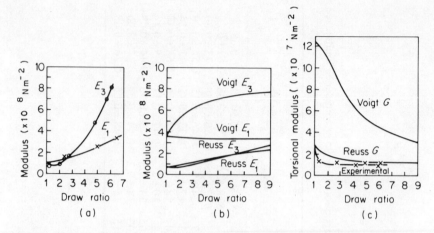

Figure 10.22. Low density polyethylene filaments: extensional (E_3), transverse (E_1) and torsional moduli (G); comparison between experimental results and simple aggregate theory for E_3 and E_1 ((a) and (b)) and for G (c).

transverse modulus E_1 with increasing draw ratio, for polypropylene and high density polyethylene E_1 remains almost constant, and for low density polyethylene E_1 increases significantly with increasing draw ratio. The overall effect is that the extensional modulus E_3 for highly oriented filaments is greater than the transverse modulus E_1 (see Table 10.3). Polyethylene terephthalate is the most anisotropic fibre in this respect, with

$$\frac{E_3}{E_1} = \frac{s_{11}}{s_{33}} \sim 27.$$

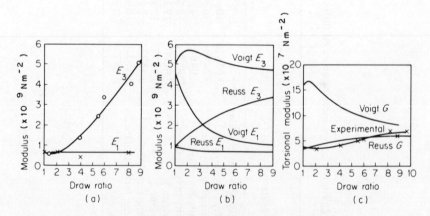

Figure 10.23. High density polyethylene filaments: extensional (E_3), transverse (E_1) and torsional moduli (G); comparison between experimental results and simple aggregate theory for E_3 and E_1 ((a) and (b)) and for G (c).

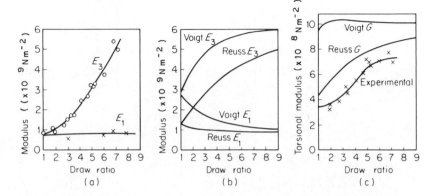

Figure 10.24. Polypropylene filaments: extensional (E_3), transverse (E_1) and torsional moduli (G); comparison between experimental results and simple aggregate theory for E_3 and E_1 ((a) and (b)) and for G (c).

Note the anomalous behaviour of low density polyethylene at draw ratios less than 2, which is in agreement with the film data of Raumann and Saunders.

A further striking difference between low density polyethylene and other fibres is shown by the change of the shear modulus G with draw ratio, a decrease greater than three times occurring over the range of molecular orientations examined, compared with only small changes for the other materials. For the other fibres s_{44} lay close in value to s_{11}: in polyethylene terephthalate, high density polyethylene and polypropylene $s_{44}/s_{11} \sim 1$; in nylon $s_{44}/s_{11} \sim 2$. Low density polyethylene appears, as far as the present room temperature measurements are concerned, to be an exceptional polymer with the extensional compliance s_{33} having the same order of magnitude as

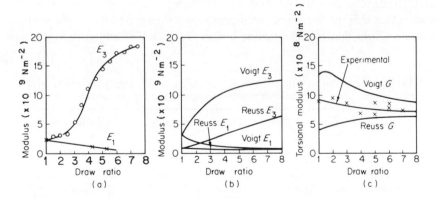

Figure 10.25. Polyethylene terephthalate filaments: extensional E_3, transverse E_1 and torsional moduli (G); comparison between experimental results and simple aggregate theory for E_3 and E_1 ((a) and (b)) and for G (c).

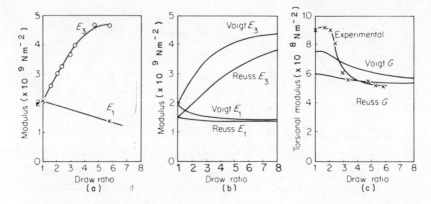

Figure 10.26. Nylon filaments: extensional (E_3), transverse (E_1) and torsional moduli (G); comparison between experimental results and simple aggregate theory for E_3 and E_1 ((a) and (b)) and for G (c).

the transverse compliance s_{11}, and the shear compliance s_{44} being more than an order of magnitude greater than either s_{33} or s_{11}. This exceptional behaviour has been exemplified by a detailed analysis of the anisotropy in this polymer (see Section 8.4.4).

It is interesting to note the similarity between the elastic constants of high density polyethylene and polypropylene: the relative values of the elastic compliances show very similar trends (see Table 10.2) and the major differences can be expressed as a marginally greater stiffness for polypropylene as might be expected from its higher melting point.

The compliance s_{13} is in all cases low, and appears to decrease rapidly with increasing draw ratio, in a similar manner to s_{33}. Thus the extensional Poisson's ratio $\nu_{13} = -s_{13}/s_{33}$ is rather insensitive to draw ratio and, with the exception of high density polyethylene, does not differ significantly from 0.5. The assumption that the fibres are incompressible is thus generally a valid approximation. (Note that for anisotropic bodies, ν_{13} is not confined to values less

Table 10.3. Comparison of elastic compliances of highly oriented and unoriented fibres (units of compliance are square metres per newton $\times 10^{-10}$).

	Highly oriented			Unoriented	
	s_{11}	s_{33}	s_{44}	$\overline{s_{11}} = \overline{s_{33}}$	$\overline{s_{44}}$
Low density polyethylene	30	12	917	81	238
High density polyethylene	15	2.3	17	17	26
Polypropylene	12	1.6	10	14	27
Polyethylene terephthalate	16	0.71	14	4.4	11
Nylon	7.3	2.4	15	4.8	12

than 0.5, but is limited solely by the inequalities necessary for a positive strain energy:

$$s_{12}^2 < s_{11}^2; \qquad s_{13}^2 < \tfrac{1}{2}s_{33}(s_{11} + s_{12})$$

(see for example Reference 3)).

The transverse Poisson's ratio $\nu_{12} = -s_{12}/s_{11}$ was subject to large experimental errors but even so the values can be seen to range widely, with those for polypropylene and high density polyethylene being considerably higher than for the other materials.

Two small unusual features of the mechanical anisotropy can be noted. There is a small minimum in the extensional modulus E_3 of high density polyethylene at low draw ratios, and a very small maximum in the torsional modulus of nylon at a draw ratio of 1.5.

10.6.4 Correlation of the Elastic Constants of an Oriented Polymer with those of an Isotropic Polymer: the Aggregate Model

It is to be expected that the mechanical properties of polymers will depend on the exact details of the molecular arrangements, i.e. both the crystalline morphology and the molecular orientation, these being intimately related so that any attempt to separate their influence must be an artificial one to a greater or lesser degree. In the case of polyethylene terephthalate it was found that the degree of molecular orientation (as measured from birefringence, for example) was the primary factor in determining the mechanical anisotropy. Table 10.4 shows some extensional and torsional modulus results for a number

Table 10.4. Physical properties of polyethylene terephthalate fibres at room temperature[77].

Birefringence	X-ray crystallinity	Extensional modulus $(\times 10^9 \, \text{N m}^{-2})$	Torsional modulus $(\times 10^9 \, \text{N m}^{-2})$
0	0	2.0	0.77
0	33	2.2	0.89
0.142	31	9.8	0.81
0.159	30	11.4	0.62
0.190	29	15.7	0.79

of polyethylene terephthalate fibres measured at room temperature. It can be seen that the influence of crystallinity on these moduli is small compared with the effect of molecular orientation on the extensional modulus. It has therefore been proposed that to a first approximation the unoriented fibre or polymer can be regarded as an aggregate of anisotropic elastic units whose elastic properties are those of the highly oriented fibre or polymer[63,84]. The average elastic constants for the aggregate can be obtained in two ways, either by assuming uniform stress throughout the aggregate (which will imply a summation of compliance constants) or uniform strain (which will imply a summation of stiffness constants). Because in general the principal axes of stress and strain do not coincide for an anisotropic solid these two approaches both involve an approximation. With the first assumption of uniform stress, the strains throughout the aggregate are not uniform; with the alternative assumption of uniform strain, non-uniformity of stress occurs. It was shown by Bishop and Hill[85] that for a random aggregate the correct value lies between the two extreme values predicted by these alternative schemes.

Consider the case of uniform stress. This can be imagined as a system of N elemental cubes arranged end-to-end forming a 'series' model (Figure 10.27(a)). Assume that each elemental cube is a transversely isotropic elastic

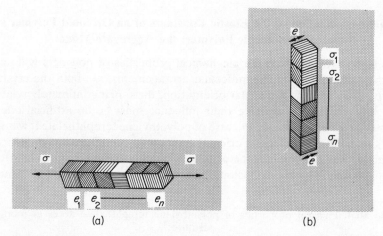

Figure 10.27. The aggregate model (a) for uniform stress; (b) for uniform strain.

solid, the direction of elastic symmetry being defined by the angle θ which its axis makes with the direction of applied external stress σ. The strain in each cube e_1 is then given by the compliance formula

$$e_1 = [s_{11} \sin^4 \theta + s_{33} \cos^4 \theta + (2s_{13} + s_{44}) \sin^2 \theta \cos^2 \theta]\sigma,$$

where s_{11}, s_{33}, etc., are the compliance constants of the cube. We ignore the fact that the cubes in general distort under the applied stress and do not satisfy compatibility of strain throughout the aggregate. Then the average strain e is

$$e = \frac{\Sigma e_1}{N} = [s_{11} \overline{\sin^4 \theta} + s_{33} \overline{\cos^4 \theta} + (2s_{13} + s_{44}) \overline{\sin^2 \theta \cos^2 \theta}]\sigma,$$

where $\overline{\sin^4 \theta}$, etc., now define the average values of $\sin^4 \theta$, etc., for the aggregate of units. For a random aggregate it is found that

$$e/\sigma = \text{average extensional compliance}$$
$$= \overline{s'_{33}} = \tfrac{8}{15}s_{11} + \tfrac{1}{5}s_{33} + \tfrac{2}{15}(2s_{13} + s_{44}). \tag{10.9}$$

In a similar manner the case of uniform strain can be imagined as a system of N elemental cubes stacked in a 'parallel' model (Figure 10.27(b)). For this case the stress in each cube σ_1 is given by the stiffness formula:

$$\sigma_1 = [c_{11} \sin^4 \theta + c_{33} \cos^4 \theta + 2(c_{13} + 2c_{44}) \sin^2 \theta \cos^2 \theta]e,$$

where c_{11}, c_{33}, etc., are the stiffness constants of the cube. The average stress σ is then

$$\sigma = \frac{\Sigma \sigma_1}{N} = [c_{11} \overline{\sin^4 \theta} + c_{33} \overline{\cos^4 \theta} + 2(c_{13} + 2c_{44}) \overline{\sin^2 \theta \cos^2 \theta}]e,$$

where $\overline{\sin^4 \theta}$, etc., are the average values of $\sin^4 \theta$. For a random aggregate

$$\frac{\sigma}{e} = \overline{c'_{33}} = \tfrac{8}{15}c_{11} + \tfrac{1}{5}c_{33} + \tfrac{4}{15}(c_{13} + 2c_{44}). \tag{10.10}$$

Equations (10.9) and (10.10) define one compliance constant and one stiffness constant for the isotropic polymer. For an isotropic polymer there are two independent elastic constants, and these two schemes predict a value for the isotropic shear compliance $\overline{s'_{44}}$ and the isotropic shear stiffness $\overline{c'_{44}}$ respectively. These are

$$\overline{s'_{44}} = \tfrac{14}{15}s_{11} - \tfrac{2}{3}s_{12} - \tfrac{8}{15}s_{13} + \tfrac{4}{15}s_{33} + \tfrac{2}{5}s_{44}, \tag{10.11}$$

$$\overline{c'_{44}} = \tfrac{7}{30}c_{11} - \tfrac{1}{6}c_{12} - \tfrac{2}{15}c_{13} + \tfrac{1}{15}c_{33} + \tfrac{2}{5}c_{44}. \tag{10.12}$$

Averaging the compliance constants defines the elastic properties of the isotropic aggregate in terms of $\overline{s'_{33}}$ and $\overline{s'_{44}}$. This is called the 'Reuss average'[86]. Averaging the stiffness constants defines the elastic properties of the aggregate in terms of $\overline{c'_{33}}$ and $\overline{c'_{44}}$. This is called the 'Voigt average'[87]. In the latter case it is desirable to invert the matrix and obtain the $\overline{s'_{33}}$ and $\overline{s'_{44}}$ corresponding to these values of $\overline{c'_{33}}$ and $\overline{c'_{44}}$ in order to compare directly the values obtained by the two averaging procedures.

The results of such a comparison are summarized in Table 10.5 for five polymers. For polyethylene terephthalate and low density polyethylene the measured isotropic compliances lie between the calculated bounds, suggesting that in these polymers the molecular orientation is indeed the primary factor determining the mechanical anisotropy. In nylon the measured compliances lie just outside the bounds, suggesting that although molecular orientation is important in determining the mechanical anisotropy, other structural factors also play an important part. Finally, in high density polyethylene and polypropylene the measured values for the isotropic compliances $\overline{s'_{11}} = \overline{s'_{33}}$ lie well outside the calculated bounds, suggesting that factors other than orientation

Table 10.5. Comparison of calculated and measured extensional and torsional compliances (in square metres per newton $\times 10^{10}$) for unoriented fibres.

| | Extensional compliance $(s'_{11} = s'_{33})$ | | | Torsional compliance (s'_{44}) | | |
| | Calculated | | | Calculated | | |
	Reuss average	Voigt average	Measured	Reuss average	Voigt average	Measured
Low density polyethylene	139	26	81	416	80	238
High density polyethylene	10	2.1	17	30	6	26
Polypropylene	7.7	3.8	14	23	11	2.7
Polyethylene terephthalate	10.4	3.0	4.4	25	7.6	11
Nylon	6.6	5.2	4.8	17	13	12

play a major role in the mechanical anisotropy. In polypropylene Pinnock and Ward[88] suggested that simultaneous changes occur in morphology and molecular mobility, both of which affect the mechanical properties.

10.6.5 Mechanical Anistropy of Fibres and Films of Intermediate Molecular Orientation

Fibres and films of intermediate molecular orientation are often produced by a two-stage process in which the first stage consists of making an approximately isotropic specimen which is then uniaxially stretched or drawn. The aggregate model can be extended to determine the mechanical anisotropy as a function of the draw ratio.

The starting point for such a theory was the observation that in general terms the birefringence–draw ratio curves for several crystalline polymers take a similar form, as noted previously by several workers (Crawford and Kolsky for low density polyethylene[89] and Cannon and Chappel for nylon[90]), with the birefringence increasing rapidly at low draw ratios, but approaching the maximum value asymptotically at draw ratios greater than about five. Results for low density polyethylene are shown in Figure 10.28(a).

Crawford and Kolsky concluded that the birefringence was directly related to the permanent strain, and they proposed a model of rod-like units rotating towards the draw direction on drawing. The essential mathematical step in the theory is illustrated in Figure 10.29. Each unit is considered to be transversely isotropic. The orientation of a single unit is therefore defined by the angle θ between its symmetry axis and the draw direction, and the angle ϕ which is the angle between the projection of the symmetry axis on a plane perpendicular to the draw direction and any direction in this plane. It is

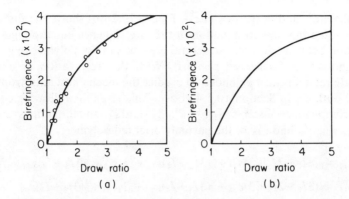

Figure 10.28. (a) Experimental and (b) theoretical curves for the birefringence of low density polyethylene as a function of draw ratio.

assumed that the symmetry axes of the anisotropic units rotate in the same manner as lines joining pairs of points in the macroscopic body, which deforms uniaxially at constant volume. This assumption is similar to the 'affine' deformation scheme of Kuhn and Grün for the optical anisotropy of rubbers[91] (see Section 4.1.2 above for a definition of 'affine'), but ignores the required change in length of the units on deformation. We will therefore call it the 'pseudo-affine' deformation scheme. Kuhn and Grün did in fact consider this scheme and reject it in their discussion of rubber-like behaviour. The angle θ in Figure 10.29 thus changes to θ', $\phi = \phi'$, and it can be shown that

$$\tan \theta' = \frac{\tan \theta}{\lambda^{3/2}}$$

where λ is the draw ratio. This relationship can be used to calculate the orientation distribution function for the units in terms of the draw ratio.

On this model the birefringence Δn of a uniaxially oriented polymer is given by

$$\Delta n = \Delta n_{max}(1 - \tfrac{3}{2}\overline{\sin^2 \theta}),$$

where $\overline{\sin^2 \theta}$ is the average value of $\sin^2 \theta$ for the aggregate of units and Δn_{max} is the maximum birefringence observed for full orientation.

This pseudo-affine deformation scheme gives a reasonable first-order fit to the birefringence data for low density polyethylene[89], nylon[90], polyethylene

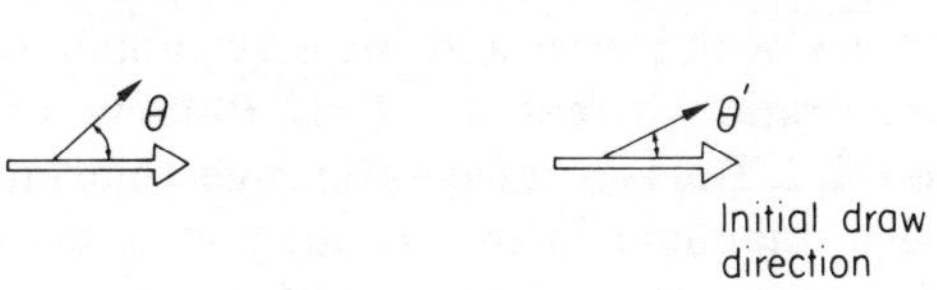

Figure 10.29. The pseudo-affine deformation.

terephthalate[62] and polypropylene[88]. Figure 10.28(b) shows the first case. It is to be noted that this formulation of the birefringence equation ignores the distinction between different structural elements in the polymer (e.g. crystalline regions and disordered regions). With this reservation in mind, the aggregate model is now extended to predict the mechanical anisotropy in the manner outlined in Section 10.5.4 above. This gives the following equations for the compliance constants s'_{11}, s'_{12}, s'_{13}, s'_{33} and s'_{44} and the stiffness constants c'_{11}, c'_{12}, c'_{13}, c'_{33} and c'_{44} of the partially oriented polymer:

$$s'_{11} = \tfrac{1}{8}(3I_2 + 2I_5 + 3)s_{11} + \tfrac{1}{4}(3I_3 + I_4)s_{13} + \tfrac{3}{8}I_1 s_{33} + \tfrac{1}{8}(3I_3 + I_4)s_{44},$$

$$c'_{11} = \tfrac{1}{8}(3I_2 + 2I_5 + 3)c_{11} + \tfrac{1}{4}(3I_3 + I_4)c_{13} + \tfrac{3}{8}I_1 c_{33} + \tfrac{1}{2}(3I_3 + I_4)c_{44},$$

$$s'_{12} = \tfrac{1}{8}(I_2 - 2I_5 + 1)s_{11} + I_5 s_{12} + \tfrac{1}{4}(I_3 + 3I_4)s_{13} + \tfrac{1}{8}I_1 s_{33} + \tfrac{1}{8}(I_3 - I_4)s_{44},$$

$$c'_{12} = \tfrac{1}{8}(I_2 - 2I_5 + 1)c_{11} + I_5 c_{12} + \tfrac{1}{4}(I_3 + 3I_4)c_{13} + \tfrac{1}{8}I_1 c_{33} + \tfrac{1}{2}(I_3 - I_4)c_{44},$$

$$s'_{13} = \tfrac{1}{2}I_3 s_{11} + \tfrac{1}{2}I_4 s_{12} + \tfrac{1}{2}(I_1 + I_2 + I_5)s_{13} + \tfrac{1}{2}I_3 s_{33} - \tfrac{1}{2}I_3 s_{44},$$

$$c'_{13} = \tfrac{1}{2}I_3 c_{11} + \tfrac{1}{2}I_4 c_{12} + \tfrac{1}{2}(I_1 + I_2 + I_5)c_{13} + \tfrac{1}{2}I_3 c_{33} - 2I_3 c_{44},$$

$$s'_{33} = I_1 s_{11} + I_2 s_{33} + I_3(2s_{13} + s_{44}),$$

$$c'_{33} = I_1 c_{11} + I_2 c_{33} + 2I_3(c_{13} + 2c_{44}),$$

$$s'_{44} = (2I_3 + I_4)s_{11} - I_4 s_{12} - 4I_3 s_{13} + 2I_3 s_{33} + \tfrac{1}{2}(I_1 + I_2 - 2I_3 + I_5)s_{44},$$

$$c'_{44} = \tfrac{1}{4}(2I_3 + I_4)c_{11} - \tfrac{1}{4}I_4 c_{12} - I_3 c_{13} + \tfrac{1}{2}I_3 c_{33} + \tfrac{1}{2}(I_1 + I_2 - 2I_3 + I_5)c_{44}.$$

$$(10.13)$$

In these equations s_{11}, s_{12}, etc., are the compliance constants and c_{11}, c_{12}, etc., are the stiffness constants for the anisotropic elastic unit, which in practice means those of the most highly oriented specimen obtained. The terms I_1, I_2, I_3, I_4, I_5 are the orientation functions, defining the average values of $\sin^4 \theta (I_1)$, $\cos^4 \theta (I_2)$, $\cos^2 \theta \sin^2 \theta (I_3)$, $\sin^2 \theta (I_4)$ and $\cos^2 \theta (I_5)$ for the aggregate. Note that only two of these orientation functions are independent parameters (e.g. $I_4 = I_1 + I_3$, $I_5 = I_2 + I_3$, $I_4 + I_5 = 1$).

The orientation functions can be calculated on the pseudo-affine deformation scheme and Figures 10.22–10.26 show that the aggregate model then predicts the general form of the mechanical anisotropy. It is particularly interesting that the predicted Reuss average curves for low density polyethylene show the correct overall pattern, including the minimum in the extensional modulus. This arises as follows. On the pseudo-affine deformation scheme $\sin^4 \theta$ and $\cos^4 \theta$ decrease and increase monotonically respectively with increasing draw ratio whereas $\sin^2 \theta \cos^2 \theta$ shows a maximum value at a draw ratio of about 1.2. Thus s'_{33} can pass through a maximum with increasing draw ratio (giving a minimum in the Young's modulus E_0) provided that $(2s_{13} + s_{44})$ is sufficiently large compared with s_{11} and s_{33} which should be approximately equal. The theory assumes elastic constants for the units which are identical with those measured for the highly oriented polymer. In low density polyethylene s_{44} is much larger than s_{11} and s_{33}, which are fairly close in value; hence these conditions are fulfilled and the anomalous mechanical anisotropy is predicted.

At low temperatures, as discussed above (see Figure 10.18), a more conventional pattern of mechanical anisotropy is observed for low density polyethylene. At the same time the polar diagram of the modulus changes (Figure 10.19) and s_{44} is no longer very much greater than the other elastic constants. These results are thus consistent with the aggregate model.

The theoretical curves of Figures 10.22–10.26 differ from those obtained experimentally in two ways. First there are features of detail (a small miminum in the transverse modulus of low density polyethylene; a small minimum in the extensional modulus of high density polyethylene) which are not predicted at all. It has been shown elsewhere[92] that such effects may be associated with mechanical twinning. Secondly, the predicted development of mechanical anisotropy with increasing draw ratio is much less rapid than is observed in practice. Deficiences in the pseudo-affine deformation scheme are not unexpected due to the simplifying nature of the assumptions made. The quantities $\overline{\sin^4 \theta}$, $\overline{\cos^4 \theta}$, and $\overline{\sin^2 \theta \cos^2 \theta}$ can also be determined experimentally by wide angle X-ray diffraction and nuclear magnetic resonance[93,94,95]. In low density polyethylene a considerably improved fit was obtained in this manner (Figure 10.30). The conclusion from these results is that the mechanical

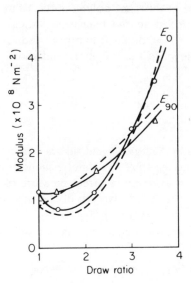

Fʼgure 10.30. Comparison of experimental (——) variation of E_0 and E_{90} for cold-drawn low density polyethylene with those predicted by the aggregate model using orientation functions from nuclear magnetic resonance (– – –).

anisotropy of low density polyethylene relates to the orientation of the crystalline regions and that it is predicted to a very good degree of approximation by the Reuss averaging scheme.

The aggregate model predicts only that the elastic constants should lie between the Reuss and Voigt average values. In polyethylene terephthalate it is clear that the experimental compliances lie approximately midway between the two bounds. For cold-drawn fibres it has been shown that this median condition applies almost exactly[64].

For low density polyethylene, the Voigt averaging scheme does not predict the anomalous behaviour. However, the Reuss average does, and therefore appears to describe the physical situation more closely. A similar conclusion was reached by Odajima and Maeda[48] who compared theoretical estimates of the Reuss and Voigt averages of single crystals of polyethylene with experimental values.

In nylon the Voigt average is closest to the experimentally observed data. It is interesting to note that both averaging schemes predict a maximum in the torsional modulus as a function of draw ratio.

The aggregate model would not appear to be generally applicable to high density polyethylene and polypropylene. It appears that for polypropylene the aggregate model is applicable only at low draw ratios[88]. As discussed above, there are simultaneous changes in morphology and molecular mobility at higher draw ratios.

It is interesting that in different polymers the Reuss or Voigt averages or a mean of these is closest to the measured values. It is likely that these conclusions will relate to the detailed nature of the stress and strain distributions at a molecular level in the polymers and should in turn be related to the structure.

10.6.6 The Sonic Velocity

It has been suggested by Morgan[30] and others[28] that the sonic modulus (i.e. the extensional modulus measured at high frequencies by a wave-propagation technique) can be used to obtain a direct measure of molecular orientation in a manner analogous to the derivation of the so-called optical orientation function $f_0 = (1 - \frac{3}{2} \sin^2 \theta)$ from the birefringence.

Consider the equations for the extensional modulus of the aggregate

$$s'_{33} = \overline{\sin^4 \theta} s_{11} + \overline{\cos^4 \theta} s_{33} + \overline{\sin^2 \theta \cos^2 \theta} (2s_{13} + s_{44}).$$

Table 10.3 summarizes the measured values of s_{11}, s_{33} and s_{44} for a number of polymers, as obtained from the monofilament data of Hadley, Pinnock and Ward[79]. It can be seen that in all cases except that of low density polyethylene s_{11} and s_{44} are of approximately the same value, and that s_{33} is comparatively small. Remembering that Poisson's ratio is usually close to 0.5, this implies that s_{13} will also be comparatively small.

This suggests that, except for high degrees of orientation, both the terms $\overline{\cos^4 \theta s_{33}}$ and $\overline{\sin^2 \theta \cos^2 \theta s_{13}}$ will be small and we can approximate to

$$s'_{33} = \overline{\sin^4 \theta s_{11}} + \overline{\sin^2 \theta \cos^2 \theta s_{44}} \tag{10.14}$$

$$= \overline{(\sin^4 \theta + \sin^2 \theta \cos^2 \theta)}s_{11}$$

$$= \overline{\sin^2 \theta s_{11}}. \tag{10.15}$$

Remembering that the birefringence is given by $\Delta n = \Delta n_{max}(1 - \frac{3}{2}\overline{\sin^2 \theta})$ it can be seen that the extensional compliance, the reciprocal of the extensional modulus, should be directly related to the birefringence through $\overline{\sin^2 \theta}$ independent of the mechanism of molecular orientation[96]. To this degree of approximation it then follows that

$$s'_{33} = \tfrac{2}{3}s_{11}(\Delta n_{max} - \Delta n). \tag{10.16}$$

We would therefore predict a linear relationship between the extensional compliance s'_{33} and the birefringence Δn, which extrapolates to zero extensional compliance at the maximum birefringence value.

Figure 10.31 shows results for polyethylene terephthalate and polypropylene which suggest that this is a reasonable approximation. But the values of s_{11} obtained from these plots do not agree with that measured experimentally for the most highly oriented fibre monofilament, suggesting that this approximate treatment is not very soundly based.

Samuels[68] has carried the sonic velocity analysis one stage further by recognizing the two-phase nature of a crystalline polymer. The natural extension of equation (10.15) would then be

$$\frac{1}{E} = s'_{33} = \frac{\beta}{E^0_{t,c}} \overline{\sin^2 \theta_c} + \frac{(1-\beta)}{E^0_{t,am}} \overline{\sin^2 \theta_{am}}, \tag{10.17}$$

where E is the sonic modulus of the sample: $E^0_{t,c}$, $E^0_{t,am}$ are the lateral moduli of the crystalline and amorphous regions, respectively; $\overline{\sin^2 \theta_c}$, $\overline{\sin^2 \theta_{am}}$ are orientation functions for the crystalline and amorphous regions respectively and β is the fraction of crystalline material.

For an isotropic sample $\overline{\sin^2 \theta_c} = \overline{\sin^2 \theta_{am}} = \frac{2}{3}$ and the isotropic sonic modulus E_u is given by

$$\frac{3}{2E_u} = \frac{\beta}{E^0_{t,c}} + \frac{1-\beta}{E^0_{t,am}}. \tag{10.18}$$

If we define orientation averages

$$f_c = \tfrac{1}{2}(3\overline{\cos^2 \theta_c} - 1), \qquad f_{am} = \tfrac{1}{2}(3\overline{\cos^2 \theta_{am}} - 1)$$

292

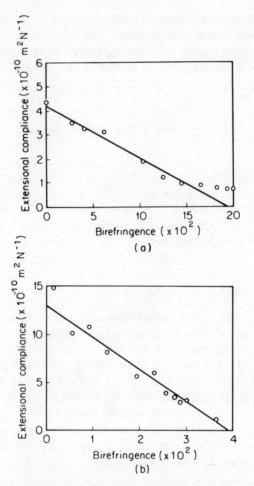

Figure 10.31. Experimental curves showing the relationship between the extensional compliance and birefringence for fibres of (a) polyethylene terephthalate; (b) polypropylene.

for the crystalline and amorphous regions respectively we can combine equations (10.17) and (10.18) to give

$$\frac{3}{2}\left\{\frac{1}{E_u} - \frac{1}{E}\right\} = \frac{\beta f_c}{E_{t,c}^0} + \frac{(1-\beta)f_{am}}{E_{t,am}^0}. \qquad (10.19)$$

Measurement of the dependence of the sonic modulus on crystallinity in isotropic samples, gives through equation (10.18) a method of determining $E_{t,c}^0$ and $E_{t,am}^0$. Measurements of the sonic modulus on oriented samples then gives, through equation (10.19), a method of determining the orientation function of the amorphous regions f_{am}, providing that f_c can be obtained from another technique, i.e. wide angle X-ray diffraction.

Samuels[68] obtained justification for this argument by combining such sonic modulus and X-ray diffraction measurements with the measurement of birefringence. Now the birefringence of a polymer on the two-phase model (ignoring form birefringence) is given by

$$\Delta n = \Delta n_c^0 f_c + (1 - \beta) \Delta n_{am}^0 f_{am} \qquad (10.20)$$

or equivalently

$$\frac{\Delta n}{\beta f_c} = \Delta n_c^0 + \Delta n_{am}^0 \left(\frac{1 - \beta}{\beta}\right) \frac{f_{am}}{f_c}. \qquad (10.21)$$

Samuels[68] showed that plots of $\Delta n / \beta f_c$ as a function of

$$\left(\frac{1 - \beta}{\beta}\right) \frac{f_{am}}{f_c}$$

gave good straight line fits for a range of polypropylene samples. This provided support for his analysis and enabled values to be deduced for Δn_c^0, Δn_{am}^0, the maximum birefringence (i.e. for a completely oriented phase) of the crystalline and amorphous regions respectively. However, it should be noted that this treatment involves several approximations, as well as the basic assumption of homogeneous stress.

10.6.7 Amorphous Polymers

There are relatively few measurements on amorphous polymers, where the degree of mechanical anisotropy is much less than in crystalline polymers. Early studies include those of Hennig[97] on polyvinyl chloride, polymethyl methacrylate and polystyrene and Robertson and Buenker[98] on bisphenol A polycarbonate. The results are summarized in Table 10.6. Hennig's measurements on s_{33} and s_{11} were obtained from dynamic testing at 320 Hz, and the s_{44} measurements at 1 Hz. Robertson and Buenker used the vibrating reed technique to obtain values in the range 100–400 Hz.

A more comprehensive investigation on uniaxially oriented sheets of polymethyl methacrylate and polystyrene was undertaken by White, Treloar and colleagues[47] using ultrasonic measurements. The results are summarized in Figure 10.32(a) and (b) where the stiffness constants are shown as a function of the birefringence. Rawson and Rider[46] have also reported ultrasonic data for oriented polyvinyl chloride and observed a similar degree of anisotropy to that seen in Table 10.6 from Hennig's work.

For amorphous polymers Ward[61] and Kausch[85,99,100] and later Rawson and Rider[46] are in agreement that the mechanical anisotropy can be discussed very satisfactorily by the aggregate model. Moreover the development of anisotropy with draw ratio can often be described by the pseudo-affine deformation scheme[46].

Table 10.6. Elastic compliances of oriented amorphous polymers (units of compliance are 10^{-9} m^2 N^{-1}).

Material	Draw ratio	s_{33}	s_{11}	s_{44}
Polyvinyl chloride	1	0.313	0.313	0.820
	1.5	0.276	0.319	0.794
	2.0	0.255	0.328	0.781
	2.5	0.243	0.337	0.769
	2.8	0.238	0.341	0.763
	∞	0.204	0.379	0.730
Polymethyl methacrylate	1	0.214	0.214	0.532
	1.5	0.208	0.215	0.524
	2.0	0.204	0.215	0.518
	2.5	0.200	0.216	0.510
	3.0	0.196	0.217	0.505
Polystyrene	1	0.303	0.303	0.769
	2.0	0.296	0.304	0.769
	3.0	0.289	0.305	0.769
Polycarbonate	1	0.376	0.376	1.05
	1.3	0.314	0.408	0.980
	1.6	0.268	0.431	0.926

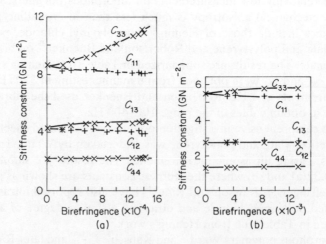

Figure 10.32. Stiffness constants of uniaxially drawn amorphous polymers, measured at room temperatures, as a function of birefringence: (a) polymethylmethacrylate, (b) polystyrene. [Redrawn with permission from Wright, Faraday, White and Treloar, *J. Phys. D*, **4**, 2002 (1971).]

10.6.8 Oriented Polyethylene Terephthalate Sheet with Orthorhombic Symmetry

All nine independent elastic constants have been determined for one-way drawn oriented polyethylene terephthalate sheet. The sheet was prepared by drawing isotropic sheet at constant width. It has been shown that there is then both a high degree of chain orientation in the draw direction and that the (100) crystal planes (which mainly reflect preferential orientation of the terephthalate residues in the chain) are preferentially oriented in the plane of the sheet. This type of orientation has been termed uniplanar axial. From the viewpoint of elastic anisotropy the sheet possesses three orthogonal planes of symmetry and can be described as possessing orthorhombic symmetry.

Collected results for the nine compliance constants are shown in Table 10.7. The z axis is the initial draw direction and the x axis lies in the plane of the sheet, following the convention indicated in Figure 10.2. s_{11} and s_{33} were obtained from measurements of extensional creep in a dead loading creep machine and refer to the 10 s response at 0.1% strain. s_{13} was obtained from the deformation of an electron microscope grid printed on the surface of the sample[101], and s_{12} and s_{23} by the Michelson interferometer method[12]. s_{22} was determined by increasing the compressional strain of strips under load in a compressional creep apparatus[7]. s_{55} was determined by the torsion of rectangular samples cut with their long axes parallel to z and x respectively[18]. s_{44} and s_{66} were also determined in this way by making measurements on samples of different aspect ratio[18]. In addition, s_{44} and s_{66} were determined by the simple shear technique[18,27] and the values quoted in Table 10.7 are weighted means for the two methods.

It can be seen from Table 10.7 that the sheet shows a very high degree of mechanical anisotropy, and it is interesting to consider how this relates to the two major structural features, the high chain axis orientation and the preferen-

Table 10.7. Full set of compliances for the oriented polyethylene terephthalate sheet with orthorhombic symmetry[18]

Compliance	Value ($\times 10^{-10}\,\mathrm{m^2\,N^{-1}}$)
s_{11}	3.61 ± 0.12
s_{22}	9.0 ± 1.6
s_{33}	0.66 ± 0.01
s_{12}	-3.8 ± 0.4
s_{13}	-0.18 ± 0.01
s_{23}	-0.37 ± 0.05
s_{44}	97 ± 3
s_{55}	5.64 ± 0.25
s_{66}	141 ± 8

tial orientation of the terephthalate residues. Infrared measurements have shown that the high degree of orientation in the crystalline regions is accompanied by a high proportion of glycol residues in the extended chain *trans* conformation and that these are also highly oriented along the draw direction[102]. The low value of the extensional compliance s_{33} can then be explained by supposing that the deformation involves bond stretching and bending in these extended chain molecules. These molecules could be the taut tie-molecules proposed by Peterlin[59]. The transverse compliances s_{11} and s_{22} are approximately an order of magnitude greater, which is consistent with these relating primarily to dispersion forces. It also follows from such considerations that if the polymer is stressed in a direction perpendicular to the draw direction the major contraction is likely to take place in the direction perpendicular to the draw direction (the 2 direction) rather than parallel to it (the 3 direction). Thus the magnitude of s_{12} would be expected to be much greater than that of s_{23}, as is observed. The value of s_{13} is similar to that of s_{23} which is consistent with this line of argument, namely the comparative difficulty of deformation in the 3 direction compared with the 1 and 2 directions. The Poisson ratios reflect the same argument and in particular ν_{31} has the very small value of 0.05.

The anisotropy of the shear compliances is also very remarkable. Both s_{44} and s_{66} are large compared with s_{55} and reflect easy shear in the 23 and 12 planes respectively, presumably where the planar terephthalate chains are sliding over each other constrained only by weak dispersion forces. The compliance s_{55} which geometrically involves distortion of the plane of the polyester molecule, is of a similar order of magnitude to s_{11} and s_{22}.

It is interesting to apply the aggregate model to these data, calculating bounds for the elastic constants of an 'equivalent fibre' by averaging the sheet constants in the plane normal to the sheet draw direction. This requires an extension of the mathematical treatment of Section 10.5.4 to deal with the case of a transversely isotropic aggregate of orthorhombic units. The basic equations have been given in detail elsewhere[103] so only the key results will be summarized here. If the orthorhombic unit constants are $s_{11}, s_{13}, \ldots, s_{66}$ (Section 10.2.2 above), the Reuss average fibre constants $s'_{33}, s'_{13}, \ldots, s'_{44}$ obtained by averaging in the 12 plane are given by

$$s'_{33} = s_{33},$$

$$s'_{11} = \tfrac{3}{8}s_{11} + \tfrac{1}{4}s_{12} + \tfrac{3}{8}s_{22} + \tfrac{1}{8}s_{66},$$

$$s'_{12} = \tfrac{1}{8}s_{11} + \tfrac{3}{4}s_{12} + \tfrac{1}{8}s_{22} - \tfrac{1}{8}s_{66},$$

$$s'_{13} = \tfrac{1}{2}(s_{13} + s_{23}),$$

$$s'_{44} = \tfrac{1}{2}(s_{44} + s_{55})$$

and the Voigt average fibre constants in similar terms are $c'_{33}, c'_{13}, \ldots, c'_{44}$ where

$$c'_{33} = c_{33},$$

$$c'_{11} = \tfrac{3}{8}c_{11} + \tfrac{1}{4}c_{12} + \tfrac{3}{8}c_{22} + \tfrac{1}{2}c_{66},$$

$$c'_{12} = \tfrac{1}{8}c_{11} + \tfrac{3}{4}c_{12} + \tfrac{1}{8}c_{22} - \tfrac{1}{2}c_{66},$$

$$c'_{13} = \tfrac{1}{2}(c_{13} + c_{23}),$$

$$c'_{44} = \tfrac{1}{2}(c_{44} + c_{55}).$$

The results of this calculation are shown in Table 10.8 together with the experimental value obtained for a highly oriented fibre monofilament. Although the experimental values do not always lie exactly within the predicted bounds, they are always in the correct range. In Table 10.8 a comparison

Table 10.8. Comparison of calculated and measured compliance constants ($\times 10^{-10}$ m^2 N^{-1}) for polyethylene terephthalate fibres based on the sheet compliances.

Compliance constant	Calculated bounds		Experimental value
	Reuss	Voigt	
Highly oriented fibres			
s_{11}	21	7.3	16.1
s_{12}	−19	−5.5	−5.8
s_{13}	−0.28	−0.25	−0.31
s_{33}	0.66	0.66	0.71
s_{44}	51	10.7	13.6
Isotropic fibres			
s_{33}	18	2.4	4.4
s_{44}	53	6.4	11

is given between the calculated and measured compliance constants for isotropic polyethylene terephthalate based on the sheet data. Again the measured values lie between the Reuss and Voigt bounds. Taking into account the very large degree of anisotropy and the very simplistic nature of these calculations, it is considered that these results afford good support for the contention that to a first approximation the mechanical anisotropy can be considered in terms of the single phase aggregate model.

10.7 EXPERIMENTAL STUDIES: ORIENTED CRYSTALLINE POLYMERS WITH LAMELLAR TEXTURES, HIGH MODULUS ORIENTED POLYMERS: THE TAKAYANAGI MODEL AND COMPOSITE MODELS

10.7.1 Oriented Polyethylene Sheets with Clear Lamellar Textures

Crystalline polymers, especially polyethylene, can be produced as oriented structures with a clear lamellar texture in a number of ways[104,105]. These include drawing and annealing; drawing, rolling and annealing, and crystallization under strain. These structures can be considered as composite solids, and it will be shown that the orientation of the lamellae, as distinct from the molecular orientation, plays a dominant role in their mechanical anisotropy.

For oriented polymers, the analogy with composite materials was first recognized by Takayanagi[69,106], who examined the dynamic mechanical behaviour of several polymers including high density polyethylene and polypropylene. Uniaxial orientation (i.e. fibre symmetry) was assumed and the dynamic extensional moduli E_1 and E_2 measured along the draw direction ($\parallel^l$ direction in Figure 10.33) and perpendicular to the draw direction ($\perp^r$ in Figure 10.33). For these drawn and annealed samples it is extremely interesting to observe that the parallel modulus (E_0 in our previous nomenclature) crosses the perpendicular modulus (E_{90} previously) at high temperatures. Thus, although $E_0 > E_{90}$ at low temperatures, $E_0 < E_{90}$ at high temperatures.

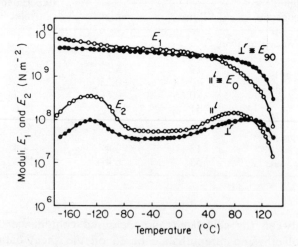

Figure 10.33. Temperature dependence of E_1 and E_2, the components of the dynamic modulus, in directions parallel ($\parallel^l$) and perpendicular ($\perp^r$) to the initial draw direction for annealed samples of high density polyethylene. [Redrawn with permission from Takayanagi, Imada and Kajiyama, *J. Polymer Sci. C*, **15**, 263 (1966).]

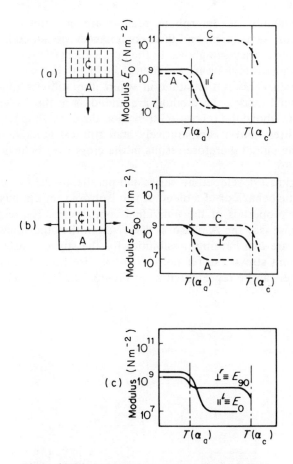

Figure 10.34. Schematic representations of change in modulus E with temperature on the Takayanagi model for (a) the $\|^l$ and (b) the $\perp^r$ situations corresponding to E_0 and E_{90} respectively. Calculations assume amorphous relaxation at temperature $T(\alpha_a)$ and crystalline relaxation at temperature $T(\alpha_c)$ and (c) shows combined results. C, crystalline phase; A, amorphous phase. [Redrawn with permission from Takayanagi, Ímada and Kajiyama, *J. Polymer Sci. C,* **15**, 263 (1966).]

Takayanagi proposed a simple model to explain this behaviour, which is shown in Figure 10.34. The basic feature of this model is illustrated by graphs of the variation of modulus with temperature (which Takayanagi terms 'dispersion curves'). Application of stress in the parallel direction (Figure 10.34(a)) gives a large fall in modulus with increasing temperature, the stiffness at high temperatures being primarily determined by the amorphous regions which are very compliant above the relaxation transition. The basic assumption is

300

that the crystalline and amorphous regions are in series, so that each is subjected to the same stress and their compliances are added like the Reuss averaging scheme discussed above. In the perpendicular direction the crystalline and amorphous regions are assumed to be in parallel, so that each suffers the same strain. This is the situation where the stiffness (or in a simple one-dimensional model, the moduli) are added as in the Voigt scheme. In physical terms, above the relaxation transition the crystalline regions support the applied stress and a comparatively high stiffness is maintained (Figure 10.34(b)). The model therefore results in the cross-over behaviour shown in Figure 10.34(c).

Takayanagi first developed his series and parallel model for understanding the viscoelastic behaviour of a blend of two isotropic amorphous polymers in terms of the properties of the individual components. He recognized that when one phase (say the A phase) is dispersed in the other phase (say the B phase) there are two limiting possibilities for the stress transfer: there may be very efficient stress transfer normal to the direction of tensile stress, so that the model can be represented as in Figure 10.35(a) which we term the

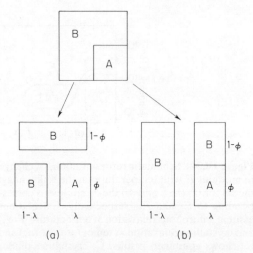

Figure 10.35. The series–parallel (a) and parallel–series (b) Takayanagi models.

series–parallel model; or else there is weak stress transfer across planes containing the tensile stress so that we have the parallel–series model of Figure 10.35(b). In later sections we will discuss the relevance of these Takayanagi models to oriented polyethylenes and nylons, where the continuous phase can be either the crystalline phase or the amorphous phase. At this point we merely wish to point out that this comparatively simple development is sufficient to give a quantitative fit to experimental data, although the main pattern of anisotropy is obtained from the simpler model of Figure 10.34.

In this section we wish to develop the theme that these annealed oriented sheets are composite systems to examine the influence of the lamellar orientation on the mechanical anisotropy. Rolling and annealing processes established by Hay and Keller[104] (see also Point[105]) enabled the production of oriented low density polyethylene sheets with very well defined crystallographic and lamellar orientations. Three types of sheet were selected by Ward and co-workers[71,80,107,108] for intensive study by mechanical and other techniques. The structures of these three sheets are shown schematically in Figure 10.36(a), (b) and (c). The 'b–c sheet' in Figure 10.36(a) shows the c axes of the polyethylene crystallites lying along the initial draw direction, the b axes in the plane of the sheet and the a axes normal to the plane of the sheet. The 'a–b' sheet in Figure 10.36(c) shows the a axes lying along the draw direction, the b axes again in the plane of the sheet and the c axes normal to the plane of the sheet. In both these samples a four-point low angle X-ray diffraction pattern is observed. This is interpreted as showing the presence of lamellae inclined at about 45° to the direction of the c axis. The morphology of the b–c and a–b sheets is represented schematically by the model shown in Figure 10.37. The solid blocks represent the crystalline lamellae and the space between them is occupied with disordered material and interlamellar tie molecules which are relaxed by the annealing treatments. The third type of sheet is the 'parallel lamellae' sheet of Figure 10.36(b). Here the lamellar planes are normal to the initial draw direction and the c axes now make an angle of about 45° with this direction. This gives a twinned structure with respect to the crystallographic orientation, but a single texture structure (a two-point low angle X-ray diffraction pattern) as far as the lamellar orientation is concerned.

The dynamic mechanical loss spectra for these three sheets are shown in Figure 10.36(d), (e) and (f) and the 10 s isochronal moduli as a function of temperature in Figure 10.36(g), (h) and (i). Consider first the loss spectra. Both the high temperature α-relaxation and the lower temperature β-relaxation process show considerable anisotropy. It has been proposed that the α-process involves shear in the c axis direction on planes containing the c axis of the crystallites (the c-shear process). The results for the three sheets are consistent with this proposal. In the b–c sheet (Figure 10.36(d)) tan δ_{45} is very much greater than tan δ_0 or tan δ_{90}. This is the situation in which there is maximum resolved shear stress parallel to the c axis directions. The parallel lamellae sheets on the other hand (Figure 10.36(e)) show greatest α-process losses when the stress direction is parallel to the initial draw directions, because this now gives maximum resolved shear stress parallel to the c axes. In the a–b sheet (Figure 10.36(f)) the α-relaxation is barely detectable because there are now no planes containing the c axis which would shear in the c direction by application of tensile stress in the plane of the sheet.

The β-relaxation has been assigned to interlamellar shear, and this is where there is the link with the Takayanagi model and the extension of his ideas to include the orientation of the lamellae as dominating the mechanical

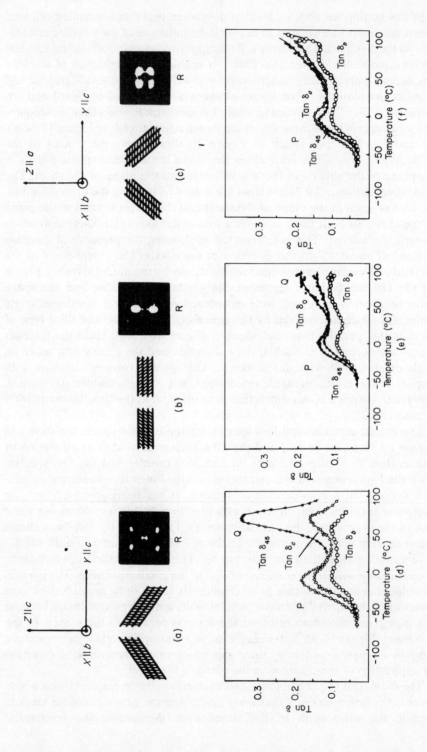

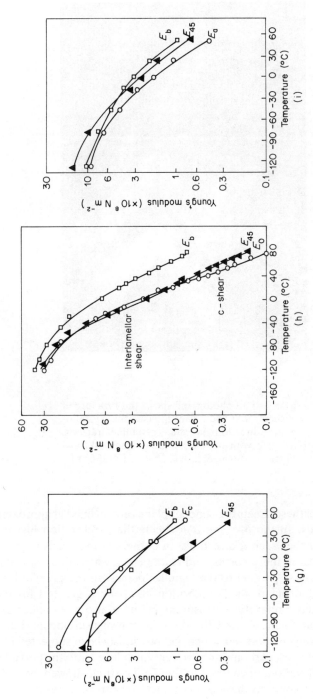

Figure 10.36. Schematic structure diagrams, mechanical loss spectra and 10 s isochronal creep moduli. (a), (d) and (g) for *b–c* sheet; (b), (e) and (h) for parallel lamellae sheet; (c), (f) and (i) for *a–b* sheet. P, interlamellar shear process; Q, c-shear process (note absence of *c*-shear process in (f)); R, small angle X-ray diagram, beam along X.

ethylene. This photograph shows the structure of
b–c sheet; a, b and c axes indicate the crystallographic direction in the crystalline regions.
[Redrawn with permission from Stachurski and
Ward, *J. Polym. Sci.*, **A2**, 161, 1817 (1968).]

anisotropy. In all these sheets the lamellar planes are inclined at approximately
45° to the c axis. In the b–c and a–b sheets the β-relaxation loss peak is
now greatest for the c and a directions (in both cases the z axis in the sheet)
respectively because these correspond to the cases where there is maximum
resolved shear stress parallel to the lamellar planes. In the parallel lamellae
sheet, on the other hand, the β-relaxation loss peak is greatest for the 45°
direction. Dielectric relaxation measurements have also been undertaken[108]
on sheets similar to those used for the dynamic measurements, but using
branched polyethylene polymer which had previously been subjected to some
oxidation so that a few carbonyl groups were introduced into the polyethylene
chain. The β-relaxation, in contrast to the α-relaxation, showed no dielectric
anisotropy, which confirms that the mechanical anisotropy arises entirely from
the geometrical arrangement of the lamellae.

Gupta and Ward[17] observed similar cross-over points in the extensional moduli to Takayanagi, in drawn and annealed sheets of low density polyethylene and in the $b-c$ and $a-b$ sheets (Figure 10.36(g) and (i) respectively). The fall in modulus in the parallel direction, i.e. in E_0 or in the c and a directions in the $b-c$ and $a-b$ sheets respectively can be attributed to the interlamellar shear process. A tensile stress in the b direction, on the other hand, will not favour interlamellar shear, as the lamellar planes are approximately parallel to the b axis. This gives $E_b > E_a \sim E_c$ above the relaxation transitions, which is the observed behaviour. Similar results for fibre symmetry sheets[107] can be explained by considering that the lamellar planes make angles of 35–40° to the initial draw direction (the fibre axis) and follow a conical distribution around this axis. Application of a tensile stress parallel to the initial draw direction results in a maximum shear stress nearly parallel to all the lamellar planes. Application of the tensile stress in the 90° direction gives maximum shear stress parallel to only some lamellar planes. This is consistent with the observation that $E_0 < E_{90}$ above the relaxation transition.

The parallel lamellae sheet shows results of exceptional interest (Figure 10.36(h)). First, there are two cross-over points in the modulus curves; one at about −40 °C corresponding to the onset of interlamellar shear, and another at about +20 °C where the c-shear relaxation is the predominant process. These results are, of course, fully consistent with the loss data of Figure 10.36(e). Secondly, the modulus in the direction normal to the lamellar planes (E_0) does fall appreciably due to the interlamellar shear relaxation. This led to the proposal[73] that *pure* shear occurs as well as *simple* shear and that the increase in compliance due to both types of shear must be added to explain the anisotropy in a range where interlamellar shear is the predominant process.

In an extensive series of experiments[13,109], the extensional, lateral and shear compliances of these parallel lamellae sheets were measured. Results showing the temperature depenence of the 10 s compliance constants are summarized in Table 10.9. At −70 °C there is very little anisotropy, $s_{33} \doteq s_{11}$ and $s_{44} \doteq s_{55} \doteq$

Table 10.9. Temperature dependence of 10 s compliance constants ($\times 10^{-9} \, \mathrm{m^2 \, N^{-1}}$) for parallel lamellae sheet.

	20 °C	−30 °C	−70 °C	Pure shear
s_{33}	9.76	1.09	0.46	
s_{23}	−7.8	−0.60		
ν_{23}	0.8	0.55		1
s_{13}	−0.82	−0.32		
ν_{13}	0.08	0.29		0
s_{11}	1.77	0.63	0.38	
s_{44}	26.7	3.25	1.12	
s_{55}	21.4	3.40	1.17	
s_{66}	10.5	1.95	0.96	

s_{66}. Although the Poisson's ratios were not determined at this low temperature these results are consistent with an isotropic material with an extensional compliance of $0.42 \times 10^{-9}\,\mathrm{m^2\,N^{-1}}$, a shear compliance of $1.08 \times 10^{-9}\,\mathrm{m^2\,N^{-1}}$ and hence a Poisson's ratio of 0.3. With increasing temperature marked anisotropy develops, and there is a quite remarkable anisotropy in the Poisson's ratios. On the simple model of plank-like parallel lamellae the compliance s_{33} relates to pure shear of the interlamellar amorphous material in the YZ (23) plane, because there is full constraint in the Z (1) direction which is the b axis of the lamellae and assumed to be of infinite length. In this case $s_{23} = -s_{33}$, $\nu_{23} = 1$ and $\nu_{13} = 0$. The 10 s compliances shown in Table 10.9 confirm that this situation is clearly approached at 20 °C. It can also be seen that s_{11} is comparatively small, and that $s_{44} \sim s_{55} > s_{66}$ again consistent with interlamellar shear being an important mechanism of deformation.

Study of the time dependence of the extensional and lateral compliances does however suggest that the c-shear relaxation is also playing an important role in the mechanical behaviour. From Table 10.10 it can be seen that the

Table 10.10. Time dependence of compliance constants $(\times 10^{-9}\,\mathrm{m^2\,N^{-1}})$ for parallel lamellae sheet at 20 °C.

	10 s	1000 s
s_{33}	9.76	22
s_{23}	−7.8	−19.2
ν_{23}	0.8	0.87
s_{13}	−0.82	−0.57
ν_{13}	0.08	0.03
s_{11}	1.77	2.36
s_{44}	26.7	36.3
s_{55}	21.4	29.5
s_{66}	10.5	24.0

20 °C behaviour approximates even more closely to pure shear at 1000 s, i.e. as the c-shear relaxation contributes more to the deformation. It was therefore concluded[13] that pure shear in the YZ plane does not arise solely because of the plank-like nature of the structure, but because intralamellar c-shear allows pure shear of the composite system in the YZ plane. It is possible that at lower temperatures (-30 °C in Table 10.9) the situation is weighted towards pure shear of the amorphous material, because the ν_{13} values are still comparatively small compared with the values of ν_{23}. Table 10.10 shows that s_{44} and s_{55} which primarily involve interlamellar shear, are much less time dependent than s_{33}, which has an appreciable component of intralamellar c-shear. s_{66}, which involves shear of the lamellar planes shows a similar time dependence to s_{33}, which is consistent with c-shear being activated.

The mechanical anisotropy has also been studied in oriented linear polyethylene structures[110], including these produced by crystallization under strain[111]. In Figure 10.38 results are shown for a cold-drawn linear polyethylene, cross-linked by radiation and crystallized by slow cooling from the melt at permanent strains of ~1000%. Following the work of Stein and Judge[112] and Keller and Machin[113], it was anticipated that a shish-kebab structure would be produced, with a core of extended chain crystallites

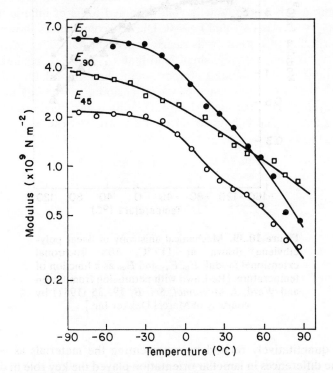

Figure 10.38. Mechanical anisotropy of strain-crystallized radiation cross-linked linear polyethylene: 10 s isochronal extensional moduli E_0, E_{45} and E_{90} as a function of temperature. [Redrawn with permission from Kapuscinski, Ward and Scanlan, *J. Macromol. Sci. B*, **11**, 475 (1975) by courtesy of Marcel Dekker Inc.]

surrounded by a parallel lamellae texture. Such materials would be expected to be appreciably stiffer than the parallel lamellae textures in the same polymer produced by hot drawing. The comparison with hot-drawn linear polyethylene is shown in Figure 10.39. The two materials do not differ markedly, suggesting that there is no significant proportion of extended chain crystallites in the strain-crystallized polymer. In fact, detailed modelling of the mechanical results for both parallel lamellae and roof-top structures[111] showed that these

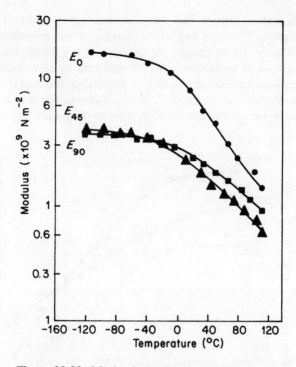

Figure 10.39. Mechanical anistropy of linear polyethylene drawn at $117\,°C$: $10\,s$ isochronal
extensional moduli E_0, E_{45} and E_{90} as a function of
temperature. [Redrawn with permission from Owen
and Ward, *J. Macromol. Sci. B*, **19**, 35 (1981) by
courtesy of Marcel Dekker Inc.]

could be quantitatively related by representing the materials as composite
solids. The differences in lamellar orientation played the key role in determining the mechanical anisotropy and a series–parallel Takayanagi model with
a continuous fraction of amorphous material provided a satisfactory basis for
a consistent interpretation of the data.

10.7.2 Specially Oriented Nylon Sheets

In nylon 6 it is possible to prepare two types of specially oriented sheet, both
with orthorhombic elastic symmetry, but with significantly different structures.
These are shown schematically in Figure 10.40. In the α-form sheet there is
a parallel lamellae type morphology with the crystalline regions in the α-form.
The molecular chain axes (the basis of the monoclinic structure) are aligned
parallel to the 3 direction, and the hydrogen bonds form layers in the (001)
planes, i.e. they are in the plane of the sheet parallel to the 1 direction. The
γ-form sheet also possesses a parallel lamellae texture but with the crystalline

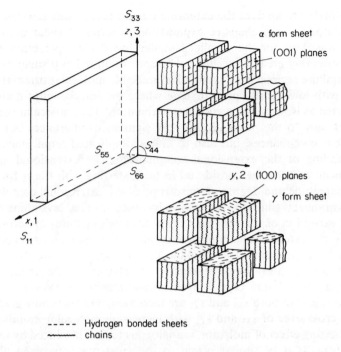

----- Hydrogen bonded sheets
〜〜 chains

Figure 10.40. Schematic diagram showing the morphologies of the α- and γ-form nylon 6 sheets in relation to the principal axes of the sheets.

regions in the γ-form. Again the molecular chain axes are aligned parallel to the 3 direction but the hydrogen bonds now form layers in the (100) planes, so that the hydrogen-bonded sheets make an angle of about 60° with the plane of the sheet.

The extensional and shear compliances were determined for both forms of sheet in the dry and wet states[82,83]. The collected results are shown in Table 10.11. It may be noted that the extensional compliances are in general of smaller magnitude than the shear compliances. This suggests that it will be

Table 10.11. Ten second compliances for wet and dry nylon 6. ($\times 10^{-10}$ m^2 N^{-1})

	α-form		γ-form	
Compliance	0% RH	65% RH	0% RH	65% RH
S_{33}	1.70 ± 0.04	3.65 ± 0.05	1.80 ± 0.02	3.10 ± 0.04
S_{11}	1.79 ± 0.01	2.60 ± 0.03	3.39 ± 0.03	4.83 ± 0.06
S_{44}	26.0 ± 2.9	30.9 ± 3.5	12.1 ± 2.3	20.1 ± 2.3
S_{55}	7.1 ± 0.3	19.0 ± 1.6	7.7 ± 0.4	15.6 ± 0.4
S_{66}	19.1 ± 1.6	6.0 ± 3.0	6.2 ± 1.4	7.2 ± 1.3

most profitable to analyze the extensional and shear results separately rather than attempt a comprehensive explanation in terms of shear deformations only, as was done for the parallel lamellae low density polyethylene sheet. The considerably greater mechanical anisotropy in nylon 6 suggests that the non-crystalline regions have retained appreciable molecular orientation, contrasting with low density polyethylene where the interlamellar material can be regarded as isotropic to a good approximation. This molecular orientation leads not only to the possibility of very significant differences between the three shear compliances, but also to more complicated requirements for an understanding of the extensional compliances. The extensional and shear compliances will now be considered in turn, starting with the α-form sheet.

The magnitude and degree of anisotropy of the extensional compliances in the α-form sheet is similar to that shown by other oriented crystalline polymers with the exception of the very high draw ultrahigh modulus materials. In the dry sheet, the extensional compliance s_{33} along the chain direction is appreciably less than the transverse compliance, although it is more than an order of magnitude greater than the crystal compliance in the chain direction determined from X-ray crystal strain measurements. However, in the wet sheet the values of both s_{33} and s_{11} are increased, and s_{33} is now greater than s_{11}. The cross-over of s_{33} and s_{11} with increasing molecular mobility due to the plasticizing effect of moisture, is analogous to that produced by increasing temperature. It is of similar origin to the cross-over discussed above for polyethylene, and can be most simply explained in terms of the Takayanagi model. There are, however, some interesting structural implications which follow from quantifying the Takayanagi model further, which will now be discussed.

The simple Takayanagi model (Figure 10.34) has the crystalline and amorphous regions in series when the stress is parallel to the draw direction. The extensional compliance s_{33} is therefore given by the Reuss average

$$s_{33} = \chi s_{33}^{c} + (1-\chi)s_{33}^{a}, \tag{10.22}$$

where s_{33}^{c}, s_{33}^{a} are the crystalline and amorphous compliances respectively in the draw direction, and χ is the volume fraction of crystalline material. Substituting the results for dry nylon into equation (10.22) by combining the crystal strain data, with experimental values of the amorphous modulus and the observed s_{33} for these sheets (see Tables 10.11 and 10.12) gives a value for χ of 0.61 (Table 10.13). This is very close to that estimated by other techniques. With this value of χ the measured wet value for s_{33} gives a value for s_{33}^{a} (wet) of $9.3 \times 10^{-10}\, \text{m}^2\, \text{N}^{-1}$. For stresses perpendicular to the draw direction the crystalline and amorphous regions are in parallel and the extensional compliance s_{11} is given by the Voigt average:

$$\frac{1}{s_{11}} = \frac{\chi}{s_{11}^{c}} + \frac{1-\chi}{s_{11}^{a}}.$$

Table 10.12. Modulus and compliance values for the crystalline and amorphous regions used in Takayanagi model calculations for the α- and γ-form structures.

Moduli ($\times 10^9$ N m^{-2})		Compliances ($\times 10^{-10}$ m^2 N^{-1})		
α	γ		α	γ	Reference

	α	γ		α	γ	Reference
Crystalline (dry and wet)						
E_{33}^c	165	21	s_{33}^c	0.061	0.49⎱	Sakurada and
E_{11}^c	11	4.2	s_{11}^c	0.88	2.4 ⎰	Kaji[114]
Amorphous (dry)						
E_{33}^a	2.3	2.3	s_{33}^a	4.3	4.3⎱	Prevorsek
E_{11}^a	0.70	0.70	s_{11}^a	14	14 ⎰	et al.[115]

Table 10.13. Results from the Takayanagi model calculations for the extensional compliances of oriented α nylon.

Model	Compliance	χ	ϕ	λ	Amorphous phase wet compliance ($\times 10^{-10}$ m^2 N^{-1})
Series	s_{33}	0.61			9.3
Parallel	s_{11}	0.46			<0
PS$_3$/SP$_1$	s_{33}	0.60	0.62	0.97	9.3
	s_{11}	0.60	0.97	0.62	37
SP$_3$/PS$_1$	s_{33}	0.60	0.61	0.98	9.3
	s_{11}	0.60	0.98	0.61	38

Method: Use dry data to calculate χ, ϕ, λ; wet data to calculate amorphous phase wet compliance.

Substitution of available data (Tables 10.11 and 10.12 again) gives a much lower value for χ of 0.46. Moreover, using this value for χ, the wet value of s_{11} implies that s_{11}^a (wet) is negative. These results, taken together, suggest that the oriented nylon cannot be represented by the simplest Takayanagi model. Instead we must consider the more general series–parallel and parallel–series forms shown in Figure 10.41. In principle either the crystalline or the amorphous phase could be the continuous phase. For these nylons continuous crystalline phase models gave physically unreasonable answers, as noted for nylon 6 by Prevorsek[116]. We therefore consider the models of Figure 10.41, where the amorphous phase is continuous. The relevant equations for the overall compliance s are

$$s = (1-\phi)s^a + \frac{\phi s^a s^c}{\lambda s^a + (1-\lambda)s^c} \tag{10.23}$$

312

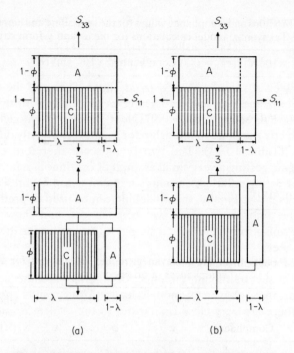

Figure 10.41. Generalized Takayanagi models with continuous amorphous phase A: (a) the series–parallel model; (b) the parallel–series model. The lower half of the diagram shows how the models break down into their components. [Redrawn with permission from Lewis and Ward, *J. Macromol. Sci. B,* **18**, 1 (1980) by courtesy of Marcel Dekker Inc.]

for the series–parallel model and

$$s = \frac{[(1-\phi)s^a + \phi s^c]s^a}{(1-\phi+\phi\lambda)s^a + \phi(1-\lambda)s^c} \tag{10.24}$$

for the parallel–series model, where ϕ and λ are the volume fractions of crystalline phase parallel and perpendicular respectively to the applied tensile stress.

It is physically reasonable to assume that if the parallel–series model represents the behaviour of tensile stress applied in the 3 direction, then the appropriate model for the 1 direction should be the series–parallel model, and vice versa. There are therefore two bounds for this Takayanagi model, the PS_3/SP_1 model and the SP_3/PS_1 model (the subscripts indicating the relevant directions of applied stress). The procedure adopted was to take first the PS_3/SP_1 model and calculate internally consistent values for χ, ϕ and λ.

It can be shown that

$$\chi = \frac{(s_{33}^a - s_{33})(s_{11}^a - s_{11})(s_{33}^a s_{11}^a - s_{33}^c s_{11}^c)}{(s_{33}^a - s_{33}^c)(s_{11}^a - s_{11}^c)(s_{33}^a s_{11}^a - s_{33} s_{11})}.$$

Using the values of s_{11}^a, s_{33}^a, s_{11}^c, s_{33}^c from Table 10.11 and the experimental dry values of s_{11} and s_{33} (Table 10.11) gives $\chi = 0.60$. With this value of χ, $\phi = 0.62$ in the 3 direction and $\lambda = 0.97$. Next, the SP_3/PS_1 model is evaluated. This gives exactly the same relationship for χ, but very slightly different values of ϕ and λ (Table 10.13). The two bounds are therefore close in their prediction of suggesting a very small amount of continuous amorphous phase, in agreement with similar conclusions on drawn nylon 6 fibres by Prevorsek and coworkers[115,117]. Finally, the modelling can be used to predict consistent values for the amorphous phase wet compliances, and show that there is considerable anisotropy here with a similar s_{11}^a/s_{33}^a ratio for the wet sheet as for the dry sheet.

The extensional compliances for the γ-form sheet are rather similar to those for the α-form sheet although the effect of humidity is rather less. A similar analysis using the Takayanagi models leads to the results shown in Table 10.14, and there are very close similarities to the α-form results. Again the

Table 10.14. Results from the Takayanagi model calculations for the extensional compliances of oriented γ-form.

Model	Compliance	χ	ϕ	λ	Amorphous phase wet compliance ($\times 10^{-10}$ m^2 N^{-1})
Series	s_{33}	0.65			8.0
Parallel	s_{11}	0.64			<0
PS_3/SP_1	s_{33}	0.65	0.65	0.996	8.0
	s_{11}	0.65	0.996	0.65	295
SP_3/PS_1	s_{33}	0.65	0.65	0.997	8.0
	s_{11}	0.65	0.997	0.65	305

Method: Use dry data to calculate χ, ϕ, λ; wet data to calculate amorphous phase wet compliance.

simple Takayanagi model gives reasonable values for χ but leads to a negative value for the wet amorphous s_{11}^a. Introducing a very small degree of amorphous continuity, even smaller than for the α-form sheet, produces very close bounds for ϕ and λ. The amorphous phase wet compliances were also calculated but the values for the 1 direction must be very uncertain because λ is so close to unity.

Next the shear compliances will be considered, again starting with the α-form sheet and the dry results in the first instance. Note that s_{55} is appreciably smaller than s_{44} or s_{66}. These results reflect the orientation of the molecular chains and the arrangement of hydrogen bonds, rather than the parallel lamellae morphology. If interlamellar shear were the dominant mechanism, s_{66} would be the smallest compliance, and *both* s_{44} and s_{55} would be large, as for the parallel lamellae low density polyethylene sheet. For the α-form nylon sheet s_{44} is the largest compliance, and this corresponds to crystallite shear parallel to the chain axis. In polyethylene this is also an easy shear mechanism, termed c-shear. For the present monoclinic structure it will be termed b-shear. Such shear within the lamellae has been observed in nylon 6 and 11, and we may imagine the hydrogen-bonded (001) planes sliding over each other in the chain direction. s_{66} is also quite large for the dry sample, suggesting that a-shear also occurs, i.e. sliding of (001) planes over each other in the a direction. It is not unreasonable to expect that this process, which involves motion parallel to the hydrogen bonds, is more difficult to activate than shear parallel to the shear axis.

The values of s_{44} for the wet and dry sheet are so close as to suggest that b-shear is the dominant mechanism in wet nylon also. However, the pattern of anisotropy now shows s_{55} to have risen in value almost to s_{44} so that we have $s_{44} \sim s_{55} > s_{66}$. This is reminiscent of the parallel lamellae low density polyethylene and suggests that interlamellar shear can occur in the wet α-form sheet.

Finally, we consider the shear compliances in the γ-form sheet. Here the values of s_{44} for both the dry and wet sheet are smaller than for the α-form sheet. This can be attributed to the fact that the hydrogen bonds now form sheets at some 60° from the sample plane, making it harder for b-shear to occur. It may also be associated with the kinking of the molecular chains in the γ-form. The value of s_{66} does not change appreciably with relative humidity, but s_{55} increases, again probably due to interlamellar shear, so that we have $s_{44} \sim s_{55} > s_{66}$ in the wet sheet.

In conclusion, it can be said that the mechanical anisotropy of these nylon sheets reflects both the composite phase morphology, so that Takayanagi modelling is appropriate for the extensional compliance, but that the shear measurements reflect the importance of the molecular structure as well. The overall anisotropy is less in the γ-form than the α-form partly due to the different arrangement of the hydrogen-bonded sheets, but also due to the reduction in the intralamellar shear in the γ-form due to the kinking of the molecular chains.

10.7.3 Ultrahigh Modulus Polyethylene

In the case of polyethylene it is possible to produce an oriented polymer with a Young's modulus approaching the theoretical crystal chain modulus of about 300 GN m^{-2}. This can be done either by solution spinning techniques[118,119] or by drawing isotropic solid polymer to very high draw ratios[120]. Figure 10.42 shows the room temperature Young's modulus as a function of draw ratio

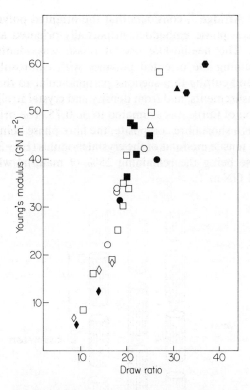

Figure 10.42. 10 s isochronal creep modulus, measured at room temperature, as a function of draw ratio for a range of quenched (open symbols) and slowly cooled (closed symbols) samples of linear polyethylene drawn at 75 °C. ●, Rigidex 140–60; △, ▲, Rigidex 25; □, ■, Rigidex 50; ○, ●, P40; ◇, ◆, H020-54P. [Redrawn with permisssion from Capaccio, Crompton and Ward, *J. Polymer Sci., Polymer Phys. Ed.*, **14**, 1641 (1976).]

for the drawing of a range of isotropic polyethylenes at 75 °C[121]. It can be seen that the modulus is dependent only on the draw ratio for starting materials of different molecular weight and initial morphology as produced by different thermal treatments. Even at room temperature the creep moduli reach 60 GN m^{-2} which is an appreciable fraction of the crystal modulus. At low temperatures dynamic moduli ~150–200 GN m^{-2} can be achieved, as will be discussed later.

There are two very different approaches to the interpretation of the mechanical behaviour of highly oriented linear polyethylene. The first approach, proposed by Arridge, Barham and Keller[122], and developed in detail by Barham and Arridge[74], considers that the oriented polymer consists of a needle-like crystal phase embedded in partially oriented amorphous phase (Figure 10.43). The needle-like crystal phase was identified with fibrils observed by staining the oriented polymer with chlorosulphonic acid and uranyl acetate and cutting thin sections perpendicular to the draw direction. From these measurements, and from density and crystal strain measurements, the concentration of fibrils was estimated to be 0.75. The oriented polymer is then modelled as a short fibre composite, the fibre phase being the needle-like crystals with the tensile modulus of the crystal modulus ($E_c \sim 300 \text{ GN m}^{-2}$) and the matrix phase being the remaining 25% of material which has a shear modulus $G_m \sim 1 \text{ GN m}^{-2}$.

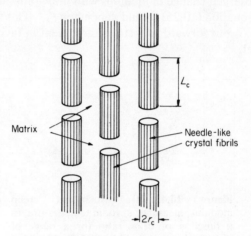

Figure 10.43. Schematic diagram of Barham and Arridge model for ultrahigh modulus polyethylene.

Following the Cox model for a fibre composite[123], the modulus E of the oriented polymer is given by

$$E = 0.75E_c\left\{1 - \frac{\tanh x}{x}\right\},\tag{10.25}$$

where

$$x = \frac{L_c}{r_c}\left(\frac{G_m}{E_c \ln 2\pi/\sqrt{2.25}}\right)^{1/2},$$

L_c and r_c being the length and radius of the fibrils respectively. The tensile modulus of the matrix is considered to be so small as to make a negligible contribution to the modulus of the oriented polymer.

The very high draw ratios required to produce ultrahigh modulus materials are obtained by first drawing the polymer through a neck to a comparatively low draw ratio (~8) and then continuing the extension so that this drawn material thins down further to achieve a final total draw ratio of 30 or more. Barham and Arridge[74] postulate that the increase in modulus on post neck drawing is due to the increase in the aspect ratio $(L_c/2r_c)$ of the fibrils, and hence their increased efficiency as reinforcing elements. Moreover, it is assumed that the drawing process is homogeneous at a structural level so that the initial aspect ratio $(L_0/2r_0)$ of the fibrils transforms affinely to $L_c/2r_c = t^{3/2}(L_0/2r_0)$, where t is the draw ratio in the post neck region. Final aspect ratios ~10 are estimated on the basis of equation (10.25). In terms of the structure this implies that 75% of the material then consists of needle-shaped crystal fibrils with lengths in the range 100–1000 nm. Barham and Arridge show that the observed change in modulus with draw ratio implies that $L/2r_c$ and hence x in equation (10.25) should depend on $t^{3/2}$. This result, illustrated in Figure 10.44, is put forward as a strong argument in favour of their fibre composite model.

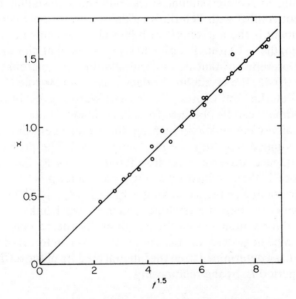

Figure 10.44. Parameter x in equation (10.25) as a function of the taper draw ratio t to the 3/2 power. [Redrawn with permission from Barham and Arridge, *J. Polymer Sci., Polymer Phys. Ed.*, **15**, 1177 (1977).]

The second approach to the interpretation of the mechanical behaviour of ultrahigh modulus polyethylene is a natural development of our previous understanding of oriented polyethylenes of lower stiffness in the traditions of Takayanagi, Peterlin and others[106,59,124]. Material of draw ratio ~10 shows a clear two-point small angle X-ray diffraction pattern and a wide angle X-ray pattern indicative of high crystallite orientation. This is consistent with a regular stacking of crystal blocks whose length, to accommodate the non-crystalline regions, is less than the long period. With increasing draw ratio the small angle pattern retains the same periodicity but diminishes in intensity[125]. At the same time there is an appreciable increase in the orientation of the non-crystalline material seen in the broad line NMR spectrum[126], and an increase in crystal length as revealed by wide angle X-ray diffraction[127] (002 reflection), dark field transmission electron microscopy[128] and nitric acid etching followed by gel permeation chromatography[129]. The average crystal length increases to about 50 nm, which compares with the constant long period of about 20 nm. Although there are some long crystals (>100 nm) the concentration of these is quite small. It was therefore proposed by Gibson, Davies and Ward[60] that the large increase in stiffness relates to the linking of adjacent crystal sequences by crystalline bridges. In the first instance the modelling invoked is just that of the Takayanagi parallel–series model (Figure 10.35(a) above). A schematic model of the structure is shown in Figure 10.45, where the lines joining adjacent crystalline stacks indicate a crystalline bridge. This model does not conflict with Peterlin's model for drawn polymers where the structural element is the microfibril with lateral dimensions between 10 and 20 nm and lengths ~10 μm. Peterlin has proposed that the axial stiffness relates to the presence of taut tie molecules within the microfibril[106]. In the model of Fig. 10.45, the crystalline bridges play a similar role to the taut tie molecules of Peterlin, and Gibson, Davies and Ward suggest how the degree of crystal continuity can be obtained from experimental data.

In absence of any information regarding the arrangement of the crystalline bridges it is assumed that they are randomly placed. The probability that a crystalline sequence traverses the disordered regions to link an adjacent crystalline block is then defined by a single parameter p. The probability of a particular crystalline sequence linking n blocks is $f_n = p^{n-1}(1-p)$, and this is also the number fraction of crystalline sequences which link n crystal blocks. If it is assumed for simplicity that the length of disordered material between the crystals can be neglected, the parameter p is given in terms of the average crystal length $\bar{L}_{002}$, determined from the integral breadth of the (002) reflection, and the long period L, by the relationship

$$p = \frac{\bar{L}_{002} - L}{\bar{L}_{002} + L}.$$

The crystal continuity is produced by all these sequences which link two or more lamellar units. On the random bridge model the weight fraction of crystalline sequences linking n crystal blocks is given by

$$F_n = np^{n-1}(1-p)^2.$$

It follows that the volume fraction of continuous phase v_f is given by

$$v_f = \chi \sum_{n=2}^{\infty} F_n = \chi p(2-p),$$

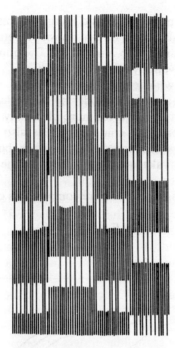

Figure 10.45. Schematic representation of the structure of the crystalline phase in ultrahigh modulus polyethylene (constructed for $p = 0.4$). [Redrawn with permission from Gibson, Davies and Ward, *Polymer*, **19**, 683 (1978). © IPC Business Press Ltd.]

where χ is the crystallinity. The Young's modulus is then

$$E = E_c \chi p(2-p) + E_a \frac{\{1-\chi+\chi(1-p)^2\}^2}{1-\chi+\chi(1-p)^2 E_a/E_c},$$ (10.26)

where the first term is the contribution of the crystalline bridge sequences and the second term is the contribution of the amorphous material and the remaining lamellar material which are considered to act in series. (Figure 10.35(a)).

Figure 10.46(a) and (b) show the dynamic mechanical tensile behaviour of a series of drawn polyethylenes. The α- and γ-relaxation regions are clearly visible, with the moduli showing a plateau region at about $-50\,°C$ between the two relaxations and approaching a second plateau at the lowest temperatures where the modulus reaches about one-half of the crystal modulus of $300\ \mathrm{GN\ m^{-2}}$. On the Takayanagi model the $-50\,°C$ plateau modulus marks

320

the region where E_a can be neglected and E is given simply by $E = \chi p (2-p)E_c$. Figure 10.47 shows results for a range of drawn polyethylenes where dynamic results have been combined with X-ray data. The good correlation provides support for the general validity of this approach.

The fall in modulus with temperature above the $-50\,°C$ plateau can be accounted for on the Takayanagi model by the reduction in E_c associated with the α-relaxation. It is, however, more physically instructive to consider the analogy with a short fibre composite. This also forms the basis for a theoretical treatment which can bring together both tensile and shear behaviour, whereas the Takayanagi modelling has the disadvantage of being a one-dimensional theory.

The modulus of an aligned short fibre composite is given by

$$E = E_f v_f \Phi + E_m V_m,$$

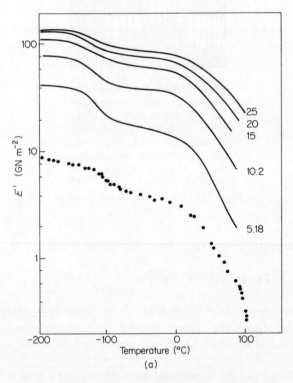

Figure 10.46. (a) The storage modulus E' and (b) the loss factor tan δ as a function of temperature for drawn linear polyethylene. Numbers on curves refer to the deformation ratio. [Redrawn with permission from Gibson, Jawad, Davies and Ward, *Polymer*, **23**, 349 (1982). © IPC Business Press Ltd.]

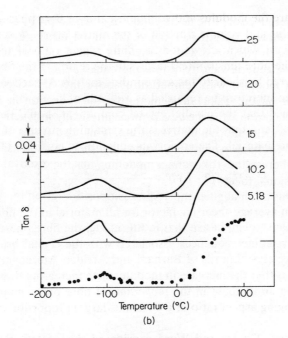

Figure 10.46. (b)

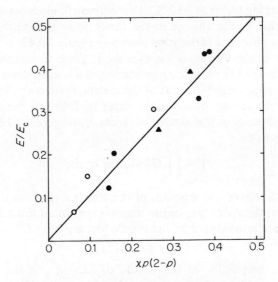

Figure 10.47. −50 °C plateau modulus E divided by the crystal modulus E_c as a function of the parameter $\chi p(2\text{-}p)$. ●, R50 extruded; ○, R50 drawn; ▲, H020 drawn.

where E_f, v_f are the modulus and volume fraction of fibre phase, and E_m, V_m are the modulus and volume fraction of the matrix phase. Φ is the so-called 'shear lag' factor which allows for the finite aspect ratio of the fibres. For short fibres, the fibre tensile stress decays in the region of the fibre ends, load being transferred by shear to the surrounding matrix. A fraction $1 - \Phi$ of the fibre phase can therefore be regarded as ineffective.

It is proposed that the parallel crystal component should be regarded as the fibre phase, and that the matrix is the remaining mixture of lamellar and non-crystalline material. The analogous equation to equation (10.26) is then

$$E = E_c \chi p (2-p) \Phi' + \frac{E_a\{1 - \chi + \chi(1-p)^2\}^2}{1 - \chi + \chi(1-p)^2 E_a / E_c}, \tag{10.27}$$

where Φ' is an average shear lag factor for all material in the fibre phase.

It is important to emphasize that in this model the fibre phase is identified with those crystalline sequences associated with the crystal bridges and not the needle-like crystal fibrils of Barham and Arridge. Moreover, the essence of this model is that the increase in modulus with increasing draw ratio arises primarily from an increase in the *proportion* of fibre phase material and not from the changing aspect ratio of a constant (large) proportion of fibre phase material.

When Gibson, Davies and Ward considered the $-50\,°C$ plateau moduli data in the light of the fibre composite model they concluded that the behaviour lay between the two extremes: (1) $\chi \sim 0.6$ and $\Phi' \sim 1$ (i.e. no shear lag effect); (2) $\chi \sim 0.8$ and $\Phi' \sim 0.8$; i.e. it is not possible to discount the presence of a significant shear lag factor at $-50\,°C$. The increase in modulus with draw ratio is essentially due to the change in the factor $p(2-p)$ and this is the basic meaning of the observed correlation shown in Figure 10.47.

The essence of the Cox model is that the stress is transmitted from fibres to matrix by shear of the matrix. Application of a tensile stress to the overall composite therefore result in shear of the matrix. Following Cox, we consider the composite to consist of a parallel array of fibres of radius r_f regularly arranged in cylinders of the matrix of radius r_m (Figure 10.48). The energy loss cycle is given by

$$W \propto \iint G_m'' \gamma_{r,x}^2 2\pi r \, dr \, dx, \tag{10.28}$$

where G_m'' is the shear loss modulus of the matrix and $\gamma_{r,x}$ is the shear strain in the matrix at the point r, x. In the tensile experiment these losses in shear of the matrix are represented by the tensile loss modulus E'' so that

$$W \propto E'' e^2 \pi r_m^2 l_f, \tag{10.29}$$

where e is the magnitude of the tensile strain and l_f is the length of the reinforcing fibres. Combining equations (10.28) and (10.29) we have

$$E'' = \frac{2G_m'}{r_m^2 l_f e^2} \iint \gamma_{r,x}^2 r \, dr \, dx. \tag{10.30}$$

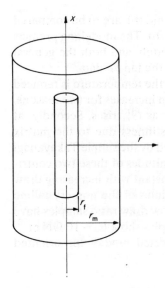

Figure 10.48. Schematic diagram of the Cox composite model. [Redrawn with permission from Gibson, Jawad, Davies and Ward, *Polymer*, **23**, 349 (1982). © IPC Business Press Ltd.]

It can be shown[127] that

$$E'' = G''_m \frac{r_f^2}{r_m^2} \frac{r_f^2}{l_f^2} \left(\frac{E_f}{G'_m}\right)^2 \frac{\ln(r_m/r_f)}{2} \frac{\beta l_f}{2} \frac{\{\sinh(\beta l_f) - \beta l_f\}}{\cosh^2(\beta l_f/2)} \quad (10.31)$$

and

$$E' = E_f v_f \left\{1 - \frac{\tanh(\beta l_f/2)}{\beta l_f/2}\right\} + E_m v_m, \quad (10.32)$$

where E_f and E_m are both considered as non-lossy.

Equation (10.31) shows that E'' depends on two geometric factors, (r_f/r_m) which relates to the volume fraction of fibre phase, and (l_f/r_f) which is the fibre aspect ratio, both of which will be constant for a given structure. It also depends on the ratio (G'_m/E_f) which is also constant for a given structure but depends on temperature through the temperature dependence of G'_m.

As discussed above $v_f = \chi p(2-p)$, and it may also be shown[130] that the average crystal length of those sequences for which $n \geqslant 2$ (the fibre phase) is

$$\bar{l}_f = \left(\frac{2-p}{1-p}\right)L.$$

Because there is no direct information on r_f, the radius of the crystalline bridge sequences, calculations were carried out for $r_f = 1$, 1.5 and 2 nm. The long period was taken as 20 nm, $\chi = 0.8$ and $E_f = 315$ GN m^{-2}. Values of G'_m and G''_m were taken from experimental data for isotropic linear polyethylene. The tensile storage modulus E' and $\tan \delta_E$ were then calculated for various values of p. It was found that the best overall match between the predicted and observed patterns of mechanical behaviour was obtained for $r_f = 1.5$ nm.

These results, which are shown in Figure 10.49(a) and (b), are to be compared with the experimental data of Figure 10.46(a) and (b). The modelling predicts the correct magnitudes for the −50 °C plateau moduli, and both the general fall in E' with temperature and the magnitudes of the tan δ_E data.

The model also throws light on the behaviour as the temperature is reduced below the −50 °C plateau region. The modulus then increases for two reasons. First, there is the increase in the shear lag factor as G'_m rises. Secondly, at low temperatures there is a contribution to the stiffness due to the matrix term in equation (10.32) as, of course, also predicted by the simple Takayanagi model. Detailed consideration of the relative magnitudes of these two contributions suggest that the matrix modulus is not constant with increasing draw ratio, but increases due to an increase in the modulus of the non-crystalline material (E_a in equation (10.27)). It appears that low draw ratio samples have values for $E_a \sim 3$ GN m^{-2}, whereas high draw samples show $E_a \sim 10$ GN m^{-2}. Such values are quite realistic in terms of expected moduli for oriented amorphous polymers (Section 10.6.7).

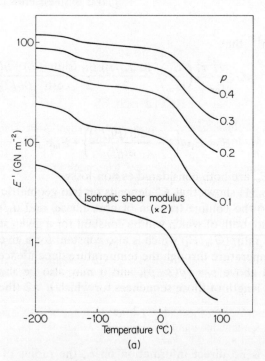

Figure 10.49. Theoretical curves for (a) the storage modulus E' and (b) the loss factor tan δ as a function of the parameter p for drawn linear polyethylene. [Redrawn with permission from Gibson, Jawad, Davies and Ward, *Polymer*, **23**, 349 (1982). © IPC Business Press Ltd.]

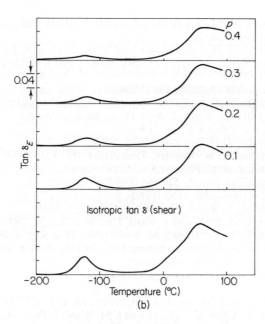

Figure 10.49. (b)

REFERENCES

1. M. A. Biot, in *International Union of Theoretical and Applied Mechanics Colloquium (Madrid)*, Springer-Verlag, Berlin, 1955, p. 251.
2. T. G. Rogers and A. C. Pipkin, *J. Appl. Maths Phys.*, **14**, 334 (1963).
3. J. F. Nye, *Physical Properties of Crystals*, Clarendon Press, Oxford, 1957.
4. C. O. Horgan, *J. Elasticity*, **2**, 169, 335 (1972).
5. C. O. Horgan, *Int. J. Solids Structure*, **10**, 837 (1974).
6. M. J. Folkes and R. G. C. Arridge, *J. Phys. D*, **8**, 1053 (1975).
7. I. Wilson, A. Cunningham, R. A. Duckett and I. M. Ward, *J. Mater. Sci.*, **11**, 2189 (1976).
8. N. H. Ladizesky and I. M. Ward, *J. Macromol. Sci. B*, **5**, 661 (1971).
9. D. Clayton, M. W. Darlington and M. M. Hall, *J. Phys. E*, **6**, 218 (1973).
10. M. W. Darlington and D. W. Saunders, in *Structure and Properties of Oriented Polymers* (ed. I. M. Ward), Applied Science Publishers, London, 1979, Chapter 10.
11. C. M. R. Dunn, W. H. Mills and S. Turner, *Brit. Plastics*, **37**, 386 (1964).
12. I. Wilson, A. Cunningham and I. M. Ward, *J. Mater. Sci.*, **11**, 2181 (1976).
13. I. D. Richardson and I. M. Ward, *J. Polymer Sci., Polymer Phys. Ed.*, **16**, 667 (1978).
14. G. Raumann, *Proc. Phys. Soc.*, **79**, 1221 (1962).
15. S. G. Lekhnitskii, *Theory of Elasticity of an Anisotropic Elastic Body*, Holden Day, San Francisco, 1964.
16. N. H. Ladizesky and I. M. Ward, *J. Macromol. Sci. B*, **5**, 759 (1971).
17. N. H. Ladizesky and I. M. Ward, *J. Macromol. Sci. B*, **9**, 565 (1974).
18. E. L. V. Lewis and I. M. Ward, *J. Mater. Sci.*, **15**, 2354 (1980).
19. M. A. Biot, *J. Appl. Phys.*, **10**, 860 (1939).

326

20. S. P. Timoshenko and J. N. Goodier, *Theory of Elasticity*, 3rd edn, McGraw-Hill, New York, 1970, Chap. 10.
21. E. L. V. Lewis, I. D. Richardson and I. M. Ward, *J. Phys. E*, **12**, 189 (1979).
22. J. H. Wakelin, E. T. L. Voong, D. J. Montgomery and J. H. Dusenbury, *J. Appl. Phys.*, **26**, 786 (1955).
23. R. Meredith, *J. Text. Inst.*, **45**, 489 (1954).
24. H. Kolsky and K. W. Hillier, *Proc. Phys. Soc. B*, **62**, 111 (1949).
25. J. W. Ballou and J. C. Smith, *J. Appl. Phys.*, **20**, 493 (1949).
26. A. W. Nolle, *J. Polymer Sci.*, **5**, 1 (1949).
27. W. J. Hamburger, *Text. Res. J.*, **18**, 705 (1948).
28. W. H. Charch and W. W. Moseley, *Text. Res. J.*, **29**, 525 (1959).
29. W. W. Moseley, *J. Appl. Polymer Sci.*, **3**, 266 (1960).
30. H. M. Morgan, *Text. Res. J.*, **32**, 866 (1962).
31. V. Davis, *J. Text. Inst.*, **50**, 1688 (1960).
32. F. I. Frank and A. L. Ruoff, *Text. Res. J.*, **28**, 213 (1958).
33. D. W. Hadley, I. M. Ward and J. Ward, *Proc. Roy. Soc. A*, **285**, 275 (1965).
34. P. R. Pinnock, I. M. Ward and J. M. Wolfe, *Proc. Roy. Soc. A*, **291**, 267 (1966).
35. H. Hertz, *Miscellaneous Papers*, Macmillan, London, 1896, p. 146.
36. E. McEwen, *Phil. Mag.*, **40**, 454 (1949).
37. S. Timoshenko and J. N. Goodier, *Theory of Elasticity*, McGraw-Hill, New York, 1951.
38. S. Abdul Jawad and I. M. Ward, *J. Mater. Sci.*, **13**, 1381 (1978).
39. O. K. Chan, F. C. Chen, C. L. Choy and I. M. Ward, *J. Phys. D*, **11**, 481 (1975).
40. M. J. P. Musgrave, *Rep. Prog. Phys.*, **22**, 77 (1959).
41. S. F. Kwan, F. C. Chen and C. L. Choy, *Polymer*, **16**, 481 (1975).
42. E. P. Papadakis, *J. Appl. Phys.*, **35**, 1474 (1964).
43. E. P. Papadakis, *J. Acoust. Soc. Amer.*, **42**, 1045 (1967).
44. R. L. Roderick and R. Truell, *J. Appl. Phys.*, **23**, 267 (1952).
54. R. F. Shaufele and T. Shimanouchi, *J. Chem. Phys.*, **47**, 3605 (1967).
46. F. F. Rawson and J. G. Rider, *J. Phys. D*, **7**, 41 (1974).
47. H. Wright, C. S. N. Faraday, E. F. T. White and L. R. G. Treloar, *J. Phys.D*, **4**, 2002 (1971).
48. A. Odajima and M. Maeda, *J. Polymer Sci. C*, **15**, 55 (1966).
49. K. Tashiro, M. Kobayashi and H. Tadokoro, *Macromolecules*, **11**, 914 (1978).
50. W. J. Lyons, *J. Appl. Phys.*, **29**, 1429 (1958).
51. L. R. G. Treloar, *Polymer*, **1**, 95, 279, 290 (1960).
52. M. A. Jaswan, P. P. Gillis and R. E. Mark, *Proc. Roy. Soc. A*, **306**, 389 (1968).
53. I. Sakurada, T. Ito and K. Nakamae, *J. Polymer Sci. C*, **15**, 75 (1966).
54. R. F. Shaufele and T. Shimanouchi, *J. Chem. Phys.*, **47**, 3605 (1967).
55. L. A. Feldkamp, G. Venkateraman and J. S. King, in *Neutron Inelastic Scattering*, Vol. II, IAEA, Vienna, 1968, p. 159.
56. L. Holliday and J. W. White, *Pure Appl. Chem.*, **26**, 545 (1971).
57. L. Holliday, in *Structure and Properties of Oriented Polymers* (I. M. Ward, ed.), Applied Science Publishers, London, 1979, Chapter 7.
58. I. M. Ward, in *Advances in Oriented Polymers* 1 (I. M. Ward, ed.), Applied Science Publishers, London, 1982, Chapter 5.
59. A. Peterlin, *Ultra-High Modulus Polymers* (A. Ciferri and I. M. Ward, eds), Applied Science Publishers, London, 1979, Chapter 10.
60. A. G. Gibson, G. R. Davies and I. M. Ward, *Polymer*, **19**, 683 (1978).
61. M. Kashiwagi, M. J. Folkes and I. M. Ward, *Polymer*, **12**, 697 (1971).
62. P. R. Pinnock and I. M. Ward, *Brit. J. Appl. Phys.*, **15**, 1559 (1964).
63. I. M. Ward, *Proc. Phys. Soc.*, **80**, 1176 (1962).
64. S. W. Allison and I. M. Ward, *Brit. J. Appl. Phys.*, **18**, 1151 (1967).

65. V. B. Gupta and I. M. Ward, *J. Macromol. Sci. B*, **1**, 373 (1967).
66. P. R. Goggin and W. N. Reynolds, *Phil. Mag.*, **16**, 317 (1967).
67. M. G. Northolt and J. J. Van Aartsen, *J. Polymer Sci. C*, **58**, 283 (1978).
68. R. J. Samuels, *Structured Polymer Properties*, Wiley, New York (1974).
69. M. Takayanagi, K. Imada and T. Kajiyama, *J. Polymer Sci. C*, **15**, 263 (1966).
70. A. Peterlin, in *Structure and Properties of Oriented Polymers* (I. M. Ward, ed.), Applied Science Publishers, London, 1975, Chapter 2.
71. V. B. Gupta and I. M. Ward, *J. Macromol. Sci. B*, **2**, 89 (1968).
72. Z. H. Stachurski and I. M. Ward, *J. Polymer Sci. A2*, **6**, 1817 (1968).
73. A. J. Owen and I. M. Ward, *J. Mater. Sci.*, **6**, 485 (1971).
74. P. J. Barham and R. G. C. Arridge, *J. Polymer Sci., Polymer Phys. Ed.*, **15**, 1177 (1977).
75. G. Raumann and D. W. Saunders, *Proc. Phys. Soc.*, **77**, 1028 (1961).
76. I. M. Ward, *Text. Res. J.*, **31**, 650 (1961).
77. P. R. Pinnock and I. M. Ward, *Proc. Phys. Soc.*, **81**, 260 (1963).
78. G. Raumann, *Brit. J. Appl. Phys.*, **14**, 795 (1963).
79. D. W. Hadley, P. R. Pinnock and I. M. Ward, *J. Mater. Sci.*, **4**, 152 (1969).
80. Z. H. Stachurski and I. M. Ward, *J. Polymer Sci. A2*, **6**, 1083 (1969); *J. Macromol. Sci. B*, **3**, 427 (1969); *J. Macromol. Sci. B*, **3**, 445 (1969).
81. G. R. Davies and I. M. Ward, *J. Polymer Sci. A2*, **10**, 1153 (1972).
82. E. L. V. Lewis and I. M. Ward, *J. Macromol. Sci. B*, **18**, 1 (1980).
83. E. L. V. Lewis and I. M. Ward, *J. Macromol. Sci. B*, **19**, 75 (1981).
84. H. H. Kausch, *Kolloidzeitschrift*, **237**, 251 (1970).
85. J. Bishop and R. Hill, *Phil. Mag.*, **42**, 414, 1248 (1951).
86. A. Reuss, *Zeit. Angew. Math. Mech.*, **9**, 49 (1929).
87. W. Voigt, *Lehrbuch der Kristallphysik*, Teubuer, Leipzig, 1928, p. 410.
88. P. R. Pinnock and I. M. Ward, *Brit. J. Appl. Phys.*, **17**, 575 (1966).
89. S. M. Crawford and H. Kolsky, *Proc. Phys. Soc. B*, **64**, 119 (1951).
90. C. G. Cannon and F. C. Chappel, *Brit. J. Appl. Phys.*, **10**, 68 (1959).
91. W. Kuhn and F. Grün, *Kolloidzeitschrift*, **101**, 248 (1942).
92. F. C. Frank, V. B. Gupta and I. M. Ward, *Phil. Mag.*, **21**, 1127 (1970).
93. V. B. Gupta, A. Keller and I. M. Ward, *J. Macromol Sci. B*, **2**, 139 (1968).
94. V. B. Gupta and I. M. Ward, *J. Macromol. Sci. B*, **4**, 453 (1970).
95. V. J. McBrierty and I. M. Ward, *Brit. J. Appl. Phys.*, **21**, 1529 (1968).
96. I. M. Ward, *Text. Res. J.*, **34**, 806 (1964).
97. J. Hennig, *Kolloidzeitschrift*, **200**, 46 (1964).
98. R. E. Robertson and R. J. Buenker, *J. Polymer Sci. A2*, **2**, 4889 (1964).
99. H. H. Kausch, *J. Appl. Phys.*, **38**, 4213 (1967).
100. H. H. Kausch, *Polymer Fracture*, Springer-Verlag, Berlin, 1978, p. 33.
101. I. Wilson, N. H. Ladizesky and I. M. Ward, *J. Mater. Sci.*, **11**, 2177 (1976).
102. A. Cunningham, I. M. Ward, H. A. Willis and V. Zichy, *Polymer*, **15**, 749 (1974).
103. A. Cunningham, Ph.D. thesis, Leeds University, 1974.
104. I. L. Hay and A. Keller, *J. Mater. Sci.*, **1**, 41 (1966).
105. J. J. Point, *J. Chim. Phys.*, **50**, 76 (1953).
106. M. Takayanagi, *Mem. Fac. Eng. Kyushu Univ.*, **23**, 41 (1963).
107. G. R. Davies, A. J. Owen, I. M. Ward and V. B. Gupta, *J. Macromol. Sci. B*, **6**, 215 (1972).
108. G. R. Davies and I. M. Ward, *J. Polymer Sci. B*, **7**, 353 (1969).
109. N. H. Ladizesky and I. M. Ward, *J. Macromol. Sci. B*, **9**, 565 (1974).
110. A. J. Owen and I. M. Ward, *J. Macromol. Sci. B*, **19**, 35 (1981).
111. M. Kapuscinski, I. M. Ward and J. Scanlan, *J. Macromol. Sci. B*, **11**, 475 (1975).
112. J. T. Judge and R. S. Stein, *J. Appl. Phys.*, **32**, 2357 (1961).
113. A. Keller and M. J. Machin, *J. Macromol. Sci. B*, **1**, 41 (1967).

328

114. K. Kaji and I. Sakurada, *J. Polymer Sci., Polymer Phys. Ed.*, **12**, 1491 (1974).

115. D. C. Prevorsek, P. J. Harget, R. K. Sharma and A. C. Reimschuessel, *J. Macromol. Sci. B*, **8**, 127 (1973).

116. D. C. Prevorsek, *J. Polymer Sci. C*, **32**, 343 (1971).

117. D. C. Prevorsek, Y. D. Kwan and R. K. Sharma, *J. Mater. Sci.*, **12**, 2310 (1977).

118. A. Zwijnenburg and A. J. Pennings, *J. Polymer Sci., Polymer Letters Ed.*, **14**, 339 (1976).

119. P. Smith and P. J. Lemstra, *J. Mater. Sci.*, **15**, 505 (1980).

120. G. Capaccio and I. M. Ward, *Nature Phys. Sci.*, **243**, 143 (1973); *Polymer*, **15**, 223 (1974).

121. G. Capaccio, T. A. Crompton and I. M. Ward, *J. Polymer Sci., Polymer Phys. Ed.*, **14**, 1641 (1976).

122. R. G. C. Arridge, P. J. Barham and A. Keller, *J. Polymer Sci., Polymer Phys. Ed.*, **15**, 389 (1977).

123. H. L. Cox, *Brit. J. Appl. Phys.*, **3**, 72 (1952).

124. E. W. Fischer, H. Goddar and W. Peisczek, *J. Polymer Sci. C*, **32**, 149 (1971).

125. J. B. Smith, G. R. Davies, G. Capaccio and I. M. Ward, *J. Polymer Sci., Polymer Phys. Ed.*, **13**, 2331 (1975).

126. J. B. Smith, A. J. Manuel and I. M. Ward, *Polymer*, **16**, 57 (1975).

127. J. Clements, R. Jakeways and I. M. Ward, *Polymer*, **19**, 639 (1978).

128. C. J. Frye, I. M. Ward, M. G. Dobb and D. J. Johnson, *Polymer*, **20**, 1309 (1979).

129. G. Capaccio and I. M. Ward, *J. Polymer Sci., Polymer Phys. Ed.*, **19**, 667 (1981).

130. A. G. Gibson, S. A. Jawad, G. R. Davies and I. M. Ward, *Polymer*, **23**, 349 (1982).

11

The Yield Behaviour of Polymers

Until comparatively recently, the yield behaviour of polymers has not received much attention. This is mainly because it was not thought profitable to treat it as a distinct mode of mechanical behaviour, different in kind from either the viscous flow processes which occur at high temperatures or the large extensions observed in the temperature range above the glass transition. The yield process in a polymer was often considered to be a softening due to a local rise in temperature and was referred to as a localized 'melting'.

A number of different factors have contributed to the appreciably greater interest in yield behaviour of polymers since about 1960. In the first instance it has been recognized that the classical concepts of plasticity are relevant to forming, rolling and drawing processes in polymers. Secondly, there have been a number of striking experimental studies of 'slip bands' and 'kink bands' in polymers which suggest that deformation processes in polymers might be similar to those in crystalline materials such as metals and ceramics. Finally, it is now evident that distinct yield points are observed and there is much interest in understanding these in the context of other ideas in polymer science.

Our first task in this chapter is to discuss the relevance of classical ideas of plasticity to the yielding of polymers. Although the yield behaviour is temperature and strain-rate dependent it will be shown that provided that the test conditions are chosen suitably, yield stresses can be measured which satisfy conventional yield criteria.

This part of the discussion is at a purely phenomenological level. Two aspects of the yield behaviour which provide information at a molecular level will also be considered. These are the temperature and strain-rate sensitivity, and the molecular reorientation associated with plastic deformation.

The temperature and time dependence often obscure some generalities of the yield behaviour. For example, it might be concluded that some polymers show necking and cold-drawing whereas others are brittle and fail catastrophically. Yet another type of polymer (a rubber) extends homogeneously to rupture. A salient point to recognize is that polymers in general show all these types of behaviour depending on the exact conditions of test (Figure 1.1). This is quite irrespective of their chemical nature and physical structure. Thus explanations of yield behaviour which involve, for example, cleavage of crystallites or lamellar slip or amorphous mobility are only relevant to specific

330

cases. As in the case of linear viscoelastic behaviour or rubber elasticity what we must first seek is an understanding of the relevant phenomenological features, decide on suitable measurable quantities and then provide a molecular interpretation of the subsequent constitutive relations.

11.1 DISCUSSION OF LOAD–ELONGATION CURVE

The most dramatic manifestation of yield is seen in tensile tests when a neck or deformation band occurs, as in Figure 11.1 In these cases the plastic deformation is concentrated either entirely or primarily in a small region of the specimen. The precise nature of the plastic deformation depends both on the geometry of the specimen and on the nature of the applied stresses. This will be discussed more fully later.

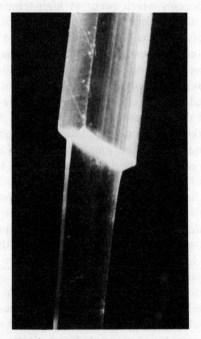

Figure 11.1. Photograph of a neck formed in the redrawing of oriented polyethylene.

The characteristic necking and cold-drawing behaviour is as follows. On the initial elongation of the specimen, homogeneous deformation occurs and the conventional load–extension curve shows a steady increase in load with increasing elongation (AB in Figure 11.2). At the point B the specimen thins to a smaller cross-section at some point, i.e. a neck is formed. Further

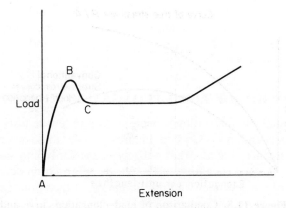

Figure 11.2. Typical load–extension curve for a
cold-drawing polymer.

elongation brings a fall in load. Continuing extension is achieved by causing
the shoulders of the neck to travel along the specimen as it thins from the
initial cross-section to the drawn cross-section. The existence of a finite or
natural draw ratio is an important aspect of polymer deformation and is
discussed in Section 11.6.3 below.

Ductile behaviour in polymers does not always give a stabilized neck. Before
analysing the requirements for necking and cold-drawing the distinction
between true and conventional stress–strain curves will be considered.

11.1.1 Necking and the Ultimate Stress

Necking and cold-drawing are accompanied by a non-uniform distribution of
stress and strain along the length of a test specimen. Let us consider these
phenomena in terms of the true stress–strain curve of a material, rather than
the conventional stress–strain curve which relates the applied tension to the
overall extension. The cross-section of the sample is decreasing with increasing
extension, so that the true stress may be increasing when the apparent or
conventional stress or load may be remaining constant or even decreasing.
This has been very well discussed by Nadai[1] and Orowan[2] and their argument
will be followed here.

Consider the conventional stress–strain curve or the load–elongation curve
for a ductile material (Figure 11.3). The ordinate is equal to the stress σ_a
obtained by dividing the load P by the original cross-sectional area A_0:

$$\sigma_a = P/A_0$$

This gives a stress–strain curve of the form shown. The load reaches its
maximum value at the instant the uniform extension of the sample stops. At
this elongation the specimen begins to neck and consequently the load falls

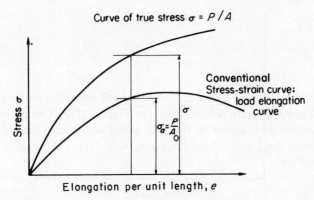

Curve of true stress $\sigma = P/A$

Stress σ

$\sigma = \dfrac{P}{A}$

$\sigma_a = \dfrac{P}{A_0}$

Conventional
Stress-strain curve:
load elongation
curve

Elongation per unit length, e

Figure 11.3. Comparison of load–elongation curve and
true stress elongation curve.

as shown by the last part of the stress–strain curve. Finally the sample fractures at the narrowest point of the neck.

It is more instructive to plot the true tensile stress at any elongation rather than the apparent stress σ_a.

The true stress $\sigma = P/A$, where A is the actual cross-section at any time.

We now assume that the deformation takes place at constant volume, this assumption being usual for plastic deformation. Then $Al = A_0 l_0$, and if we put $l/l_0 = 1 + e$ where e is the elongation per unit length,

$$A = \frac{A_0 l_0}{l} = \frac{A_0}{1+e}$$

The true stress is given by

$$\sigma = \frac{P}{A} = \frac{(1+e)P}{A_0} = (1+e)\sigma_a$$

This gives the load:

$$P = \frac{A_0 \sigma}{1+e}$$

Thus if we know σ, the true stress, as a function of e, i.e. the true stress–strain curve, P can be computed for any elongation. In particular $P_{\max}$, the maximum load, is defined by the condition $dP/de = 0$, i.e.

$$\frac{dP}{de} = \frac{A_0}{(1+e)^2}\left[(1+e)\frac{d\sigma}{de} - \sigma\right] = 0$$

or

$$\frac{d\sigma}{de} = \frac{\sigma}{1+e}$$

The measured ultimate stress can be obtained from the true stress–strain curve by the simple construction shown in Figure 11.4. The ultimate stress is obtained when the tangent to the true stress–strain curve $d\sigma/de$ is given by the line from the point -1 on the elongation axis. The angle α in Figure 11.4 is defined by

$$\tan \alpha = \frac{d\sigma}{de} = \frac{\sigma}{1+e} = \frac{\sigma}{\lambda},$$

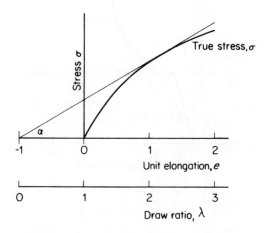

Figure 11.4. The Considère construction.

where λ is the draw ratio. This construction is called the 'Considère construction' and is useful in discussing whether a polymer will neck and cold-draw.

The significance of the argument at this stage relates to the failure of plastics in the ductile state. Orowan[2] first pointed out that for ductile materials the ultimate stress is entirely determined by the stress–strain curve, that is by the plastic behaviour of the material, without any reference to its strength properties, provided that fracture does not occur before the load maximum corresponding to $d\sigma/de = \sigma/(1+e)$ is reached. This explains why the yield stress is such an important property in many plastics. In fact it defines the practical limit of behaviour much more than the ultimate fracture, unless the plastic fails by brittle fracture.

11.1.2 Necking and Cold-drawing: A Phenomenological Discussion

Figure 11.5 shows that there are three distinct regions on the true stress–strain curve of a typical cold-drawing polymer.

(1) Initially the stress rises in an approximately linear manner as the applied strain increases.

(2) At the yield point there is a fall in true stress and a region where the stress rises less steeply with the strain. In some earlier discussions[3] it was not recognized that a fall in true stress can occur, and it was proposed only that the true stress rises less steeply with increasing strain (dotted line in Figure 11.5). This region was attributed to 'strain softening'.

(3) Finally at large extensions the slope of the true stress–strain curve increases again, i.e. a 'strain-hardening' effect occurs.

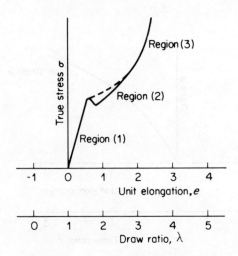

Figure 11.5. The true stress–strain curve for a cold-drawing polymer.

There are two ways in which a neck may be initiated. First, if, for a given applied load, one element is subjected to a higher true stress, because its effective cross-sectional area is smaller, that element will reach the yield point at a lower tension than any other point in the sample. Secondly, a fluctuation in material properties may cause a localized reduction of the yield stress in a given element so that this element reaches the yield point at a lower applied tension. When a particular element has reached its yield point it is easier to continue deformation entirely within this element because it has a lower effective stiffness than the surrounding material. Hence further deformation of the sample is accomplished by straining in only one region and a 'neck' is formed.

This localized deformation will continue until strain-hardening increases the effective stiffness of the element, i.e. it reaches the third part (3) of the true stress–strain curve in Figure 11.5. At this point the deformation will stabilize in the highly strained element and in order to accommodate further extension of the sample as a whole, new elements will be brought to their yield point. In this way a neck propagates along the length of a sample as successive elements are brought to a degree of strain hardening which is greater than the yield stress of the undeformed material.

11.1.3 Use of the Considère Construction

In Figure 11.6 two tangent lines have been drawn to the true stress–strain curve from the point $e = -1$ or $\lambda = 0$. In terms of the extension ratio or draw ratio λ, Considère's construction gives the conventional stress

$$\frac{P}{A_0} = \frac{\sigma}{1+e} = \frac{\sigma}{\lambda}$$

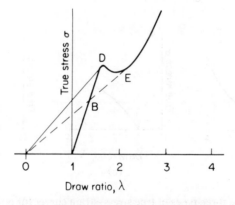

Figure 11.6. A true stress–strain curve for a cold-drawing polymer and the Considère construction.

Thus a line from the point $\lambda = 0$ to a point on the true stress–strain curve has slope σ/λ and gives us the conventional stress, i.e. the applied tension at that point.

The first tangent line has been drawn from O to D, i.e. to the yield point. At this point σ/λ and hence the conventional stress is a maximum. As further deformation takes place the slope of σ/λ decreases continuously as the true stress–strain curve is traced, until the point E is reached. This is where strain hardening occurs. The final tension settles down to a value represented by the slope of the line OE, the second tangent line, and drawing takes place

336

by deforming successive elements by an amount corresponding to this extension ratio. After the entire sample has been drawn, further deformation may take place along the steeper part of the stress–strain curve until fracture occurs.

It should be noted that these arguments do not satisfactorily explain how the sample draws at constant tension throughout its length (along the line BE in the diagram) and at the same time apparently follows the true stress–strain curve along BDE. This anomaly has been attributed by Vincent[3] to our ignorance regarding the actual stress conditions in the neck, where from the corresponding work in metals[4] it can be implied that there is a complex combination of tensile and hydrostatic stresses.

The Considère construction can be used as a criterion to decide whether a polymer will neck, or will neck and cold-draw. There are three possible situations:

Case 1: $d\sigma/d\lambda$ *is always greater than* σ/λ. This is shown in Figure 11.7(a), and it can be seen that there is no tangent line which can be drawn to the true stress–strain curve from the point $\lambda = 0$. The polymer therefore extends uniformly with increasing load, and no neck is formed.

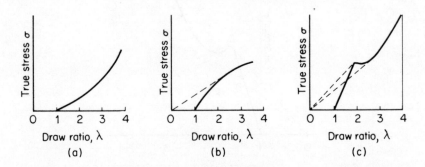

Figure 11.7. The three types of true stress–strain curves for polymers. (a) Case 1, $d\sigma/d\lambda > \sigma/\lambda$; (b) case 2, $d\sigma/d\lambda = \sigma/\lambda$ at one point; and (c) case 3, $d\sigma/d\lambda = \sigma/\lambda$ at two points. [Redrawn with permission from Vincent, *Polymer*, **1**, 7 (1960). © IPC Business Press Ltd.]

Case 2: $d\sigma/d\lambda = \sigma/\lambda$ *at one point*. In this case (Figure 11.7(b)) the polymer extends uniformly up to the point where $d\sigma/d\lambda = \sigma/\lambda$ and then necks. The neck gets steadily thinner and then the measured load decreases until fracture occurs. This case has been discussed previously, where the treatment of Nadai and Orowan[1,2] was presented (Section 11.1.1).

Case 3: $d\sigma/d\lambda = \sigma/\lambda$ *at two points*. This is the case where necking and cold-drawing occurs (Figure 11.7(c)), and it is only necessary to state that the requirement of two tangents becomes a criterion for necking *and* cold-drawing.

Note that these arguments do not take into account the strain rate dependence of the yield or flow stress σ. As will be discussed in detail below, the flow stress increases with increasing strain rate. Hence in the neck where the cross-sectional area is decreasing rapidly, the strain rate is correspondingly increasing. This leads to an increase in the flow stress which tends to stabilize the neck.

11.1.4 Definition of Yield Stress

A simple definition of the yield stress is to regard it as the minimum stress at which permanent strain is produced when the stress is subsequently removed. Although this definition is satisfactory for metals, where there is a clear distinction between elastic recoverable deformation and plastic irrecoverable deformation, in polymers the distinction is not so straightforward. In many cases, such as the tensile tests discussed above, yield coincides with the observation of a maximum load in the load–elongation curve. The yield stress can then be defined as the true stress at the maximum observed load. Because this stress is achieved at a comparatively low elongation of the sample it is often adequate to use the engineering definition of the yield stress as the maximum observed load divided by the *initial* cross-sectional area.

In some cases there is no observed load drop (e.g. shear tests in Figure 11.29) and another definition of yield stress is required. One approach is to determine the stress where the two tangents to the initial and final parts of the load–elongation curve intersect (Figure 11.33).

An alternative is to attempt to define an initial linear slope on the stress–strain curve and then to draw a line parallel to this which is offset by a specified strain, say 2%. The interception of this line with the stress–strain curve then defines a stress which is called the offset or proof stress, and is considered to be the yield stress.

11.2 IDEAL PLASTIC BEHAVIOUR

The simplest theories of plasticity do not contain time as a variable and ignore any features of the behaviour which take place below the yield point. In other words we assume a rigid plastic material whose stress–strain relationship in tension is shown in Figure 11.8. For stresses below the yield stress σ_Y there is no deformation. For stresses above the yield stress the deformation is determined by the movement of the applied loads.

The constitutive relations of elasticity and viscoelasticity relate the magnitudes of the stresses in the material to the magnitudes of the strains. In plasticity the situation is somewhat different in that, although the relationships between the different components of the plastic strain increment relate to the stresses in the material, the absolute magnitudes of the plastic strain increments

338

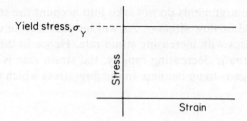

Figure 11.8. Stress–strain relationship for
an ideal rigid-plastic material.

are determined by the movement of the external applied loads and there is
no unique relationship between stress and strain.

There are two aspects to classical plasticity, or the behaviour of an ideal
rigid plastic material. First, we wish to define the stress situations in which
plastic flow can occur. For a simple tensile test this merely involves a definition
of the yield stress as in Section 11.1.4. Secondly we will seek to define the
appropriate plastic strain-increment relationships in terms of the stress situation at yield.

It should perhaps be pointed out that the idealization of a rigid plastic
material can be simply extended to the more realistic idealization of an elastic
plastic material if we wish to consider the behaviour under load. In many
cases, however, we are concerned either with the magnitudes of the stresses
causing yield or the permanent strains after unloading.

11.2.1 The Yield Criterion: General Considerations

The first question which may be asked regarding the yielding of materials
concerns the relationship between the measured yield point for different types
of testing, e.g. tension and simple shear. The aim is to find a function of all
the components of stress which reaches a critical value for all tests, i.e. for
different combinations of stress. This function is called the 'yield criterion',
and in its most general form can be written as a function of the components
of stress, $f(\sigma_{xx}, \sigma_{yy}, \sigma_{zz}, \sigma_{xy}, \sigma_{yz}, \sigma_{zx})$.

The actual form of the function f can be restricted by several considerations.
In the first instance it will be assumed that the material is isotropic. The term
f must then be a function of the invariants of the stress tensor. If we refer
our stresses to the principal axes of stress, the stress tensor

$$\begin{bmatrix} \sigma_{xx} & \sigma_{xy} & \sigma_{xz} \\ \sigma_{xy} & \sigma_{yy} & \sigma_{yz} \\ \sigma_{xz} & \sigma_{yz} & \sigma_{zz} \end{bmatrix}$$

becomes

$$\begin{bmatrix} \sigma_1 & 0 & 0 \\ 0 & \sigma_2 & 0 \\ 0 & 0 & \sigma_3 \end{bmatrix}.$$

In terms of the principal components of stress σ_1, σ_2, σ_3, the three simplest stress invariants are

$$J_1 = \sigma_1 + \sigma_2 + \sigma_3,$$

$$J_2 = \sigma_1\sigma_2 + \sigma_2\sigma_3 + \sigma_3\sigma_1,$$

$$J_3 = \sigma_1\sigma_2\sigma_3,$$

and it can be shown that all other invariants of stress can be expressed in terms of these three.

This gives the yield criterion as

$$f(J_1, J_2, J_3) = \text{constant.}$$

The apparent number of variables in our yield criterion has been reduced (by eliminating the three shear-stress components) but at the expense of defining our reference axes (which requires three direction variables).

In metals it has been established that the yield behaviour is to a first approximation independent of the hydrostatic component of stress. This is not true for polymers, but in our preliminary discussions we will consider the simplification which such an approximation allows.

The yield behaviour will now depend only on the components of the deviatoric stress tensor σ'_{ij}, obtained by subtracting the hydrostatic components of stress from the total stress tensor. We have that $\sigma'_{ij} = \sigma_{ij} - p\delta_{ij}$ where the prime indicates the deviatoric stress tensor. In Cartesian notation σ'_{ij} is given by

$$\begin{bmatrix} \sigma_{xx} - p & \sigma_{xy} & \sigma_{xz} \\ \sigma_{xy} & \sigma_{yy} - p & \sigma_{yz} \\ \sigma_{xz} & \sigma_{yz} & \sigma_{zz} - p \end{bmatrix},$$

where $p = \frac{1}{3}(\sigma_{xx} + \sigma_{yy} + \sigma_{zz})$.

In terms of principal components of stress the yield criterion is a function of

$$\sigma'_1 = \sigma_1 - p, \quad \sigma'_2 = \sigma_2 - p \quad \text{and} \quad \sigma'_3 = \sigma_3 - p.$$

Since $\sigma'_1 + \sigma'_2 + \sigma'_3 = 0$ the yield criterion reduces to

$$f(J'_2, J'_3) = 0,$$

where

$$J'_2 = -(\sigma'_1\sigma'_2 + \sigma'_2\sigma'_3 + \sigma'_3\sigma'_1) = \tfrac{1}{2}(\sigma'^2_1 + \sigma'^2_2 + \sigma'^2_3) = \tfrac{1}{2}\sigma'_{ij}\sigma'_{ij},$$

$$J'_3 = \sigma'_1\sigma'_2\sigma'_3 = \tfrac{1}{3}(\sigma'^3_1 + \sigma'^3_2 + \sigma'^3_3) = \tfrac{1}{3}\sigma'_{ij}\sigma'_{jk}\sigma'_{ki}.$$

A further simplification is obtained by assuming that if a stress σ_{ij} can induce yield, then so can a stress $-\sigma_{ij}$, e.g. it is implied that yield stresses in simple tension and compression are equal. Analytically this means either that f does not involve J_3' or that it involves only even powers of J_3'.

This simplification is not generally true for polymers, where the difference between tensile and compressive yield (the Bauschinger effect[5]) plays an important part in the yield behaviour. However, as with the hydrostatic component of stress we will develop our initial argument by considering that there is no Bauschinger effect in polymers.

11.2.2 The Tresca Yield Criterion

The earliest yield criterion to be suggested for metals was Tresca's proposal that yield occurs when the maximum shear stress reaches a critical value[6], i.e.

$$\sigma_1 - \sigma_3 = \text{constant}$$

with

$$\sigma_1 > \sigma_2 > \sigma_3.$$

This yield criterion is of a similar nature to the Schmid critical resolved shear-stress law for the yield of metal single crystals (see Section 11.2.7). We will see that the Tresca yield criterion is only of limited interest in polymers.

11.2.3 The von Mises Yield Criterion

Von Mises[7] proposed that the yield criterion did not involve J_3', but was a function of J_2' only, i.e. that yield occurs when J_2' reaches a critical value, i.e.

$$J_2' = \text{constant} = \mathbf{K}^2$$

(say). It is important to discuss three alternative ways of expressing the von Mises yield criterion.

(1) In terms of the principal components of the deviatoric stress tensor

$$J_2' = -(\sigma_1'\sigma_2' + \sigma_2'\sigma_3' + \sigma_3'\sigma_1') = \tfrac{1}{2}(\sigma_1'^2 + \sigma_2'^2 + \sigma_3'^2)$$
$$= \tfrac{1}{2}\{(\sigma_1 - p)^2 + (\sigma_2 - p)^2 + (\sigma_3 - p)^2\}.$$

This gives the yield criterion in terms of the principal components of the total stress tensor:

$$(\sigma_1 - p)^2 + (\sigma_2 - p)^2 + (\sigma_3 - p)^2 = 2\mathbf{K}^2, \tag{11.1}$$

where $p = \tfrac{1}{3}(\sigma_1 + \sigma_2 + \sigma_3)$ is the hydrostatic pressure.

(2) An equally well known representation of this criterion is

$$(\sigma_1 - \sigma_2)^2 + (\sigma_2 - \sigma_3)^2 + (\sigma_3 - \sigma_1)^2 = 6\mathbf{K}^2, \tag{11.2}$$

which is readily obtained by algebraic manipulation. This representation relates very simply to the so-called octahedral shear stress τ_{oct}. We have

$$\tau_{oct} = \tfrac{1}{3}\{(\sigma_1-\sigma_2)^2 + (\sigma_2-\sigma_3)^2 + (\sigma_3-\sigma_1)^2\}^{1/2},$$

giving the von Mises yield criterion as $\tau_{oct}^2 = \tfrac{2}{3}\mathbf{K}^2$.

(3) Finally, if we refer our stresses to a Cartesian set of axes x, y, z which are other than principal axes of stress the stress tensor,

$$\begin{bmatrix} \sigma_1 & 0 & 0 \\ 0 & \sigma_2 & 0 \\ 0 & 0 & \sigma_3 \end{bmatrix}$$

becomes

$$\begin{bmatrix} \sigma_{xx} & \sigma_{xy} & \sigma_{xz} \\ \sigma_{xy} & \sigma_{yy} & \sigma_{yz} \\ \sigma_{xz} & \sigma_{yz} & \sigma_{zz} \end{bmatrix}.$$

Equation (11.2) can be written as

$$2(\sigma_1^2 + \sigma_2^2 + \sigma_3^2) - (\sigma_1\sigma_2 + \sigma_2\sigma_3 + \sigma_3\sigma_1) = 6\mathbf{K}^2,$$

i.e.

$$3(\sigma_1^2 + \sigma_2^2 + \sigma_3^2) - (\sigma_1 + \sigma_2 + \sigma_3)^2 = 6\mathbf{K}^2.$$

Remembering that the invariants of the stress tensor are

$$\sigma_1 + \sigma_2 + \sigma_3 = \sigma_{xx} + \sigma_{yy} + \sigma_{zz}$$

and

$$\sigma_1^2 + \sigma_2^2 + \sigma_3^2 = \sigma_{xx}^2 + \sigma_{yy}^2 + \sigma_{zz}^2 + 2(\sigma_{xy}^2 + \sigma_{yz}^2 + \sigma_{xz}^2),$$

the von Mises yield criterion becomes

$$3[(\sigma_{xx}^2 + \sigma_{yy}^2 + \sigma_{zz}^2) + 2(\sigma_{xy}^2 + \sigma_{yz}^2 + \sigma_{zx}^2)] - (\sigma_{xx} + \sigma_{yy} + \sigma_{zz})^2 = 6\mathbf{K}^2,$$

i.e.

$$(\sigma_{xx} - \sigma_{yy})^2 + (\sigma_{yy} - \sigma_{zz})^2 + (\sigma_{zz} - \sigma_{xx})^2 + 6(\sigma_{xy}^2 + \sigma_{yz}^2 + \sigma_{zx}^2) = 6\mathbf{K}^2.$$

$$(11.3)$$

We will see that the von Mises yield criterion is of considerable interest in the case of polymers, but that we cannot ignore the pressure dependence of the yield behaviour. The von Mises criterion can be modified in a number of ways. One simple way is to allow $\mathbf{K}$ to be an arbitrary function of the hydrostatic pressure.

11.2.4 The Coulomb Yield Criterion

On the Tresca yield criterion, the critical shear stress for yield is independent of the normal pressure on the plane in which yield is occurring. Coulomb had previously proposed a more general yield criterion for the failure of soils[8]. This states that the critical shear stress τ for yielding to occur in any plane varies linearly with the stress normal to this plane, i.e.

$$\tau = \tau_c - \mu\sigma_N, \tag{11.4}$$

τ_c is the 'cohesion' of the material, μ is the coefficient of friction (sometimes μ is written as $\tan\phi$, for reasons which will be made evident below), and σ_N is the normal stress on the yield plane. For a compressive stress σ_N has a negative sign so that the critical shear stress τ for yielding to occur in any plane increases linearly with the pressure applied normal to this plane.

We will see that this yield criterion is of considerable interest in the case of polymers.

11.2.5 Geometrical Representations of the Tresca, von Mises and Coulomb Yield Criteria

The Tresca and von Mises Yield Criteria

The Tresca and von Mises yield criteria take very simple analytical forms when expressed in terms of the principal stresses. One reason for this is the assumption of material isotropy which implies that σ_1, σ_2 and σ_3 are interchangeable. Thus the yield criteria form simple surfaces in principal stress space, i.e. space where the three rectangular Cartesian axes are parallel to the principal stress directions.

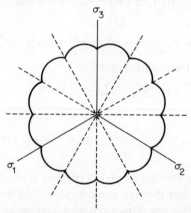

Figure 11.9. Cross-section of the yield surface normal to the [111] direction in principal stress space.

Because the yield criterion is independent of the hydrostatic component of stress we can replace σ_1, σ_2 and σ_3 by σ_1+p, σ_2+p and σ_3+p respectively. Thus if the point σ_1, σ_2, σ_3 lies on the yield surface, so does the point σ_1+p, σ_2+p, σ_3+p. This shows that the yield surface must be parallel to the [111] direction in principal stress space and can be represented geometrically by the cross-section normal to this direction (Figure 11.9). The material isotropy implies equivalence of σ_1, σ_2 and σ_3 and hence that the section has a threefold symmetry about the [111] direction. The assumption that $f(\sigma_{ij}) = f(-\sigma_{ij})$ (i.e. no Bauschinger effect) implies equivalence of σ_1 and $-\sigma_1$, and hence we have finally sixfold symmetry about the [111] direction.

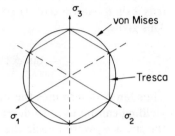

Figure 11.10. Cross-section of Tresca and von Mises yield surfaces normal to the [111] direction in principal stress space.

The cross-section normal to the [111] direction therefore consists of 12 equivalent parts (see Figure 11.9), and for the Tresca and von Mises yield criteria takes the particularly simple forms shown in Figure 11.10. The Tresca criterion gives a regular hexagon and the von Mises criterion a circle.

The Coulomb Yield Criteria

The Coulomb yield criterion is

$$\tau = \tau_c - \mu\sigma_N, \tag{11.4}$$

where $\mu = \tan\phi$, as explained above.

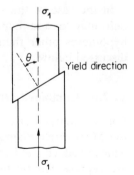

Figure 11.11. The yield direction for a material obeying the Coulomb criterion which is subjected to a compressive stress σ_1.

For uniaxial compression under a compressive stress σ_1 where yield occurs on a plane whose normal makes an angle θ with the direction of σ_1 (Figure 11.11), the shear stress is

$$\tau = \sigma_1 \sin\theta \cos\theta$$

and the normal stress $\sigma_N = -\sigma_1 \cos^2 \theta$. Yield occurs when

$$\sigma_1 \sin \theta \cos \theta = \tau_c + \sigma_1 \tan \phi \cos^2 \theta,$$

i.e. when

$$\sigma_1(\cos \theta \sin \theta - \tan \phi \cos^2 \theta) = \tau_c.$$

In practice this is achieved by yield occurring in the plane which maximizes $(\cos \theta \sin \theta - \tan \phi \cos^2 \theta)$ so that yield occurs for the smallest value of σ_1. This gives

$$\tan \phi \tan 2\theta = -1 \quad \text{or} \quad \theta = \frac{\pi}{4} + \frac{\phi}{2} \tag{11.5}$$

Thus $\tan \phi$ determines the direction of yield and conversely the direction of yielding can be used to define ϕ, where $\tan \phi$ is the coefficient of friction. If the stress σ_1 were tensile the angle θ would be given by

$$\theta = \frac{\pi}{4} - \frac{\phi}{2}.$$

We see that the Coulomb yield criterion therefore defines both the stress condition required for yielding to occur *and* the directions in which the material will deform. Where a deformation band forms the direction of the deformation band is the direction which is neither rotated nor distorted by the plastic deformation. This follows because the band direction marks the direction which establishes material continuity between the deformed material in the deformation band and the undistorted material in the rest of the specimen. If volume is conserved the band direction therefore denotes the direction of shear in a simple shear (by the definition of a shear strain). Thus for a Coulomb yield criterion the band direction is defined by equation (11.5).

In the case of the von Mises yield criterion the prediction of the plastic deformation and hence the deformation-band direction requires further hypotheses other than those contained in the yield criterion. This will be discussed below.

11.2.6 Combined Stress States

For the analysis of combined stress states in the two-dimensional situation the Mohr circle diagram is of value. In Figure 11.12(a) two states of stress which produce yield with principal stresses σ_1 and σ_2, σ_3 and σ_4 respectively are represented by two circles of identical radius, tangential to the yield surface. The yield criterion in this case is assumed to be that of Tresca and the yield surface degenerates for the two-dimensional case to two lines parallel to the normal stress axis.

In Figure 11.12(b) two states of stress causing yield for a material which satisfies the Coulomb criterion are shown as σ_5 and σ_6, σ_7 and σ_8, respectively.

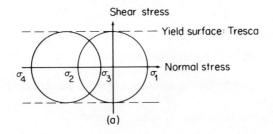

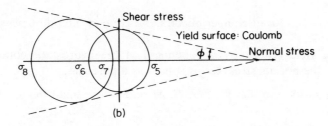

Figure 11.12. Mohr circle diagram for two states of stress
which produce yield in a material satisfying (a) the Tresca
yield criterion and (b) the Coulomb yield criterion.

In this case the yield surface is represented by two lines each making an angle
ϕ with the normal stress axis.

11.2.7 Yield Criteria for Anisotropic Materials

A very simple yield criterion for anisotropic materials is the critical resolved
shear stress law of Schmid (see Reference 9, p. 4). This law states that yield
occurs when the resolved shear stress in the slip direction in the slip plane
reaches a critical value. For a tensile stress σ making angles α and β with
the slip direction and the normal to the slip plane respectively this gives the
critical resolved shear stress τ_c as

$$\tau_c = \sigma \cos \alpha \cos \beta. \tag{11.6}$$

We will see that although this law is extensively used in metal plasticity it
is of restricted application in polymers.

For polymers, a generalization of the von Mises yield criterion by Hill[10]
has proved more useful. Hill's yield criterion applies to anisotropic materials
which possess three mutually orthogonal planes of symmetry at every point,
i.e. the material possesses at least orthorhombic symmetry. It will also apply
in a simplified form for solids possessing transverse isotropy.

The intersections of the three orthogonal planes define the principal axes
of symmetry, and these directions, rather than the principal axes of stress,
are chosen as Cartesian axes of reference. The yield criterion then takes the

form

$$F(\sigma_{yy} - \sigma_{zz})^2 + G(\sigma_{zz} - \sigma_{xx})^2 + H(\sigma_{xx} - \sigma_{yy})^2$$
$$+ 2L\sigma_{yz}^2 + 2M\sigma_{zx}^2 + 2N\sigma_{xy}^2 = 1, \tag{11.7}$$

where F, G, H, L, M and N are parameters which characterize the anisotropy of the yield behaviour.

This yield criterion satisfies several requirements:

(1) It reduces to the von Mises yield criterion for vanishingly small anisotropy.

(2) There is no Bauschinger effect, i.e. it contains no odd powers of stress components.

(3) The yield criterion is independent of the hydrostatic component of stress, i.e. the normal stress terms appear as differences.

We may also note that if F, G, H, M and N are small the yield criterion reduces to

$$2L\sigma_{yz}^2 = 1,$$

which is equivalent to the Tresca yield criterion. In physical terms this means that a single yield process represented by the constant L dominates the yield behaviour.

11.2.8 The Stress–Strain Relations for Isotropic Materials

Once we achieve the combination of stresses required to produce yield in an idealized rigid plastic material, deformation can proceed without altering the stresses, and is determined by the movements of the external constraints, e.g. the displacement of the jaws of the tensometer in a tensile test. This means that there is no unique relationship between the stresses and the *total* plastic deformation. Instead the relationships which do exist relate the stresses and the *incremental* plastic deformation. This was first recognized by St Venant, who proposed that for an isotropic material the principal axes of the strain *increment* are parallel to the principal axes of stress.

If the material is assumed to remain isotropic after yield there is no dependence on the deformation or stress history. Furthermore, if we assume that the yield behaviour is independent of the hydrostatic component of stress then the principal axes of the strain increment are parallel to the principal axes of the stress deviator.

Lévy[11] and von Mises[7] independently proposed that the principal components of the strain-increment tensor

$$\begin{bmatrix} de_1 & 0 & 0 \\ 0 & de_2 & 0 \\ 0 & 0 & de_3 \end{bmatrix}$$

and the deviatoric stress tensor

$$\begin{bmatrix} \sigma_1' & 0 & 0 \\ 0 & \sigma_2' & 0 \\ 0 & 0 & \sigma_3' \end{bmatrix}$$

are proportional, i.e.

$$\frac{de_1}{\sigma_1'} = \frac{de_2}{\sigma_2'} = \frac{de_3}{\sigma_3'} = d\lambda, \qquad (11.8)$$

where $d\lambda$ is *not* a material constant but is determined by our choice of the extent of deformation of the material, e.g. by the displacement of the jaws of the tensometer. Since

$$\sigma_1' + \sigma_2' + \sigma_3' = 0$$

and

$$d\lambda = \frac{de_1 + de_2 + de_3}{\sigma_1' + \sigma_2' + \sigma_3'}$$

it is implied that $de_1 + de_2 + de_3 = 0$, i.e. that the deformation takes place at constant volume, or that the material is incompressible.

If the stress–strain relations are referred to other than principal axes we have

$$de_{ij} = \sigma_{ij}' \, d\lambda,$$

i.e.

$$\frac{de_{xx}}{\sigma_{xx}'} = \frac{de_{yy}}{\sigma_{yy}'} = \frac{de_{zz}}{\sigma_{zz}'} = \frac{de_{yz}}{\sigma_{yz}'} = \frac{de_{zx}}{\sigma_{zx}'} = \frac{de_{xy}}{\sigma_{zy}'}. \qquad (11.9)$$

These equations are called the Lévy–Mises equations.

11.2.9 The Plastic Potential

The Lévy–Mises equations can also be considered to arise from more sophisticated considerations based on the concept of the plastic potential and the 'normality' rule for ideal plastic behaviour.

These ideas follow from a general representation of plastic behaviour, extensively discussed by Hill[10], which assumes that the components of the plastic strain increment tensor are proportional to the partial derivatives of a function called the 'plastic potential', which is a scalar function of stress. This is analogous to the use of a strain-energy function to derive stresses in elastostatics, but it is important to emphasize that whereas the strain-energy function relates to the total strains in an elastic material, the plastic potential relates to the strain increments.

A particularly simple situation is obtained if the plastic potential is assumed to be a surface in stress space of the same shape as the yield surface. The St

Venant principle is now seen as the mathematical expression of the fact that the plastic strain increments occur in directions normal to the yield surface. This is sometimes called the 'normality' condition of ideal plasticity, and some workers (notably Drucker[12]) have attempted to justify this flow rule in terms of a maximum work criterion.

11.2.10   The Stress–Strain Relations for an Anisotropic Material of Orthorhombic Symmetry

By analogy with the Lévy–Mises equations for an isotropic plastic material, Hill has proposed the following relations for the tensor components of plastic strains for an anisotropic material referred to the principal axes of anisotropy.

$$\left.\begin{aligned}
d\varepsilon_{xx} &= d\lambda\left[H(\sigma_{xx}-\sigma_{yy})+G(\sigma_{xx}-\sigma_{zz})\right], & d\varepsilon_{yz} &= d\lambda L\sigma_{yz}, \\
d\varepsilon_{yy} &= d\lambda\left[F(\sigma_{yy}-\sigma_{zz})+H(\sigma_{yy}-\sigma_{xx})\right], & d\varepsilon_{zx} &= d\lambda M\sigma_{zx}, \\
d\varepsilon_{zz} &= d\lambda\left[G(\sigma_{zz}-\sigma_{xx})+F(\sigma_{zz}-\sigma_{yy})\right], & d\varepsilon_{xy} &= d\lambda N\sigma_{xy}.
\end{aligned}\right\} \quad (11.10)$$

It is to be noted that in this case the principal axes of the plastic strain increment only coincide with the axes of anisotropy when the principal axes of stress coincide with the latter. It is also to be noted that the proportionality factor $d\lambda$ is dimensionally different from that in the Lévy–Mises equations for an isotropic plastic material.

11.3   THE YIELD PROCESS

We have seen that yield is often associated with a load drop on the load–extension curve, and always involves a change in slope on the true stress–strain curve. This load drop has sometimes been attributed either to adiabatic heating of the specimen or to the geometrical reduction in cross-sectional area on the formation of a neck. It is necessary to examine these explanations in detail, determine their shortcomings and establish the test conditions under which true yield behaviour can be observed. With this information it is then possible to consider the relevance of experimental data on the yield behaviour of polymers to our discussions of an ideal rigid plastic material.

11.3.1   The Adiabatic Heating Explanation

It was soon recognized that under conventional conditions of cold-drawing, where the specimen is extended at strain rates of the order of $10^{-2}\,\mathrm{s}^{-1}$, a considerable rise of temperature occurs in the region of the neck. Marshall and Thompson[13], following Müller[14], proposed that cold-drawing involves a local temperature rise and that necking occurs because of strain softening produced by the consequent fall in stiffness with rising temperature. The stability of the drawing process was then attributed to the stability of an adiabatic process of heat transfer through the shoulders of the neck, with extension taking place at constant tension throughout the neck.

Let us consider Marshall and Thompson's argument as they presented it. The starting point was to assume that the isothermal load–extension curves for polyethylene terephthalate were of the form shown in Figure 11.13. It was then argued that under their conditions of drawing (between rollers at high speeds) the drawing process takes place under adiabatic conditions. The 'adiabatic' load–extension curve was then calculated by assuming that all the work done in stretching appears as heat within the filament, i.e. no account was taken either of increases in elastic potential energy or of heat liberated due to crystallization. This calculation was performed by estimating the work done in successive 10% extensions, adjusting the temperature level at each 10% step, and finally checking the consistency of the integrated area under the resultant curve with the corresponding total temperature rise.

This produces a load–extension curve where the load falls with extension (curve A in Figure 11.13), which is an unstable situation. Marshall and Thompson suggested that instead of drawing each element unstably under variable tension with the temperature varying as for these adiabatic conditions, a condition of constant tension is obtained with heat being transferred through the neck to maintain the differences in temperature required to achieve this.

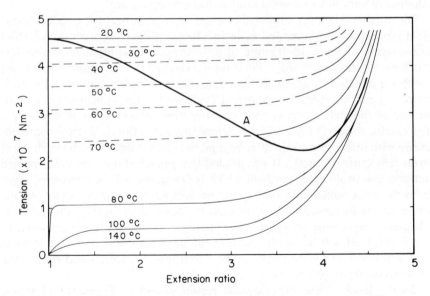

Figure 11.13. Tension–elongation–temperature properties of amorphous polyethylene terephthalate of various temperatures. Dashed curves are interpolated data, not obtainable in practice because necking occurred. A is estimated curve from 20 °C. [Redrawn with permission from Marshall and Thompson, *Proc. Roy. Soc. A*, **221**, 541 (1954).]

A heat balance equation was proposed which predicted to a reasonable approximation the measured length of the shoulder of the neck in the cold-drawing of polyethylene terephthalate.

Hookway[15] later attempted to explain the cold-drawing of Nylon 66 on somewhat similar grounds, suggesting that there is actually a possibility of local melting in the neck due to a combination of hydrostatic tension and temperature. The idea of local melting raises an important point with regard to the structure of the cold-drawn polymer. Nylon is crystalline in the undrawn, unoriented state. If the crystals do not break down in the necking and drawing process, the morphological state of the undrawn polymer will be particularly important in determining the structure of the drawn polymer.

There is no doubt that an appreciable rise in temperature does occur at conventional drawing speeds, and the ideas of Marshall and Thompson are very relevant to an understanding of the complex situation. Calorimetric measurements by Brauer and Müller[16] have, however, shown that at slow rates of extension the increase in temperature is quite small ($\sim 10\,^{\circ}$C) and not sufficient to give an explanation for necking and cold-drawing in terms of adiabatic heating. More recently, it has been clearly demonstrated that necking can still take place under quasi-static conditions. This was noted by Lazurkin[17], who observed cold-drawing in elastomers at very low speeds of drawing. (The drawing took place at temperatures below the glass transition temperature.) Vincent confirmed this result by showing that cold-drawing occurs in polyethylene at very low extension rates at room temperature[3].

The adiabatic heating explanation arose at least in part because the initial yield process was not regarded as distinct from the drawing process. A further examination of the cold-drawing of polyethylene terephthalate was undertaken by Allison and Ward[18]. Their results showed that although the *drawing* process is affected by adiabatic heat generation at high strain rates, the *yield* process is not affected. The principal evidence for this conclusion is a comparison of the yield stress and the drawing stress as a function of strain rate. The results, shown in Figure 11.14, demonstrate that the yield stress continues to rise with increasing strain rate, beyond the strain rate at which the drawing stress falls quite distinctly. It was argued that provided the drawing is carried out at a low strain rate, any heat which is generated will be conducted away from the neck sufficiently rapidly for no temperature rise to occur. As the strain rate is increased and the process becomes more nearly adiabatic, the effective temperature at which the drawing is taking place is increased. In particular, heat will be conducted into the unyielded portion of the sample. This will lower the yield stress of the undeformed material and reduce the force necessary to propagate the neck.

On the basis of the experimental results shown in Figure 11.14 it was possible to make an approximate calculation of the temperature rise in a sample caused by the drawing process. Two sets of information were required for this calculation: (1) the effect of increasing strain rate on the yield and drawing stresses and (2) the effect of temperature on the yield stress. It was

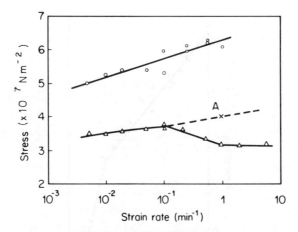

Figure 11.14. Comparison of yield stress (○) and drawing stress (△) as a function of strain rate for polyethylene terephthalate.

further assumed that to a first approximation both the yield stress and the drawing stresses are measures of the force necessary to initiate large scale molecular motion and that similar mechanisms are operative in both initial yielding and the subsequent propagation of the neck.

The data shown in Figure 11.14 indicate that initially both the yield and drawing stresses increase by similar amounts when the strain rate is increased. It would be expected that, if the drawing process were isothermal, the drawing stress would continue to increase with increasing strain rate in a manner similar to the yield stress. The difference between the drawing stress predicted on the assumption of an isothermal process and that measured experimentally can therefore be attributed to an increase in the sample temperature.

Subsidiary experiments showed that over a wide range the yield stress is a linear function of temperature. The data are shown in Figure 11.15, and it was calculated that an increase of 10 °C in temperature will cause the stress to decrease by about $0.48 \times 10^7 \, \text{N m}^{-2}$. A typical temperature rise in the neck during drawing was then calculated as follows. From Figure 11.14 we obtain by extrapolation a drawing stress of $3.9 \times 10^7 \, \text{N m}^{-2}$ at a strain rate of $1 \, \text{min}^{-1}$ in the absence of any heating effects (point A in Figure 11.14). The measured drawing stress is $3.0 \times 10^7 \, \text{N m}^{-2}$ and the yield stress $6.0 \times 10^7 \, \text{N m}^{-2}$. Assuming a similar dependence on temperature for the drawing stress and the yield stress, the fall in the drawing stress from the predicted value of $3.9 \times 10^7 - 3.0 \times 10^7 \, \text{N m}^{-2}$ corresponds to a temperature rise of $0.9 \times 6/3.9 \times 10/0.48 = 29 \, °C$. This gives the point B in Figure 11.16.

The calculation was performed for polyethylene terephthalate and nylon (the latter based on the experimental results of Hookway[15]). The results are shown in Figure 11.16, together with experimentally measured temperature

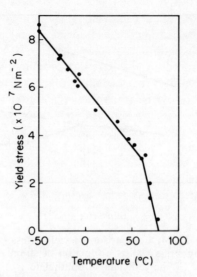

Figure 11.15. The yield stress of polyethylene terephthalate as a function of temperature.

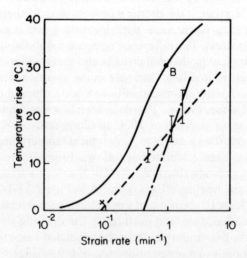

Figure 11.16. The temperature rise as a function of strain rate: ——, calculated from Figures 11.14 and 11.15; — · —, calculated from Hookway's (1958) data[15]; ----, reported by Vincent (1960)[3].

rises reported by Vincent[3] for polyethylene and polyvinyl chloride. It is apparent that adiabatic heating effects become important as the strain rate is raised above about $10^{-1} \, \text{min}^{-1}$.

This calculated temperature rise agrees approximately with that calculated from the work done in drawing, assuming that no heat is generated due to crystallization. In polyethylene terephthalate, X-ray diffraction diagrams of cold-drawn fibres show that very little crystallization has occurred.

The work done per unit volume is then given by $W = \sigma_D(D_r - 1)$ where σ_D is the drawing stress and D_r the natural draw ratio. From the results obtained $\sigma_D = 2.3 \times 10^7 \, \text{N m}^{-2}$ when $D_r = 3.6$, giving $W = 4.7 \, \text{Mjm}^{-3}$. For polyethylene terephthalate the specific heat is $67 \, \text{jkg}^{-1} \, \text{K}^{-1}$ and the density is $1.38 \, \text{Mgm}^{-3}$. This gives a calculated temperature rise of $57 \, ^\circ\text{C}$, compared with $42 \, ^\circ\text{C}$ obtained from Figure 11.16.

11.3.2 The Isothermal Yield Process: the Nature of the Load Drop

There is no doubt that a temperature rise does occur in cold-drawing under many conditions of test. We have shown, however, that there is very good evidence to support the view that necking can still take place under quasi-static conditions where there is no appreciable temperature rise. Vincent[3] therefore proposed that the observed fall in load is a geometrical effect due to the fact that the fall in cross-sectional area during stretching is not compensated by an adequate degree of strain hardening. This effect was called strain softening and attributed to the reduction in the slope of the stress–strain curve with increasing strain.

Contrary to this latter explanation of the load drop in terms of geometric softening, results reported by Andrews and Whitney[19] showed a yield drop in compression for polystyrene and polymethyl methacrylate. This led Brown and Ward[20] to make a detailed investigation of yield drops in polyethylene terephthalate. They studied isotropic and oriented specimens, under a variety of test conditions (tension, shear and compression) and concluded that in most cases there is clear evidence for the existence of an intrinsic yield drop, i.e. that a fall in true stress can occur in polymers, as in metals.

There is, however, a significant difference between polymers and many metals with regard to yield behaviour. In a polymer, as shown in Figure 11.2, it is to be noted that only one maximum is observed on the load–extension curve. This contrasts with metals (illustrated by mild steel in Figure 11.17) where often two maxima are observed on a typical load–extension curve. The first maximum (point A in Figure 11.17) is the upper yield point. This represents a fall in true stress, an intrinsic load drop, and corresponds to a sudden increase in the amount of plastic strain which relaxes the stress. From B to C Lüders bands propagate throughout the specimen. Lüders bands have also been observed in polymers[21]. At C the specimen is homogeneously strained and the stress begins to rise as the material work hardens uniformly. A second maximum is observed at point D. This second type of maximum is

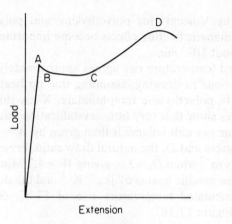

Figure 11.17. Load–extension curve
in tension of mild steel.

always associated with the beginning of necking in the specimen. Necking occurs when the strain-hardening of the metal is exceeded by the geometrical softening due to the reduction in the cross-sectional area of the specimen as it is strained, i.e. the Orowan–Vincent explanation, discussed in Section 11.1.1 above.

The second maximum, as we have seen previously, is not observed if the true stress–strain curve is plotted instead of the load–extension curve. The first maximum, on the other hand, would exist on the true stress–strain curve. It is called an intrinsic yield point, because it relates to the intrinsic behaviour of the material.

In polymers, as we have emphasized, only one maximum is observed in the load–extension curve. The investigations of Andrews and Whitney[19] and Brown and Ward[20], show that this maximum combines the effect of the geometrical changes and an intrinsic load drop, and cannot be attributed to the geometrical changes alone. In particular, the cold-drawing results are not accounted for by a decrease in the slope of the true stress–strain curve, as suggested in the explanation of Vincent. It is important to note that every element of the material does not follow the same true stress–strain curve, since the stress for initiation is greater than that for propagation of yielding. This confirms that it will not be possible to give a complete explanation of necking and cold-drawing in terms of the Considère construction on a true stress–strain curve as has already been remarked (Section 11.1.3).

11.4 EXPERIMENTAL EVIDENCE FOR YIELD CRITERIA IN POLYMERS

Studies of the yield behaviour of polymers have often bypassed the question of strain rate and temperature dependence and sought to establish a yield criterion as discussed in Section 11.2 above. This approach will be followed in the first instance, with a historical review of the subject.

11.4.1 Isotropic Glassy Polymers

Polystyrene and Other Polymers

Whitney and Andrews[19] studied the behaviour of polystyrene, polymethyl methacrylate, polycarbonate and polyvinyl formal. They were particularly concerned with the effect of the hydrostatic component of stress on the yield behaviour and the volume changes occurring before and during yielding. Their results for polystyrene are summarized in Figure 11.18 which shows the section of the yield surface normal to the principal stress σ_3. Tensile stress in the $\sigma_1\sigma_2$ plane is positive and compressive stress negative. The numbered circles represent the following tests: (1) uniaxial tension, (2) uniaxial compression, (3) torsional shear, (4) biaxial tension and (5) biaxial compression.

Solid cylinders were used for torsional shear. The biaxial tensile stress was produced by straining, in a single direction, a square film sample (length = width ≫ thickness) held in wide grips. The biaxial compression was produced

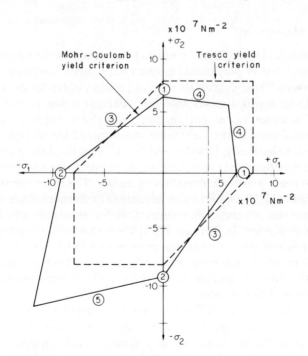

Figure 11.18. Yield and fracture envelopes for polystyrene in the $\sigma_1\sigma_2$ stress plane. [Redrawn with permission from Whitney and Andrews, *J. Polymer Sci. C*, **16**, 2981 (1967).]

by using a so-called 'double punch' test in which the sample is squeezed between two rectangular bars lying across the rectangular sample, under conditions of plane strain. It is not clear how point (5) is obtained in the $\sigma_1\sigma_2$ plane plot of Whitney and Andrews as the σ_3 stress is not zero in this case.

With this reservation regarding point (5) it is still clear that the results do not fit either the Tresca or the von Mises yield criterion. There is a definite asymmetry between the uniaxial tensile and compressive yield stresses and this asymmetry is confirmed by the torsion and combined stress measurements. Whitney and Andrews propose that the yield behaviour is affected either by the mean normal stress or by the hydrostatic component of the stress (these are not, in fact, identical). They point out that the observed data will fit a Coulomb criterion where the yield stress is assumed to depend linearly on the mean normal stress as has been discussed in Section 11.2.4 above.

The Coulomb criterion also specifies the direction in which yielding will occur. It states that the material yields in shear in the plane where the shear stress reaches the critical value $\tau = \tau_c - \mu\sigma_N$, where σ_N is the normal stress on the plane. Whitney and Andrews suggested that the direction of yield, i.e. of the deformation bands, was consistent with their assumption of the Coulomb yield criterion.

Polymethyl Methacrylate

In a review article, Thorkildsen[22] quotes a few measurements on the behaviour of thin-walled tubes of polymethyl methacrylate under combined tension and internal pressure. These suggest that the von Mises yield criterion is applicable. From the discussion it appears that the yield stress was considered as the stress at 0.2% strain, i.e. an engineering proof stress, but it is not clear that a consistent definition of the yield point was assumed for all experiments.

In a more detailed study, Bowden and Jukes[23] examined the yield behaviour of polymethyl methacrylate in a plane-strain compression test first developed by Ford[24] to study plastic deformation of metals. The experimental set-up is shown in Figure 11.19. A particular advantage of this technique is that yield behaviour can be observed in compression for materials which normally fracture in a tensile test. In this case PMMA was studied at room temperature, i.e. below its brittle–ductile transition temperature in tension.

The yield point in compression σ_1 was measured for various values of applied tensile stress σ_2, and the results (Figure 11.20) were analysed in terms of the Coulomb yield criterion.

The data gave $\tau = 4.66 - 0.258\sigma_N$ ($\times 10^7$ N m^{-2}) but it is to be noted that σ_1 is plotted as a true stress (applied compressive load divided by cross-sectional area of the dies), and σ_2 as a nominal stress (applied tensile load divided by initial cross-sectional area of sheet). A more consistent representation in terms of true stress for both σ_1 and σ_2 gives

$$\tau = 4.74 - 0.158\sigma_N.$$

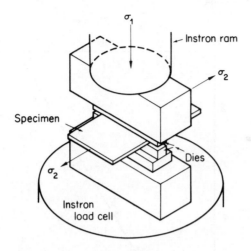

Figure 11.19. The plane-strain compression test. [Redrawn with permission from Bowden and Jukes, *J. Mater. Sci.*, **3**, 183 (1968).]

The raw data of Figure 11.20 give $\sigma_1 = -11.1 + 1.365\sigma_2$, when both σ_1 and σ_2 are expressed as true stresses in units of newtons per square metre $\times 10^7$. This illustrates directly the divergence from the Tresca criterion where $\sigma_1 - \sigma_2 = $ constant at yield. For a von Mises yield criterion the third stress $\sigma_3 = \nu(\sigma_1 + \sigma_2)$ is involved (ν = Poisson's ratio). It can also be shown that Bowden and Jukes' data do not fit a von Mises criterion. (As they remark, for $\nu = \frac{1}{2}$, the von Mises criterion for a plane strain test degenerates to $\sigma_1 - \sigma_2 = $ constant.)

Attempts to fit the direction of the deformation band to the Coulomb criterion (see Section 11.2.5) in these tests were not very satisfactory. The discrepancies were attributed to the complications of establishing general strain increment laws for materials where the yield criterion is not independent of the hydrostatic component of stress and where volume need not be conserved, and also to difficulties in taking into account elastic recovery effects.

11.4.2 Influence of Hydrostatic Pressure on Yield Behaviour

There is also a considerable body of direct evidence for the effect on the yield behaviour of polymers of the hydrostatic component of stress, as was more indirectly indicated by the experiments described above. Early studies by Ainbinder, Laka and Maiors[25] examined the tensile behaviour of polymethyl methacrylate, polystyrene, Kapron (Nylon 6), polyethylene and several other polymeric materials under superposed hydrostatic pressure. In every case both the modulus and yield strength increased with increasing hydrostatic pressure.

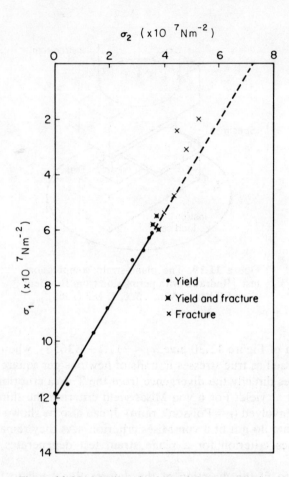

Figure 11.20. Measured values of the compressive yield stress σ_1 (true stress) plotted against applied tensile stress σ_2 (nominal stress). The full circles denote ductile yield, the crosses, brittle fracture, and the combined points, tests where ductile yielding occurred, followed immediately by brittle fracture. [Redrawn with permission from Bowden and Jukes, *J. Mater. Sci.*, **3**, 183 (1968).]

Recently there have been detailed studies by Baer, Radcliffe and coworkers[26–28] by Mears, Pae and Sauer[29,30], and by Ward, Parry and coworkers[31,32]. Figure 11.21 shows how the peak yield stress of high density polyethylene increases linearly with hydrostatic pressure to a very good approximation. Biglione, Baer and Radcliffe determined the load–extension behaviour in tension for polystyrene as a function of hydrostatic pressure from atmosphere pressure to $600 \, \text{MN m}^{-2}$. At atmospheric pressure this polymer shows brittle behaviour in tension. It was found that at pressures of

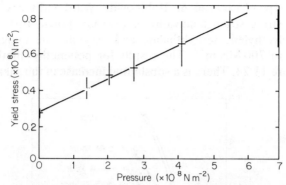

Figure 11.21. Yield stress as a function of hydro-static pressure for isotropic high density polyethyl-ene. [Redrawn with permission from Mears, Pae and Sauer, *J. Appl. Phys.*, **40**, 4229 (1969).]

$300 \, MN \, m^{-2}$ and greater the polymer became ductile, and necked with the formation of deformation bands. In this region of ductility the yield stress increased linearly with increased applied pressure, as shown in Figure 11.22.

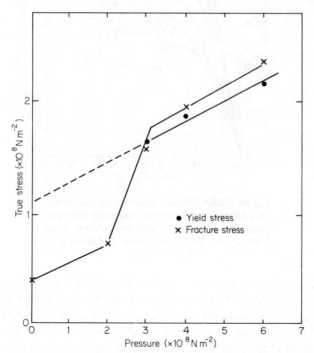

Figure 11.22. Effect of hydrostatic pressure on the yield and fracture stress of isotropic polystyrene. [Redrawn with permission from Biglione, Baer and Radcliffe, in *Fracture 1969* (P. L. Pratt, ed.), Chapman and Hall, London (1969), p. 520.]

In another type of experiment, Rabinowitz, Ward and Parry[31] determined the torsional stress–strain behaviour of isotropic polymethyl methacrylate, crystalline polyethylene terephthalate and polyethylene under hydrostatic pressures up to $700\,\text{MN m}^{-2}$. The results for polymethyl methacrylate are shown in Figure 11.23. There is a substantial increase in the shear yield stress

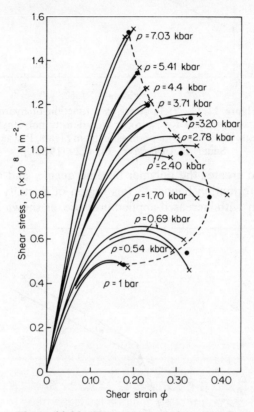

Figure 11.23. Shear stress–strain curves for polymethyl methacrylate showing fracture envelope. [Redrawn with permission from Rabinowitz, Ward and Parry, *J. Mater. Sci.*, **5**, 29 (1970).]

up to a hydrostatic pressure of about $300\,\text{MN m}^{-2}$. After this pressure brittle failure occurs, which can be prevented by protecting the specimens from the hydraulic fluid[32] (for example by coating with a layer of solidified rubber solution). The strain at which yield occurs also increases with increasing pressure, similar to the results of other workers for tensile tests under pressure. The shear yield stress increases linearly with pressure to an excellent approximation (Figure 11.24).

There are two ways in which these results can be represented. First, recalling Section 11.2.6 and Figure 11.12, the Mohr circle diagram can be constructed

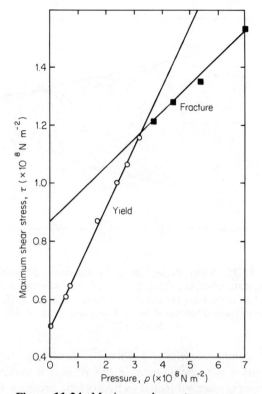

Figure 11.24. Maximum shear stress τ as a function of hydrostatic pressure p for polymethyl methacrylate. $\bigcirc$, Yield; $\blacksquare$, fracture. [Redrawn with permission from Rabinowitz, Ward and Parry, *J. Mater. Sci.*, **5**, 29 (1970).]

from the data. This is shown in Figure 11.25 where Bowden and Jukes' results are shown as crossed points. This diagram leads naturally to a Coulomb yield criterion.

It is, however, equally reasonable to interpret Figure 11.24 directly in terms of the equation

$$\tau = \tau_0 + \alpha p, \tag{11.11}$$

where τ is the shear yield stress at pressure p, τ_0 is the shear yield stress at atmospheric pressure and α is the coefficient of increase of shear yield stress with hydrostatic pressure.

We will see that this simple form of pressure-dependent yield criterion is more satisfactory than the Coulomb criterion when it is attempted to develop a representation which includes the effects of temperature and strain rate on the yield behaviour. In physical terms, the hydrostatic pressure can be seen as changing the state of the polymer by compressing the polymer significantly, unlike the situation in metals where the bulk moduli are much larger

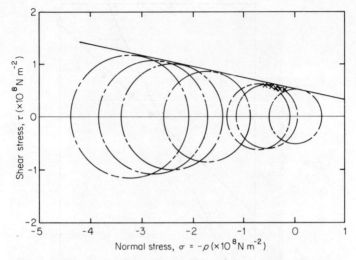

Figure 11.25. Mohr circles for yield behaviour of polymethyl methacrylate obtained from results of Rabinowitz, Ward and Parry. The crosses are the results of Bowden and Jukes. [Redrawn with permission from Rabinowitz, Ward and Parry, *J. Mater. Sci.*, **5**, 29 (1970).]

($\sim 100 \, \text{GN m}^{-2}$ compared with $\sim 5 \, \text{GN m}^{-2}$). Although such experimental evidence as exists is not unequivocal in this respect it seems likely that the flow rules for the polymer subjected to hydrostatic pressure are still given by equation (11.9), i.e. pressure increases the magnitude of the yield stresses but any volume changes are comparatively small.

Recent studies of yield behaviour, using a variety of multiaxial stressing experiments, can all be adequately described by a generalization of equation (11.11), i.e. a generalized von Mises equation where τ is replaced by the octahedral shear stress. This is developed in Section 11.6.1 below.

Finally it can be noted that the coefficient α in equation (11.11) depends on the temperature of measurement and increases markedly near a viscoelastic transition. Briscoe and Tabor[33] have pointed out that α is equivalent to the coefficient of friction μ in sliding friction, and shown that there is good numerical agreement between values of μ and the values of α obtained from yield stress/pressure measurements.

11.4.3 Anisotropic Polymers

If an oriented polymer is subjected to further extension in a direction not parallel to the initial draw direction, the deformation is sometimes concentrated into a narrow deformation band, as shown in Figure 11.26. As previously mentioned, this is analogous to the formation of a neck in an unoriented polymer, and the development of the deformation band is associated with a yield drop, which cannot be accounted for by a geometrical softening[20].

Figure 11.26. Photograph of a deformation band in an oriented sheet of polyethylene terephthalate.

The deformation bands which form are of two types. One type is nearly parallel to the initial draw direction and has the appearance of a slip band in metals; the other type is more diffuse, makes a larger angle with the draw direction, and has the general appearance of a kink band in metals.

It has been suggested by F. C. Frank that the first type of band be called a 'slippy' band and the second a 'kinky' band to indicate their partial but incomplete kinship with these two modes of deformation observed in metal crystals.

A detailed examination of the plastic deformation in such specimens is especially rewarding for two major reasons. First, the yield criterion will be more complex than for unoriented polymers but clearly gives greater scope for an understanding of yield processes in molecular terms. Secondly, when we consider the molecular changes associated with the plastic deformation, we are now measuring changes in an already highly ordered structure, using techniques such as X-ray diffraction which are particularly suited for this purpose. Even macroscopic properties such as birefringence can be more precisely tested when there is a high degree of initial orientation.

In this chapter the yield behaviour of oriented polymers will be discussed first, and the molecular aspects of the plastic deformation deferred to a later section.

Nylon

Slip bands or deformation bands were first observed in oriented nylons by Zaukelies[34]. Singly or doubly oriented bristles of Nylon 66 and Nylon 610 were prepared by drawing, and drawing and rolling, respectively. The bristles were compressed in a tensile testing machine and kink bands were observed. Examination of the deformed samples showed that there was very considerable reorientation of the material within the kink band.

Zaukelies explained his results in terms of crystal plasticity theory. In particular he suggested that the angle θ between the kink band plane and the slip plane was consistent with the Orowan equation:

$$\cot \theta = 2md/nc,$$

where c is the crystal lattice spacing along the slip plane, d is the crystal lattice spacing between adjacent slip planes, n is the number of c-axis spacings per slip plane, and m is the number of slip planes acting in unison. The kink-band angles at 25 and 100 °C were 43.5° and 35.5° respectively, which Zaukelies proposed was consistent with the Orowan equation for slip on 010 planes putting $m = 2$ and 3, respectively. He also proposed various dislocation motions as explanations for the slip processes.

Polyethylene

In a subsequent publication, Kurakawa and Ban[35] (and later Keller and Rider[36]) describe the observation of deformation bands in tensile tests on oriented high density polyethylene sheets. Kurakawa and Ban concluded that in some cases, when there was only a small angle between the initial draw direction (the IDD) and the tensile axis in the redrawing, the band direction was in the c-axis direction (i.e. the (001) direction in the crystalline regions of the polymer) and coincided with the c-axis direction of the deformed material in the band. In other cases, the deformation band was noted to be a little inclined to the (001) direction. They therefore suggested that the basic deformation process was not simple slip in the (001) direction, but a combination of (001) slip and twinning. Mechanical twinning in polyethylene had been proposed as an explanation of crystal reorientation during rolling by Frank, Keller and O'Connor[37].

When the tensile axis made a large angle with the IDD, kink bands were observed. Wide angle X-ray diffraction measurements showed that gross reorientation occurred in the kink bands, as in the case of nylon.

Keller and Rider, in a more detailed examination[36], reported the observation of similar deformation bands to those described by Kurakawa and Ban. A variety of drawn; drawn and rolled; drawn, rolled and annealed sheets were examined. In each case the alignment of the c axis, as determined by wide angle X-ray diffraction data, was concluded to be an important factor in determining the nature of the deformation bands.

It was noted that the band boundary was generally not parallel to the c-axis direction, and that the boundary of the kink bands did not bisect the angle between the c-axis directions on either side of the boundary. In spite of these anomalies it was concluded that the ductile deformation approximated to slip in the c direction within the crystalline regions and the similarity to the plastic behaviour of hexagonal metal single crystals was emphasized.

Keller and Rider[36] measured the yield stress as a function of the angle θ between the tensile testing direction and the initial draw direction. Their results are shown in Figure 11.27. They proposed that the data can be fitted to the Coulomb yield criterion.

In this case $\tau_c = \sigma(\sin\theta\cos\theta + k\sin^2\theta)$ where θ is the angle between the tensile stress direction and the initial draw direction and k is a constant.

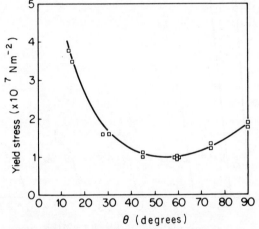

Figure 11.27. Yield stress plotted against θ, the angle between the tensile axis and the initial draw direction, for drawn high density polyethylene sheets, measured at room temperature. [Redrawn with permission from Keller and Rider, *J. Mater. Sci.*, **1**, 389 (1966).]

The Coulomb yield criterion also defines the plane in which yield will occur (Section 11.2.5 above). In this respect it did not satisfactorily describe the high density polyethylene data, as the deformation band did not form in the direction predicted from the yield criterion.

Polyethylene Terephthalate

A further polymer to be studied in this area is polyethylene terephthalate, which is of much lower crystallinity than high density polyethylene (~30% as against ~85%)[20,38–40]. It was found that the direction of the deformation band differed significantly from the initial drawing direction for most directions of test.

A typical set of tensile tests for different angles θ between the test direction and the initial draw direction is shown in Figure 11.28. The corresponding

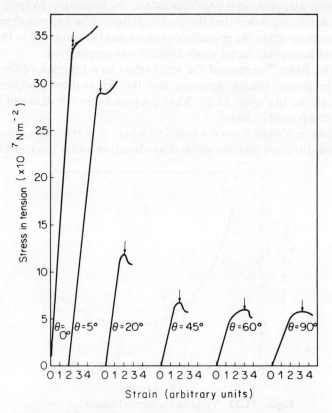

Figure 11.28. Stress–strain curves for various angles θ between the tensile axis and the initial draw direction for tensile tests on drawn polyethylene terephthalate sheets. The yield points are marked by arrows. [Redrawn with permission from Brown, Duckett and Ward, *Phil. Mag.*, **18**, 483 (1968).]

variation in yield stress with angle θ between the tensile testing direction and the initial draw direction of the film is plotted in Figure 11.29.

We have seen that the simplest yield criterion for an anisotropic solid would be the critical resolved shear-stress law of Schmid. For a tensile stress σ, $\tau_c = \sigma \sin \theta \cos \theta$. Figure 11.29 shows that it is not possible to fit the PET data to this equation. Attempts to fit the data to a Coulomb criterion were also unrewarding. Although the critical resolved shear-stress laws were moderately successful for angles θ between 10° and 60°, where the strain observed is basically shear parallel to the deformation band directions, agreement broke down at higher values of θ.

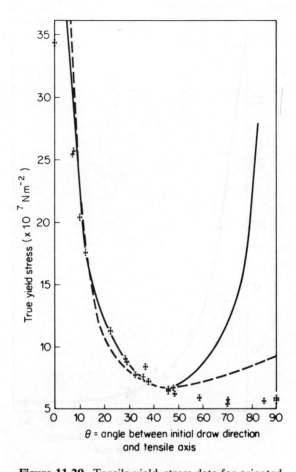

Figure 11.29. Tensile yield–stress data for oriented polyethylene terephthalate sheets showing the prediction of a critical resolved shear stress law. ——, Resolving shear stress parallel to the initial draw direction; – – – –, resolving shear stress parallel to the deformation band direction. [Redrawn with permission from Brown, Duckett and Ward, *Phil. Mag.*, **18**, 483 (1968).]

The yield stress data can, however, be shown to be consistent with the Hill anisotropic yield criterion, as shown in Figure 11.30.

For the plane stress situation of the tensile tests the Hill yield criterion for a material of orthorhombic symmetry (Section 11.2.7) reduces to

$$\sigma^2\{(G+H)\cos^4\theta + (H+F)\sin^4\theta$$
$$+ 2(N-H)\sin^2\theta\cos^2\theta\} = 1. \qquad (11.11)$$

We have chosen rectangular Cartesian axes with x parallel to the initial draw direction, y in the plane of the sheet and z normal to the sheet.

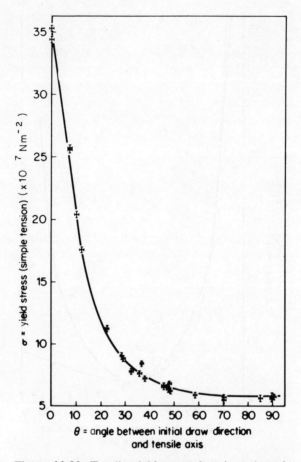

Figure 11.30. Tensile yield stress data for oriented polyethylene terephthalate sheets showing the best fit obtained from the Hill theory for anisotropic sheets of orthorhombic symmetry. [Redrawn with permission from Brown, Duckett and Ward, *Phil. Mag.*, **18**, 483 (1968).]

This gives the yield stress σ as

$$\sigma = \{(G+H)\cos^4\theta + (H+F)\sin^4\theta + 2(N-H)\sin^2\theta\cos^2\theta\}^{-1/2}.$$

In Figure 11.30 we see a curve following this equation with constants chosen to give an exact fit to data for PET sheet at 0°, 45° and 90°. It can be seen that very good agreement with the experimental data is obtained.

11.4.4 The Band Angle in the Tensile Test

In Section 11.2.10 above we discussed the modified Lévy–Mises equations relating the plastic strain increments to the applied stresses for a material of orthorhombic symmetry.

In the simple tensile tests $\sigma_{zz} = 0$ and these equations reduce to

$$\left. \begin{aligned}
d\varepsilon_{xx} &= d\lambda [(G+H)\sigma_{xx} - H\sigma_{yy}], \\
d\varepsilon_{yy} &= d\lambda [(H+F)\sigma_{yy} - H\sigma_{xx}], \\
d\varepsilon_{zz} &= -d\lambda [G\sigma_{xx} + F\sigma_{yy}], \\
d\varepsilon_{xy} &= d\lambda N\sigma_{xy}.
\end{aligned} \right\} \tag{11.12}$$

As discussed in Section 11.2.5, the deformation-band direction is the direction which is common to the deformed and the undeformed material, and must therefore define a direction which is neither rotated nor distorted by the plastic deformation. This means that the plastic strain increment must be zero in the band direction. There are two such directions in the material; one defines the direction of the 'slippy' band and the other the 'kinky' band.

We have chosen a rectangular set of axes x, y, z with x parallel to the initial draw direction, y in the plane of the sheet. Now consider a set of axes x', y', z', produced by a rotation about the z axis through an angle β. We refer the plastic strain increments to the new frame of reference by using the following transformations:

$$\begin{aligned}
d\varepsilon'_{xx} &= d\varepsilon_{xx} \cos^2 \beta + d\varepsilon_{yy} \sin^2 \beta + 2d\varepsilon_{xy} \sin \beta \cos \beta, \\
d\varepsilon'_{yy} &= d\varepsilon_{xx} \sin^2 \beta + d\varepsilon_{yy} \cos^2 \beta - 2d\varepsilon_{xy} \sin \beta \cos \beta, \\
d\varepsilon'_{zz} &= d\varepsilon_{zz}, \\
d\varepsilon'_{xy} &= -(d\varepsilon_{xx} - d\varepsilon_{yy}) \sin \beta \cos \beta + d\varepsilon_{xy}(\cos^2 \beta - \sin^2 \beta). \tag{11.13}
\end{aligned}$$

The condition for a band forming at an angle β to the initial draw direction is therefore the $d\varepsilon'_{xx} = 0$, i.e.

$$d\varepsilon_{xx} \cos^2 \beta + d\varepsilon_{yy} \sin^2 \beta + 2 d\varepsilon_{xy} \sin \beta \cos \beta = 0$$

or

$$d\varepsilon_{yy} \tan^2 \beta + 2d\varepsilon_{xy} \tan \beta + d\varepsilon_{xx} = 0.$$

From equations (11.12) above we have $d\varepsilon_{xx}$, $d\varepsilon_{yy}$, and $d\varepsilon_{xy}$. Now put

$$\begin{aligned}
\sigma_{xx} &= \sigma \cos^2 \theta, \\
\sigma_{yy} &= \sigma \sin^2 \theta, \\
\sigma_{xy} &= \sigma \sin \theta \cos \theta.
\end{aligned}$$

On substitution we have a quadratic in $\tan \beta$. Thus for each value of θ the theory predicts two possible band directions depending on θ, F, G, H and N.

370

Figure 11.31 shows the fit obtained to the 'slippy' band angle in polyethylene terephthalate[39] using the yield stress data. Good agreement is obtained which confirms the applicability of both the modified von Mises yield criterion and the associated plastic strain-increment rules proposed by Hill.

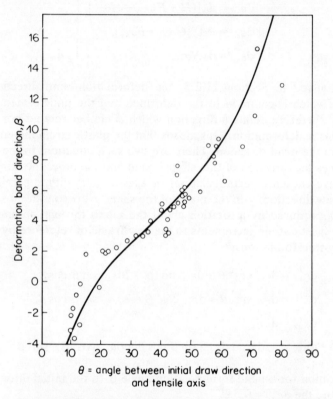

Figure 11.31. Graph of deformation band angle β against θ for tensile tests on drawn polyethylene terephthalate sheets. The curve shows the predictions of the Hill theory modified to include the Bauschinger effect using appropriate values of the constants, F, G, H, N and σ_i as described in the text. [Redrawn with permission from Brown, Duckett and Ward, *Phil. Mag.*, **18**, 483 (1968).]

11.4.5 Simple Shear–Stress Yield Tests

The yield stresses for oriented polyethylene terephthalate sheets were also measured in a series of simple shear tests[39,40], where the direction of shear

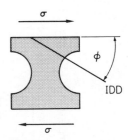

Figure 11.32. A typical specimen for the simple shear test with the direction of the applied shear stress σ making an angle ϕ with the initial draw direction.

displacement made varying angles ϕ with the initial draw direction (Figure 11.32). The stress–strain curves are shown in Figure 11.33, and the yield stress as a function of angle in Figure 11.34. It can be seen that the stress–strain curves vary markedly in form with angle ϕ. Note in particular that there is a clear yield drop when $\phi = 45°$ but no yield drop at 135°.

The applied shear stress can be resolved into a compressive stress and a tensile stress at right angles. When $\phi = 45°$, the initial draw direction is parallel to the compressive component of stress. Thus it has been argued[39] that the lower yield stress in this direction occurs because the stress situation is favourable for compressing the chains, which is a comparatively easy process. When $\phi = 135°$, the mean direction of the chains is parallel to the tensile component of stress and the chains are under tension. This is a difficult process which may, for example, involve breaking chains, so that the yield stress is

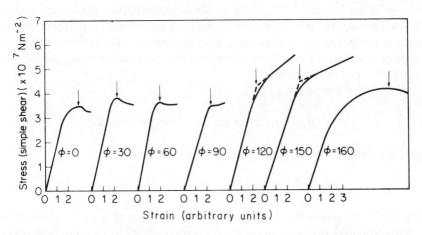

Figure 11.33. Stress–strain curves for various values of ϕ in simple shear tests on drawn polyethylene terephthalate sheets. Arrows mark the yield points. [Redrawn with permission from Brown, Duckett and Ward, *Phil. Mag.*, **18**, 483 (1968).]

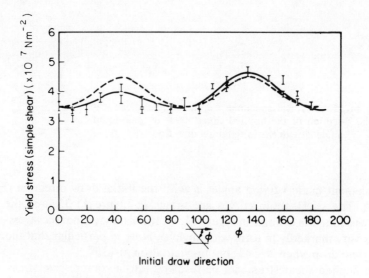

Figure 11.34. Graph of yield stress in simple shear against ϕ for drawn polyethylene terephthalate sheet – – –, Best fit obtained from Hill theory for orthorhombic sheets; ——, best fit obtained from Hill theory modified to include the Bauschinger term. [Redrawn with permission from Brown, Duckett and Ward, *Phil. Mag.*, **18**, 483 (1968).]

now much higher (see Figure 11.35). The different nature of the stress–strain curves in the 45° and 135° directions can be attributed to differences in strain hardening. At 45° the chains are disoriented by the plastic deformation, the polymer softens, and a clear yield drop is observed. At 135° the chains are further oriented by plastic deformation, the polymer stiffens and no clear yield drop can be observed.

In simple shear tests

$$-\sigma_{xx} = +\sigma_{yy} = \sigma \sin 2\phi, \qquad \sigma_{xy} = \sigma \cos 2\phi$$

and the Hill anisotropic yield criterion reduces to

$$\sigma^2\{(G+F+4H)\sin^2 2\phi + 2N\cos^2 2\phi\} = 1.$$

Figure 11.34 shows the experimental value of σ and a curve obeying the above equation with appropriately chosen constants to give the best overall fit (dashed line). Whilst the overall nature of the curve appears to be correct, in that two maxima and two minima are correctly predicted at approximately the observed values of ϕ, the theory predicts that the two maxima be equal. It can be seen that this is not true.

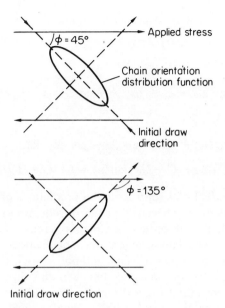

Figure 11.35. Line drawing showing the relative orientations of the initial draw direction and the applied stress for $\phi = 45°$ and $\phi = 135°$.

11.4.6 The Bauschinger Effect

We have seen that the yield stress values in the shear tests for $\phi = 45°$ and $\phi = 135°$ differ markedly. This was attributed to the difference between the situations where the compressive and tensile components of the applied stress are parallel to the initial draw direction respectively. A subsidiary experiment was performed in which specimens of oriented PET rods cut with their axes parallel to the draw direction were subjected to direct tension and compression[39]. It was found that the compressive yield stress parallel to the draw direction was appreciably less than the tensile yield stress in this direction. This gives direct evidence for a Bauschinger effect in oriented PET.

A simple method of introducing the Bauschinger effect into the yield criterion[39,40] is to include a term σ_i which represents the difference between the tensile and compressive yield points parallel to the initial drawing direction, i.e. the x direction in equation (11.7). (A more complete treatment would also add similar terms in the y and z directions.)

The Hill anisotropic yield criterion is modified to

$$F(\sigma_{yy} - \sigma_{zz})^2 + G(\sigma_{zz} - \sigma_{xx} + \sigma_i)^2 + H(\sigma_{xx} - \sigma_i - \sigma_{yy})^2$$
$$+ 2L\sigma_{yz}^2 + 2M\sigma_{zx}^2 + 2N\sigma_{xy}^2 = 1. \tag{11.14}$$

The shear tests are then described by the equation

$$(G+H)\sigma_{xx}^2 + (H+F)\sigma_{yy}^2 - 2H\sigma_{xx}\sigma_{yy} + 2N\sigma_{xy}^2 = 1, \qquad (11.15)$$

where

$$\sigma_{xx} = -\sigma \sin 2\phi - \sigma_{\mathrm{i}}, \qquad \sigma_{yy} = \sigma \sin 2\phi, \qquad \sigma_{xy} = \sigma \cos 2\phi.$$

On substitution we have

$$\sigma^2\{(G+F+4H)\sin^2 2\phi + 2N \cos^2 2\phi\}$$
$$+ 2\sigma\sigma_{\mathrm{i}}(G+2H) \sin 2\phi = 1 + (G+H)\sigma_{\mathrm{i}}^2. \qquad (11.16)$$

The full line in Figure 11.34 represents the predictions of this modified Hill theory. It is immediately apparent that the introduction of the Bauschinger term gives a much better fit to the experimental data.

Somewhat similar shear measurements have been undertaken by Robertson and Joynson[41] on oriented polyethylene and polypropylene. In both cases the plots of shear stress versus angle ϕ between the direction of the shear displacement and the initial draw direction, showed two unequal maxima. This was attributed to the ease of chain kinking (which, it is suggested, predominates when $\phi = 45°$) compared with glide between fibrils (which, it is suggested, predominates when $\phi = 135°$). This explanation is similar to the introduction of the Bauschinger effect, which implies that it is more difficult to extend an already oriented structure than to contract it.

Finally it is necessary to note that the introduction of the Bauschinger term into the yield criterion will lead to a further modification of the Lévy–Mises equations for predicting the band angle in the tensile test. To do this on the simple scheme proposed above, we merely replace $\sigma_{xx} = \sigma \cos^2 \theta$ by $\sigma_{xx} = \sigma \cos^2 \theta - \sigma_{\mathrm{i}}$. The quadratic in tan β, then predicts for each value of β two possible band directions depending on θ, F, G, H, N and σ_{i}.

Figure 11.31 also shows the fit obtained to the 'slippy' band angle in PET[39] using the appropriate constants which are consistent with the tensile and shear yield stress data. Good agreement is obtained, showing the validity of the modified von Mises equation. There is of course some latitude in the method of obtaining a fit when there are a large number of disposable constants. In particular, we have seen that the Bauschinger term σ_{i}, which is essential to describe the shear yield stress data, can be omitted and a fit to the tensile yield stress data and the band angle still be obtained, provided that appropriate changes are made in the other parameters F, G, H and N.

11.5 THE TEMPERATURE AND STRAIN-RATE DEPENDENCE OF YIELD AND DRAWING PROCESSES

It is very apparent that polymers are not ideal plastic materials. However, we have seen that provided temperature and strain rate are maintained constant, and conditions chosen so that adiabatic heating does not occur, the yield behaviour can be profitably discussed in terms of theories of ideal plasticity.

The regions where a yield drop may be observed are bounded by the brittle–ductile transition at low temperatures (Section 12.1) and, in amorphous polymers, by the glass transition at high temperatures. It is natural to ask how such situations fit into considerations of the temperature and strain rate sensitivity of the yield process. Increasing strain rate increases the yield stress without much affecting the brittle stress and hence increases the temperature of the brittle–ductile transition which determines the low boundary of yield behaviour (Section 12.1). The upper boundary has been examined by Andrews and his collaborators[42] in research which will now be discussed.

The variation of the yield stress σ_Y and the drawing stress σ_D with temperature and strain rate was studied for a number of amorphous polymers. Andrews prefers the nomenclature upper and lower yield stress for σ_Y and σ_D, respectively.

Figure 11.36 shows the variation in yield stress with temperature and strain rate for polymethyl methacrylate. It is notable that approximately straight

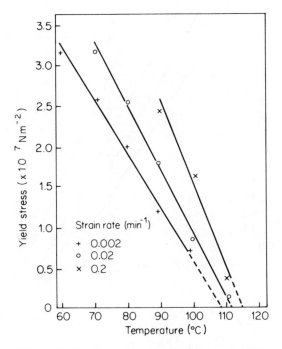

Figure 11.36. Variation in the yield stress with temperature at various strain rates for polymethyl methacrylate. [Redrawn with permission from Langford, Whitney and Andrews. *Mater. Res. Lab. Res. Rept. No. R63-49*, MIT School of Engineering, Cambridge, Mass., 1963.]

lines were obtained with higher dependence on temperature at higher strain rates. In addition the lines appear to converge and would extrapolate to zero yield stress at about 110 °C, which is close to the glass transition temperature for this polymer.

Figure 11.37 shows similar results for the drawing stress. This again shows a series of straight lines converging at a temperature close to the transition temperature, above which the polymer stretches homogeneously.

The strain rate and temperature dependence of the yield process has been the subject of much recent research. In many cases this research has been essentially of a phenomenological nature starting from the basis of yield as an activated rate process. This approach will be discussed in detail in the following section.

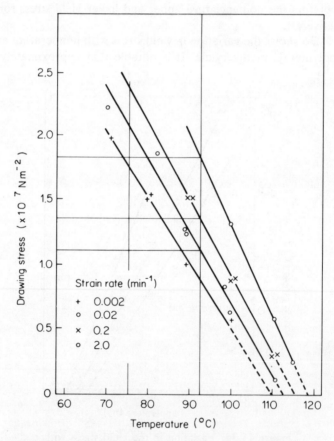

Figure 11.37. Variation in the drawing stress with temperature at various strain rates for polymethyl methacrylate. [Redrawn with permission from Langford, Whitney and Andrews, *Mater. Res. Lab. Res. Rept. No. R63-49*, MIT School of Engineering, Cambridge, Mass., 1963].

11.6 THE MOLECULAR INTERPRETATIONS OF YIELD AND COLD-DRAWING

11.6.1 Yield as an Activated Rate Process: The Eyring Equation

Many workers[17,43-50] have considered that the applied stress induces molecular flow much along the lines of the Eyring viscosity theory where internal viscosity decreases with increasing stress. On this view the yield stress denotes the point at which the internal viscosity falls to a value such that the applied strain rate is identical to the plastic strain rate $\dot{e}$ predicted by the Erying equation.

Furthermore the measurement of a yield stress in a constant strain rate test is analogous to the measurement of the creep rate at constant applied stress. We can therefore recall equation (9.22),

$$\dot{e} = \dot{e}_0 \exp - \frac{\Delta H}{RT} \sinh \frac{v\sigma}{RT}, \tag{11.17}$$

where ΔH is the activation energy, σ is the applied tensile stress and v the activation volume, which is considered to represent the volume of the polymer segment which has to move as a whole in order for plastic deformation to occur.

For high values of stress $\sinh x = \frac{1}{2}\exp x$ and

$$\dot{e} = \frac{\dot{e}_0}{2} \exp\left[-\left(\frac{\Delta H - v\sigma}{RT}\right)\right]. \tag{11.18}$$

This gives the shear yield stress τ in terms of strain rate as

$$\frac{\sigma}{T} = \frac{\Delta H}{vT} + \frac{R}{v} \ln \frac{2\dot{e}}{\dot{e}_0},$$

i.e.

$$\frac{\sigma}{T} = \frac{R}{v}\left\{\frac{\Delta H}{RT} + \ln \frac{2\dot{e}}{\dot{e}_0}\right\}. \tag{11.19}$$

This suggests that plots of yield stress/T against log(strain rate) for a series of temperatures should give a series of parallel straight lines. Figure 11.38 shows results obtained by Bauwens and coworkers[46] for the tensile yield stress of polycarbonate together with calculated lines based on equation (11.19) with constant values of ΔH and v.

In an earlier paper, Lazurkin[17] rejected a previous proposal by Hookway[15] and by Horsley and Nancarrow[51] that the molecular flow occurs because the applied stress reduced the melting point of the crystals. He remarked that similar behaviour is observed for both crystalline and non-crystalline polymers, the dependence of the yield stress on strain rate following the logarithmic form in both cases.

Haward and Thackray[49] have compared the Eyring activation volumes obtained from yield stress data with the volume of the 'statistical random

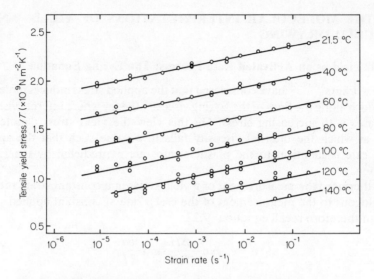

Figure 11.38. Measured ratio of yield stress to temperature as a function of the logarithm of strain rate for polycarbonate. The set of parallel straight lines is calculated from equation (11.19). [Redrawn with permission from Bauwens-Crowet, Bauwens and Homès, *J. Polymer Sci.* A2, **7**, 735 (1969).]

link'. The latter was obtained from solution studies, by assuming that the real chain can be represented by an equivalent chain with freely jointed links of a particular length. Table 11.1 is based on data collated by Haward and Thackray and shows that the activation volumes are very large in molecular terms and range from about two to 10 times that of the statistical random link. The result suggests that yield involves the cooperative movement of a

Table 11.1. A comparison of the statistical segment volume for a polymer measured in solution with the flow volumes derived from the Eyring Theory (after Haward and Thackray[49])

Polymer	Volume of statistical link in solution (nm³)	Eyring flow volume v (nm³)
Polyvinylchloride	0.38	8.6
Polycarbonate	0.48	6.4
Polymethyl methacrylate	0.91	4.6
Polystyrene	1.22	9.6
Cellulose acetate	2.06	8.8
Cellulose trinitrate	2.62	6.1
Cellulose acetate	2.05	17.4

larger number of chain segments than would be required for a conformational change in dilute solution.

Pressure Dependence

We have seen that the effect of pressure on the shear yield stress of a polymer can be very well represented by the equation (11.11):

$$\tau = \tau_0 + \alpha p.$$

This suggests that the Eyring equation may be very simply modified[52] to include the effect of the hydrostatic component of stress p to give

$$\dot{e} = \frac{\dot{e}_0}{2} \exp\left[-\left(\frac{\Delta H - \tau V + p\Omega}{RT}\right)\right], \tag{11.20}$$

where V and Ω are the shear and pressure activation volumes respectively. This modification of the Eyring equation can be considered to arise from a linear increase in the activation energy with increasing pressure, and this is the simplest approach. Alternatively, Bauwens[53] has argued that for a general state of stress we must consider the energy W_α required by the deviatoric stress τ_0 to produce an infinitesimal strain γ_0 plus the contribution of the hydrostatic stress p to the momentary formation of a hole of size Δv^*. Then

$$W_\alpha = \sqrt{\tfrac{3}{2}}\gamma_0\tau_0 + \Delta v^* p \tag{11.21}$$

and

$$\dot{e} = \dot{e}_0 \exp\left(-\frac{\Delta H}{RT}\right) \sinh \frac{W_\alpha}{2RT}. \tag{11.22}$$

For high stress, where $\sinh x = \tfrac{1}{2}\exp x$, equations (11.20) and (11.22) are of identical form.

Equation (11.20) may be conveniently expressed in terms of the octahedral yield stress

$$\tau_{\text{oct}} = \tfrac{1}{3}\{(\sigma_1 - \sigma_2)^2 + (\sigma_2 - \sigma_3)^2 + (\sigma_3 - \sigma_1)^2\}^{1/2}$$

and the octahedral strain rate

$$\dot{\gamma}_{\text{oct}} = \tfrac{2}{3}\{(\dot{e}_1 - \dot{e}_2)^2 + (\dot{e}_2 - \dot{e}_3)^2 + (\dot{e}_3 - \dot{e}_1)^2\}^{1/2}$$

to give a generalized representation suitable for all stress fields

$$\dot{\gamma}_{\text{oct}} = \frac{\dot{\gamma}_0}{2} \exp\left[-\frac{(\Delta H - \tau_{\text{oct}} V + p\Omega)}{RT}\right]. \tag{11.23}$$

For a constant strain rate test we therefore have

$$\Delta H - \tau_{\text{oct}} V + p\Omega = \text{constant},$$

from which an equation similar to equation (11.11) is obtained with

$$\tau_{\text{oct}} = (\tau_{\text{oct}})_0 + \alpha p,$$

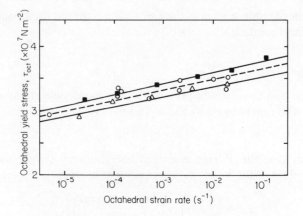

Figure 11.39. The strain rate dependence of the octahedral shear stress τ_{oct} at atmospheric pressure using data from torsion (○), tension (△), and compression (■). [Redrawn with permission from Duckett, Goswami, Smith, Ward and Zihlif, *Brit. Polymer J.* **10**, 11 (1978).]

where $\alpha = \Omega/V$. Figure 11.39 shows results for polycarbonate at atmospheric pressure[54] using data from torsion, tension and compression. It can be seen that on average the values of τ_{oct} lie in the order compression > torsion > tension. The differences are therefore consistent with the observed linear dependence of τ_{oct} on pressure shown by direct measurement of the yield stress in torsion over a range of hydrostatic pressures (Section 11.4.1 above), and there is good numerical agreement between the two sets of measurements.

The Two-stage Eyring Process Representation

Extensive studies of the yield behaviour of polymethyl methacrylate and polycarbonate over very wide ranges of strain rate and temperature by Roetling and Bauwens have shown that the yield stresses increase more rapidly with increasing strain rate and decreasing temperature at low temperatures and high strain rates than at high temperature and low strain rates. Following Ree and Eyring it has therefore been proposed that equation (11.19) should be extended by assuming that there is more than one activated rate process with all species of flow units moving at the same rate, the stresses being additive. For polymethyl methacrylate, polyvinyl chloride and polycarbonate it has been shown that the yield behaviour can be represented very satisfactorily by the addition of two activated processes. The equivalent equation to (11.19) above is then

$$\frac{\sigma}{T} = \frac{R}{v_1}\left\{\frac{\Delta H_1}{RT} + \ln 2\,\frac{\dot{e}}{\dot{e}_{01}}\right\} + \frac{R}{v_2}\sinh^{-1}\left\{\frac{\dot{e}}{\dot{e}_{02}}\exp\frac{\Delta H_2}{RT}\right\}, \qquad (11.24)$$

where the two activated processes are denoted by the subscript symbols 1 and 2 respectively. At high temperatures and low strain rates process 1 predominates and this has a comparatively low strain rate dependence (v_1 is large). We can therefore use the approximation $\sinh x = \frac{1}{2} \exp x$. Process 2 also becomes important at low temperatures and high strain rates and shows a much higher strain rate dependence (v_2 is small compared to v_1). The sinh form is retained to cover the intermediate range where process 2 is giving a smaller contribution to the magnitude of the total yield stress. Figure 11.40

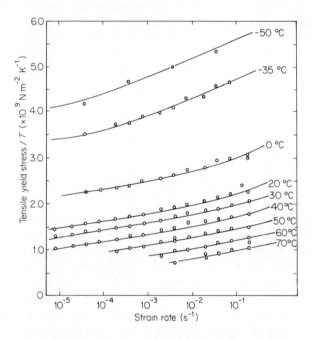

Figure 11.40. Measured ratio of yield stress to temperature as a function of logarithm of strain rate for polyvinyl chloride. The set of parallel curves is calculated from equation (11.24). [Redrawn with permission from Bauwens-Crowet, Bauwens and Homès, *J. Polymer Sci.* A2, **7**, 735 (1969).]

shows the fit obtained using equation (11.24) to experimental data for polyvinyl chloride[46]. Similar results were also obtained for polycarbonate, although in a later paper on polymethyl methacrylate[55] it was shown that the Ree–Eyring equation only fitted the data well in the region where the approximation $\sinh x = \frac{1}{2} \exp x$ is valid. It was proposed that a modification of the theory taking into account a distribution of relaxation times not only gave a much better fit to the theory but established a quantitative link between process 2 and the dynamic mechanical β-relaxation.

The Relationship of Yield to Creep

As discussed in Section 9.6, Eyring and collaborators had already considered the application of activated rate theory to the creep of polymers. For polymethyl methacrylate Sherby and Dorn[56] showed that the creep rate $\dot{e}$ could be fitted to an equation of the form

$$\dot{e} = A(e)\exp\left[-\left(\frac{\Delta H - B\sigma}{RT}\right)\right],\qquad(11.25)$$

where B is a constant (equivalent to the activation volume v of equation (11.18)) and $A(e)$ is a function of creep strain.

Mindel and Brown[57], in a later study, proposed that for the initial part of the creep curve it could be considered that the logarithmic creep rate diminishes linearly with strain. Then

$$\dot{e} = \dot{e}_0\exp\left[-\left(\frac{\Delta H - B\sigma}{RT}\right)\right]\exp(-ce_R),\qquad(11.26)$$

where we denote the strain in this initial region as e_R, recoverable strain and c is a constant. We can also write

$$\dot{e} = \dot{e}_0\exp\left(-\frac{\Delta H}{RT}\right)\exp\left[\frac{(\sigma - \sigma_{int})v}{RT}\right],\qquad(11.27)$$

where $\sigma_{int}v/RT = ce_R$ defines a rubber-like internal stress σ_{int} which is proportional to absolute temperature T. (For a further discussion see Section 9.6.)

In an earlier development Haward and Thackray[49] had proposed a very similar representation to describe the yield behaviour of polymers. Their model is shown schematically in Figure 11.41. The initial part of the stress–strain curve is modelled by the Hookean spring E and the yield point and subsequent strain hardening by the Eyring dashpot and the Langevin spring. Haward and Thackray relate the total strain e and the plastic strain e_A from the activated dashpot to the nominal stress σ_n (load applied divided by initial cross-sectional area). We have

$$e = \frac{\sigma_n(1+e)}{E} + e_A\qquad(11.28)$$

and

$$\frac{d[\ln(1+e_A)]}{dt} = \dot{e}_A\exp\left(-\frac{\Delta H}{RT}\right)\sinh\frac{v(\sigma_n - \sigma_R)}{RT},\qquad(11.29)$$

where σ_R is the internal rubber-like stress which it is proposed can be determined from rubber elasticity theory, so that

$$\sigma_R = \tfrac{1}{3}NkTn^{1/2}\left[\mathscr{L}^{-1}\left(\frac{1+e_A}{n^{1/2}}\right) - (1+e_A)^{-3/2}\mathscr{L}^{-1}\frac{1}{(1+e_A)^{1/2}n^{1/2}}\right],$$
$$(11.30)$$

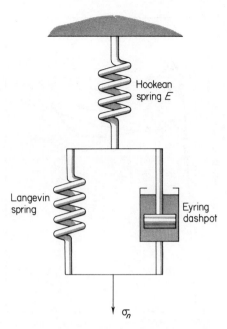

Figure 11.41. Schematic diagram of
the Haward–Thackray model.

where $\mathscr{L}^{-1}$ is the inverse Langevin function, N is the number of chains between cross-link points per unit volume and n is the average number of random links per chain.

Equation (11.29) was then integrated numerically, using equations (11.28) and (11.30), to give results like those shown in Figure 11.42. It can be seen that the Haward and Thackray model is able to reproduce the main features of the stress–strain curve and provide a semi–quantitative fit to the experimental data. However, it may be recalled that the size of the activation volumes are very large compared with the size of an individual molecular segment.

In a recent publication, Fotheringham and Cherry[58] have adopted a similar representation to Haward and Thackray and used the stress-transient dip test to determine the internal stress σ_R and hence the effective stress $\sigma_n - \sigma_R$ acting on the Eyring dashpot. Fotheringham and Cherry propose a model based on cooperative Eyring processes with the probability of a successful cooperative event involving the simultaneous occurrence of n transitions. Then

$$\dot{e} = \dot{e}_0 \exp\left[-\left(\frac{n\,\Delta H}{RT}\right)\right] \sinh^n\left(\frac{v\tau}{2RT}\right). \tag{11.31}$$

Results for linear polyethylene were fitted to give a value of about three for n and an activation volume of $0.5\,\text{nm}^3$ which is in the same range as the

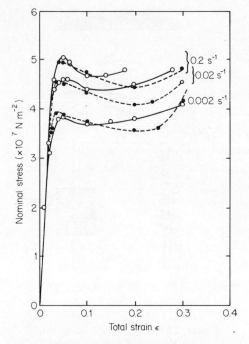

Figure 11.42. Stress–strain curves for cellulose nitrate at 23 °C. Experimental curves (○) and calculated curves, Langevin model (●). $n^{1/2} = 0.30$, $N = 1.57 \times 10^{26}$ chain m^{-3}. [Redrawn with permission from Haward and Thackray, *Proc. Roy. Soc. A*, **302**, 453 (1968).]

volume swept out by an elementary displacement of a defect moving through the crystal lattice.

11.6.2 Molecular and Structural Theories of Yield

The Robertson Theory

Robertson[59] has developed a slightly more elaborate version of the Eyring viscosity theory. For simplicity is is considered that there are only two rotational conformations, the *trans* low energy state and the *cis* high energy state, which Robertson terms the 'flexed state'. Applying a shear stress τ causes the energy difference between the two stable conformational states of each bond to change from ΔU to $(\Delta U - \tau v \cos \theta)$. $\tau v \cos \theta$ represents the work done by the shear stress in the transition between the two states and θ is the angle defining the orientation of a particular element of the structure with respect to the shear stress.

Prior to application of stress the fraction of elements in the high energy state is

$$\chi_i = \frac{\exp\{-\Delta U/k\theta_g\}}{1 + \exp\{-\Delta U/k\theta_g\}}, \tag{11.32}$$

where $\theta_g = T_g$ if the test temperature $T < T_g$ and

$$\theta_g = T \quad \text{if} \quad T > T_g,$$

i.e. below T_g the configurational state 'freezes' at that which exists at T_g. For application of a shear stress τ at a temperature T, the fraction of elements in the upper state with orientation θ is given by

$$\chi_f(\theta) = \frac{\exp\{-\Delta U - \tau v \cos \theta)/kT\}}{1 + \exp\{-(\Delta U - \tau v \cos \theta)/kT\}}. \tag{11.33}$$

Clearly the fraction of flexed elements increases for orientations such that

$$\frac{\Delta U - \tau v \cos \theta}{kT} \leq \frac{\Delta U}{k\theta_g}.$$

For one part of the distribution of structural elements, applying the stress tends to make for an equilibrium situation where there are more flexed bonds and this can be regarded as corresponding to a rise in temperature. For the other part of the distribution, the effect of stress can be regarded as tending to lower the temperature. Robertson now argues that the *rate* at which conformational changes occur is very dependent on temperature (cf. WLF equation). Hence the rate of approach to equilibrium is much faster for these elements which flex under the applied stress, so that changes in the others can be ignored in calculating the maximum flexed-bond fraction which can occur under a given applied stress. This maximum corresponds to a rise in temperature to a temperature θ_1. The strain rate $\dot{e}$ at θ_1 is calculated from the WLF equation

$$\dot{e} = \frac{\tau}{\eta_g} \exp\left\{-2.303\left[\left(\frac{C_1^g C_2^g}{\theta_1 - T_g + C_2^g}\right)\frac{\theta_1}{T} - C_1^g\right]\right\}, \tag{11.34}$$

where C_1^g, C_2^g are the universal WLF parameters (see Section 7.4.1) and η_g is the 'universal' viscosity of a glass at T_g.

Duckett, Rabinowitz and Ward[60] have modified the Robertson model to include the effect of the hydrostatic component of stress p. It was proposed that p also does work during the activation event and that the energy difference between the two states should therefore be

$$\Delta U - \tau v \cos \theta + p\Omega,$$

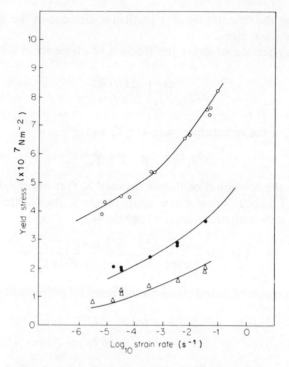

Figure 11.43. Yield stress of polymethyl methacrylate as a function of strain rate. ○, Compression at 23 °C; △, tension at 90 °C; ●, tension at 60 °C. Curves represent the best theoretical fit (see text).

where Ω is a constant with the dimensions of volume. Figure 11.43 shows that in this modified form the Robertson model can bring consistency to yield data in tension and compression for polymethyl methacrylate, together with the measured effect of hydrostatic pressure.

The Argon Theory

Argon[61] has proposed a theory of yielding for glassy polymers based on the concept that deformation at a molecular level consists in the formation of a pair of molecular kinks. The unit process of deformation is shown in Figure 11.44. The resistance to double kink formation is considered to arise from the elastic interactions between a chain molecule and its neighbour, i.e. from intermolecular forces in contrast to the Robertson theory where intramolecular forces are the primary consideration. The intermolecular energy change associated with a double kink is then calculated by modelling these as two wedge disclination loops as proposed by Li and Gilman[62] (Figure 11.45).

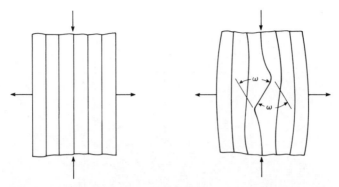

Figure 11.44. Schematic representation of the unit process of deformation consisting of unbending and bending a molecular segment. [Redrawn with permission from Argon, *Phil. Mag.*, **28**, 839 (1973).]

The activation energy (strictly enthalpy) for the formation of a pair of molecular kinks under an applied shear stress τ is given by

$$\Delta H^* = \frac{3\pi G\omega^2 a^3}{16(1-\nu)}\left[1 - 6.75(1-\nu)^{5/6}\left(\frac{\tau}{G}\right)^{5/6}\right], \qquad (11.35)$$

where G, ν are the shear modulus and Poisson's ratio respectively, a is the molecular radius and ω the angle of rotation of the molecular segment (Figure

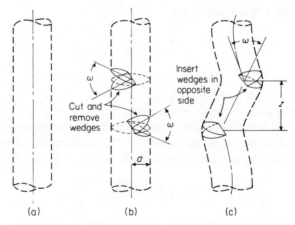

Figure 11.45. Modelling of a molecular kink pair by pair of wedge disclination loops: (a) outline of polymer molecule in an elastic surrounding made up of other neighbouring molecules; (b) make the circular cuts of radius a a distance 2 apart and cut and remove wedges of angle ω; (c) insert cut wedges into opposite side and join all parts together. [Redrawn with permission from Argon, *Phil. Mag.* **28**, 839 (1973).]

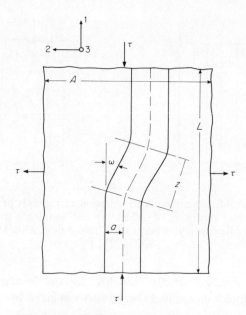

Figure 11.46. Plastic strain increment by formation of a pair of kinks in a polymer molecule. [Redrawn with permission from Argon, *Phil. Mag.*, **28**, 839 (1973).]

11.46). The shear strain rate is then

$$\dot{\gamma} = \gamma_0 \Omega C \nu_a \exp\left\{-\frac{\Delta H^*}{kT}\right\}, \tag{11.36}$$

where γ_0 is the shear strain in the local volume $\Omega = \pi a^2 z_{eq}$ (z_{eq} is the equilibrium molecular segment length), C is the total volume density of potentially rotatable segments in the polymer, and ν_a is a frequency factor of the order of (but somewhat smaller than) the atomic frequency.

It can be noted that when equation (11.35) is substituted into (11.36) the resultant equation is quite similar in form to the Eyring equation

$$\dot{\gamma} = \dot{\gamma}_0 \exp\left[-\left(\frac{\Delta H - \tau v}{RT}\right)\right].$$

The shear yield stress τ is given from equation (11.36) as

$$\tau = \frac{0.102G}{(1-\nu)}\left[1 - \frac{16(1-\nu)}{3\pi G\omega^2 a^3}kT\ln\frac{\dot{\gamma}_0}{\dot{\gamma}}\right]^{6/5}, \tag{11.37}$$

where $\dot{\gamma}_0 = \gamma_0 \nu_a \Omega C$.

Figure 11.47 shows Argon's fit to very extensive data for polyethylene terephthalate. The fit is good, but it may be noted that if we replace the factor

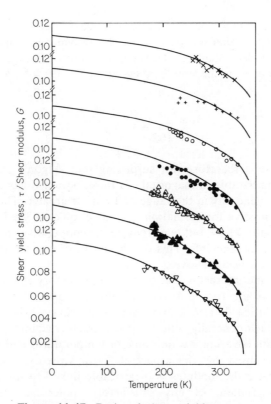

Figure 11.47. Ratio of shear yield stress to shear modulus as a function of temperature at different strain rates, for amorphous polyethylene terephthalate. Points from unpublished data of Foot and Ward, curves from Argon theory. Strain rates: $\times$, $1.02 \times 10^2 \, \text{s}^{-1}$; $+$, $21.4 \, \text{s}^{-1}$; ∇, $1.96 \, \text{s}^{-1}$; $\bigcirc$, $9 \times 10^{-2} \, \text{s}^{-1}$; $\square$, $9 \times 10^{-3} \, \text{s}^{-1}$; $\triangle$, $9 \times 10^{-4} \, \text{s}^{-1}$; $\diamond$, $9 \times 10^{-5} \, \text{s}^{-1}$.

6/5 in equation (11.35) by unity, which makes a comparatively small difference numerically, then this equation reduces to

$$\tau = \frac{0.102G}{1-\nu} - \frac{16 \times 0.102kT}{3\pi\omega^2 a^3} \ln \frac{\dot{\gamma}_0}{\gamma}, \tag{11.38}$$

which is of similar form to the Eyring equation where

$$\tau = \frac{\Delta H}{v} - \frac{kT}{v} \ln \frac{\dot{\gamma}_0}{\dot{\gamma}}. \tag{11.39}$$

The two approaches of Argon and Eyring therefore cannot be clearly distinguished at a curve-fitting level. On the Argon theory the shear yield stress at 0 K is simply a function of the shear modulus and Poisson's ratio.

This is consistent with the observation, first made by Vincent,[63] and supported by later workers[18,64] that the yield stress of a polymer is proportional to the modulus. Argon also calculates the shear activation volume from his theory as $5.3\omega^2 a^3$. Comparing the simplified equation (11.39) with equation (11.38) we have

$$v = \frac{3\pi\omega^2 a^3}{16 \times 0.102} = 5.77\omega^2 a^3 \approx 10a^3$$

(since $\omega \sim 1$). It is therefore not surprising that Argon's shear activation volumes are comparable in magnitude to those obtained from fits to Eyring theory, and are generally in the range of 1 nm^3 or greater. The yield process cannot merely involve the formation of double kinks or single molecules as envisaged in Figure 11.44, but must require the cooperative change of several adjacent molecular segments. This conclusion is, of course, identical to that reached in Section 11.6.1 above.

A point of some interest is that on the Argon theory changes in modulus are automatically incorporated. At a phenomenological level, this explains the success of fitting data to a single activated process, whereas the Eyring equation approach generally requires two processes acting in parallel when dealing with data covering a wide range of temperatures and strain rates. It could also be claimed that this aspect of Argon's theory, and other similar theories[65,66] where the modulus is inextricably linked to the yield stress, is an essential ingredient of any satisfactory molecular theory of yield behaviour.

A final consideration is that the Argon theory essentially regards yield as nucleation controlled, analogous to the stress-activated movement of dislocations in a crystal produced by the applied stress, aided by thermal fluctuations. The application of the Eyring theory, on the other hand, implies that yield is not concerned with the initiation of the deformation process, but only that the application of stress changes the rate of deformation until it equals the imposed rate of change of strain. The Eyring approach is consistent with the view that the deformation mechanisms are essentially present at zero stress, and are identical to those observed in linear viscoelastic measurements (see site model analyses in Section 7.3.1). Here a very low stress is applied merely to enable detection of the thermally activated process, without modification of the polymer structure.

At present these two approaches appear to be alternative ways of dealing with the yield behaviour of polymers. It could be argued that the Eyring equation is likely to be appropriate at high temperatures, whereas the Argon theory and similar theories are most relevant to the behaviour at very low temperatures. In this respect it is interesting to recall that as we approach absolute zero temperature the ratio of yield stress to modulus approaches a limiting value which is consistent with classical theoretical shear strength arguments.

Free-volume Theories

It has been proposed that yielding is due to the increase in free volume under stress, the free volume rising until it reaches its value at the glass transition temperature. Attempts to verify this proposition are made difficult by the large elastic volume changes arising because of the high stresses at yield. It is, however, interesting to note that Whitney and Andrews[67] have reported that a relative volume dilation does occur in compression yielding.

11.6.3 Cold-drawing

General Considerations

We have seen that strain hardening is a necessary prerequisite for cold-drawing (Section 11.13). There are two possible sources of strain hardening in polymers.

(1) Strain-induced crystallization may occur at the high degree of extension occurring in cold-drawing. This may be similar to the crystallization observed in rubbers at high degrees of stretching. It may correspond at a morphological level to extended chain crystallization and to the formation of shish-kebab structures[68]. The occurrence of crystallization on cold-drawing is probably dependent on raising the local temperature in the specimen sufficiently to permit the necessary molecular mobility for the required structural reorganization.

(2) There is a change in the directional properties as the molecular orientation changes during drawing such that the stiffness increases along the draw direction. This is a general phenomenon, true for both crystalline and amorphous polymers. (Note that the theories of mechanical anisotropy developed in Sections 10.6 and 10.7 apply to the final drawn material and do not relate directly to the strain-hardening effect.)

Cold-drawing occurs in both crystalline and amorphous polymers. Examples of the former are nylon and polyethylene[69]; examples of the latter are polymethyl methacrylate and polyethylene methylterephthalate[70-72]. It is also clear that although the general effect of drawing is to produce some degree of molecular alignment parallel to the draw direction, the morphological changes are complex. Whereas some degree of crystallization occurs when amorphous polyethylene terephthalate filaments are cold-drawn, an exactly contrary effect occurs in sodium thymonucleate where the crystalline filaments cold-draw to an amorphous drawn state[73].

Amorphous Polymers and the Natural Draw Ratio

Cold-drawing occurs at temperatures below the glass transition, sometimes as much as ~150 °C below. It has been suggested by Andrews and others that the yield process and subsequent cold-drawing do not involve long-range molecular flow but are associated with molecular rearrangements between points of entanglement and/or cross-linkage. This view is consistent with the observation of yield, necking and cold-drawing, in highly cross-linked rubbers at temperatures below their glass transition. It is evident that cross-linking does not prevent the required molecular rearrangements.

The natural draw ratio for amorphous polymers is very sensitive to the degree of pre-orientation, i.e. the molecular orientation in the polymer before cold-drawing. This was reported for polyethylene terephthalate by Marshall and Thompson[13] and for PMMA and polystyrene by Whitney and Andrews[67].

It has been proposed[74] that the sensitivity of natural draw ratio to pre-orientation arises as follows. The extension of an amorphous polymer to its natural draw ratio is regarded as equivalent to the extension of a network to a limiting extensibility. This limiting extensibility is then a function of the original geometry of the network and the nature of the links of which it is comprised.

During fibre-spinning, the network forms immediately below the point of extrusion from the small holes, and the fibre is subsequently stretched in the rubber-like state before cooling further and being collected as a frozen stretched rubber. Quantitative stress-optical measurements have confirmed this part of the hypothesis[75]. Cold-drawing then extends the network to its limiting extensibility. The ratio of the extended to unextended lengths of the network is a constant independent of the division of the extension between the spinning, hot-drawing and cold-drawing processes, providing that the junction points holding the network together are not ruptured nor the links in the chain broken.

The dimensions of the unstrained network can be measured by shrinking the pre-oriented fibres back to the state of zero strain, i.e. isotropy[75]. These results can then be combined with measurements of the natural draw ratio to give the maximum extensibility for the network.

Consider the cold-drawing of a sample of length l_1 (Figure 11.48). If the fibre were allowed to shrink back to its isotropic state, length l_0, the shrinkage s would be defined by

$$s = \frac{l_1 - l_0}{l_1} \tag{11.40}$$

Cold-drawing to a length l_2 gives a natural draw ratio

$$N = \frac{l_2}{l_1} \tag{11.41}$$

Combining Equations (11.40) and (11.41) we have

$$\frac{l_2}{l_0} = \frac{N}{1-s}. \tag{11.42}$$

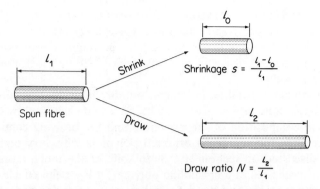

Figure 11.48. A representation of the shrinkage and drawing processes.

Table 11.2 shows collected results for a series of PET filaments. It can be seen that N varied from 4.25 to 2.58 and s from 0.042 to 0.378, but the ratio l_2/l_0 calculated from equation (11.42) remained constant at a value of about 4.0.

It is, of course, possible that the natural draw ratio is determined directly by the strain-hardening requirements. This does not invalidate the hypothesis that cold-drawing involves the extension of a molecular network, but suggests that strain hardening increases very rapidly as the network reaches its limiting extensibility.

Crystalline Polymers

The plastic deformation of crystalline polymers, in particular polyethylene, has been studied intensively from the viewpoint of changes in morphology. Notable contributions to this area have been made by Keller and his coworkers and by Peterlin, Geil and others[76–78]. It is now evident that very drastic reorganization occurs at the morphological level, with the structure changing from a spherulitic to a fibrillar type as the degree of plastic deformation

Table 11.2. Value of $l_2/l_0 = N/1 - s$ for samples of differing amounts of pre-orientation (polymer: polyethylene terephthalate, see Reference 74).

Initial birefringence $\times 10^3$	Natural draw ratio, N	Shrinkage, s	$(1-s)$	$N/(1-s)$
0.65	4.25	0.042	0.958	4.44
1.6	3.70	0.094	0.906	4.08
2.85	3.32	0.160	0.840	3.96
4.2	3.05	0.202	0.798	3.83
7.2	2.72	0.320	0.680	4.01
9.2	2.58	0.378	0.622	4.14

increases. The molecular reorientation processes are very far from being affine or pseudo-affine (p. 287) and can also involve mechanical twinning in the crystallites. It is surprising that some of the continuum ideas for mechanical anisotropy are nevertheless still relevant (see p. 289) although they must be appropriately modified.

In a few highly crystalline polymers, notably high density polyethylene, extremely large draw ratios, ~30 or more, have been achieved by optimizing the chemical composition of the polymers and the drawing conditions[79,80]. These high draw ratios lead to oriented polymers with very high Young's moduli as discussed in Section 10.7.3. In spite of the much more complex deformation processes in a crystalline polymer, it has been concluded[81] that the molecular topology and the deformation of a molecular network are still the overriding considerations in determining the strain hardening behaviour and the ultimate draw ratio achievable. For high molecular weight, high density polyethylene, the key network junction points are physical entanglements, as in amorphous polymers. For low molecular weight, high density polyethylene both physical entanglements and crystallites where more than one molecular chain is incorporated, can provide the network junction points. In the case where these are associated with the crystallites they will be of a temporary nature. The very large draw ratios involve the breakdown of the crystalline structure and the unfolding of molecules, so that the simpler ideas of a molecular network suggested for amorphous polymers have to be extended and modified.

11.7 DEFORMATION BANDS IN ORIENTED POLYMERS

As mentioned above, the molecular reorientation which occurs in a deformation band is important because it can lead to a molecular understanding of the deformation process.

With this objective in mind, Brown and Ward[82] made an exact analysis of the geometry of the deformation bands by measuring the deformation of grids scratched lightly on the surface of oriented sheets of polyethylene terephthalate. The primary deformation was a simple shear parallel to the deformation band direction. Additionally, in some cases there was a thinning of the band accompanied by a tensile strain normal to the band direction (a pure shear deformation).

The molecular reorientation within the deformation band was determined from optical measurements of the rotation of the extinction direction between the undeformed matrix and the material in the band[83] and by measurements of refractive indices[84]. The extinction direction in the band could be predicted exactly in terms of the strain in the deformation band, if it was assumed that the oriented polymer deforms as an oriented continuum. The assumption was made that the refractive index ellipsoid always has principal axes parallel to those of the strain ellipsoid. Figure 11.49 illustrates the situation. The orientation of the strain ellipsoid in the deformation band was calculated from the

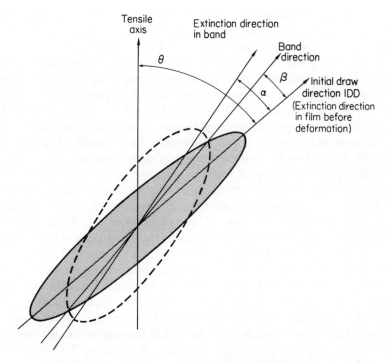

Figure 11.49. The strain ellipsoid in the undeformed and deformed material (full and dotted lines respectively) for a shear strain of 3 with no normal strain.

total strain, i.e. it is necessary to add in an appropriate manner the strain introduced by the initial drawing process and that determined from the analysis of the scratched grid. The agreement between measured and predicted extinction directions for a typical case is shown in Table 11.3.

It was later shown[84] that an extension of the aggregate theory (Section 10.6) can be used to predict the refractive indices in the deformation band

Table 11.3 Comparison of measured and predicted extinction directions for drawn polyethylene terephthalate sheet[83] (for explanation of angles see Figure 11.49).

θ (deg)	β (deg)	Shear strain in band	Extensional strain in band	α (deg) Predicted	α (deg) Measured
21	1.2	1.67	1.00	10.5	9.7
33	3.0	3.43	1.00	7.9	8.5
34	5.2	3.67	1.00	10.5	9.7
60	0	1.34	1.40	6.0	4.2
80	53	0.18	1.10	79.7	79.5

to a reasonable approximation. The total strain, as described above, is used to determine the principal axes of the strain ellipsoid. The pseudo-affine model (p. 287) can then be used to predict the refractive index ellipsoid.

This simple explanation of the molecular reorientation shows that polyethylene terephthalate undergoes deformation as a continuum. It has subsequently been shown that the same theoretical treatment is also applicable to oriented sheets of polypropylene and linear polyethylene[85,86]. In the case of both polyethylene and polyethylene terephthalate, independent measurements of the molecular reorientation in the deformation bands made using wide angle X-ray diffraction broadly confirm the optical measurements. The optical and X-ray diffraction results taken together suggest that the material within the deformation band, whether crystalline or not, becomes realigned about the new direction of maximum elongation as if controlled by the deformation of an effective molecular network. It was clear from the X-ray diffraction measurements that in polyethylene terephthalate a considerable loss in crystallinity can occur. In polyethylene the X-ray results suggest that a break-up of crystals occurs followed by a rapid recrystallization along the new direction of maximum elongation. These results are very much in line with the Peterlin theories of drawing where the isotropic lamellar texture is broken up and reformed as a fibrillar structure containing aligned crystal blocks. They are, however, difficult to reconcile with theoretical analyses where the molecular reorientation in high density polyethylene has been analysed entirely in terms of crystalline c-slip[34,36,37,87].

REFERENCES

1. A. Nadai, *Theory of Flow and Fracture of Solids*, McGraw-Hill, New York, 1950.
2. E. Orowan, *Rept. Prog. Phys.*, **12**, 185 (1949).
3. P. I. Vincent, *Polymer*, **1**, 7 (1960).
4. P. W. Bridgeman, *Studies in Large Plastic Flow and Fracture*, McGraw-Hill, New York, 1952.
5. Reference 1, p. 20.
6. H. Tresca, *C. R. Acad. Sci.* (*Paris*) **59**, 754 (1864) and **64**, 809 (1867).
7. R. von Mises, *Gottinger Nach. Math.-Phys. Kl.*, 582 (1913).
8. C. A. Coulomb, *Mem. Math. Phys.*, **7**, 343 (1773).
9. A. H. Cottrell, *Dislocations and Plastic Flow in Crystals*, Clarendon Press, Oxford, 1953.
10. R. Hill, *The Mathematical Theory of Plasticity*, Clarendon Press, Oxford, 1950, p. 50, 318.
11. M. Levy, *C. R. Acad. Sci* (*Paris*), **70**, 1323 (1870).
12. D. C. Drucker, in *Proceedings of the First US National Congress on Applied Mechanics*, ASME, New York, 1952, p. 487.
13. I. Marshall and A. B. Thompson, *Proc. Roy. Soc. A*, **221**, 541 (1954).
14. F. H. Müller, *Kolloidzeitschrift*, **114**, 59 (1949); **115**, 118 (1949) and **126**, 65 (1952).
15. D. C. Hookway, *J. Text. Inst.*, **49**, 292 (1958).
16. P. Brauer and F. H. Müller, *Kolloidzeitschrift*, **135**, 65 (1954).
17. Y. S. Lazurkin, *J. Polymer Sci.*, **30**, 595 (1958).

18. S. W. Allison and I. M. Ward, *Brit. J. Appl. Phys.*, **18**, 1151 (1967).
19. W. Whitney and R. D. Andrews, *J. Polymer Sci. C*, **16**, 2981 (1967).
20. N. Brown and I. M. Ward, *J. Polymer Sci.* **A2**, **6**, 607 (1968).
21. J. Miklowitz, *J. Colloid Sci.*, **2**, 193 (1947).
22. R. L. Thorkildsen, *Engineering Design for Plastics*, Reinhold, New York, 1964, p. 322.
23. P. B. Bowden and J. A. Jukes, *J. Mater. Sci.*, **3**, 183 (1968).
24. J. G. Williams and H. Ford, *J. Mech, Eng. Sci.*, **6**, 405 (1964).
25. S. B. Ainbinder, M. G. Laka and I. Y. Maiors, *Mekhanika Polimerov*, **1**, 65 (1965).
26. G. Biglione, E. Baer and S. V. Radcliffe, in *Fracture 1969* (P. L. Pratt, ed.) Chapman and Hall, London, 1969, p. 520.
27. A. W. Christiansen, E. Baer and S. V. Radcliffe, *Phil. Mag.*, **24**, 451 (1971).
28. K. Matsushigi, S. V. Radcliffe and E. Baer, *J. Mater, Sci.*, **10**, 833 (1975).
29. K. D. Pae, D. R. Mears and J. A. Sauer, *J. Polymer Sci. B*, **6**, 773 (1968).
30. D. R. Mears, K. D. Pae and J. A. Sauer, *J. Appl. Phys.* **40**, 4229 (1969).
31. S. Rabinowitz, I. M. Ward and J. S. C. Parry, *J. Mater. Sci.*, **5**, 29 (1970).
32. J. S. Harris, I. M. Ward and J. S. C. Parry, *J. Mater. Sci.*, **6**, 110 (1971).
33. B. J. Briscoe and D. Tabor, in *Polymer Surfaces* (D. T. Clark and W. J. Feast, eds), John Wiley, New York, 1978, Chap. 1.
34. D. A. Zaukelies, *J. Appl. Phys.*, **33**, 2797 (1961).
35. M. Kurakawa and T. Ban, *J. Appl. Polymer Sci.*, **8**, 971 (1964).
36. A. Keller and J. G. Rider, *J. Mater. Sci.*, **1**, 389 (1966).
37. F. C. Frank, A. Keller and A. O'Connor, *Phil. Mag.*, **3**, 64 (1958).
38. N. Brown and I. M. Ward, *Phil. Mag.*, **17**, 961 (1968).
39. N. Brown, R. A. Duckett and I. M. Ward, *Phil. Mag.*, **18**, 483 (1968).
40. C. Bridle, A. Buckley and J. Scanlan, *J. Mater. Sci.*, **3**, 622 (1968).
41. R. E. Robertson and C. W. Joynson, *J. Appl. Phys.*, **37**, 3969 (1966).
42. G. Langford, W. Whitney and R. D. Andrews, *Mater. Res. Lab. Res. Rept No. R63-49*, MIT School of Engineering, Cambridge, Mass., 1963.
43. Y. S. Lazurkin and R. A. Fogelson, *Zhur. Tech. Fiz.*, **21**, 267 (1951).
44. R. E. Robertson, *J. Appl. Polymer Sci.*, **7**, 443 (1963).
45. C. Crowet and G. A. Homès, *Appl. Mater. Res.*, **3**, 1 (1964).
46. C. Bauwens-Crowet, J. A. Bauwens and G. Homès, *J. Polymer Sci. A2*, **7**, 735 (1969).
47. J. C. Bauwens, C. Bauwens-Crowet and G. Homès, *J. Polymer Sci. A2*, **7**, 1745 (1969).
48. J. A. Roetling, *Polymer*, **6**, 311 (1965).
49. R. N. Haward and G. Thackray, *Proc. Roy. Soc. A*, **302**, 453 (1968).
50. D. L. Holt, *J. Appl. Polymer Sci.*, **12**, 1653 (1968).
51. R. A. Horsley and H. A. Nancarrow, *Brit. J. Appl. Phys.*, **2**, 345 (1951).
52. I. M. Ward, *J. Mater. Sci.*, **6**, 1397 (1971).
53. J. C. Bauwens, *J. Polymer Sci.*, *A2*, **5**, 1145 (1967).
54. R. A. Duckett, B. C. Goswami, L. S. A. Smith, I. M. Ward and A. M. Zihlif, *Brit. Polymer J.*, **10**, 11 (1978).
55. J. C. Bauwens, *J. Mater. Sci.*, **7**, 577 (1972).
56. O. D. Sherby and J. E. Dorn, *J. Mech. Phys. Solids*, **6**, 145 (1958).
57. M. J. Mindel and N. Brown, *J. Mater. Sci.*, **8**, 863 (1973).
58. D. G. Fotheringham and B. W. Cherry, *J. Mater. Sci.*, **13**, 951 (1978).
59. R. E. Robertson, *J. Chem. Phys.*, **44**, 3950 (1966).
60. R. A. Duckett, S. Rabinowitz and I. M. Ward, *J. Mater. Sci.*, **5**, 909 (1970).
61. A. S. Argon, *Phil. Mag.*, **28**, 839 (1973).
62. J. C. M. Li and J. J. Gilman, *J. Appl. Phys.*, **41**, 4248 (1970).
63. P. I. Vincent, in *Encyclopaedia of Polymer Science and Technology*, Wiley, New York, 1967.

398

64. N. Brown, *Mater. Sci. Eng.*, **8**, 69 (1971).
65. P. B. Bowden and S. Raha, *Phil. Mag.*, **29**, 129 (1974).
66. N. Brown, *Bull. Amer. Phys. Soc.*, **16**, 428 (1971).
67. W. Whitney and R. D. Andrews, *J. Polymer Sci. C*, **16**, 2981 (1967).
68. A. Keller, Rept. *Prog. Phys.*, **31**, 623 (1968).
69. C. W. Bunn and T. C. Alcock, *Trans. Faraday Soc.*, **41**, 317 (1945).
70. I. M. Ward, *Text. Res. J.*, **31**, 650 (1961).
71. E. A. W. Hoff, *J. Appl. Chem.* **2**, 441 (1952).
72. R. J. Curran and R. D. Andrews, *Mater. Res. Lab. Res. Rept. No. R63-55*, MIT School of Engineering, Cambridge, Mass., 1963.
73. M. F. H. Wilkins, R. G. Gosling and W. E. Seeds, *Nature*, **167**, 759 (1951).
74. S. W. Allison, P. R. Pinnock and I. M. Ward, *Polymer*, **7**, 66 (1966).
75. P. R. Pinnock and I. M. Ward, *Trans. Faraday Soc.*, **62**, 1308 (1966).
76. I. L. Hay and A. Keller, *Kolloidzeitschrift*, **204**, 43 (1965).
77. A. Peterlin, *J. Polymer Sci.*, **69**, 61 (1965).
78. P. H. Geil, *J. Polymer Sci. A*, 2, 3835 (1964).
79. G. Capaccio and I. M. Ward, *Nature Phys. Sci.*, **243**, 43 (1973).
80. G. Capaccio and I. M. Ward, *Polymer*, **15**, 233 (1974).
81. G. Capaccio, T. A. Crompton and I. M. Ward, *J. Polymer Sci.*, *Polymer Phys. Ed.*, **14**, 1641 (1976).
82. N. Brown and I. M. Ward, *Phil. Mag.*, **17**, 961 (1968).
83. N. Brown, R. A. Duckett and I. M. Ward, *J. Phys. D*, **1**, 1369 (1968).
84. I. D. Richardson, R. A. Duckett and I. M. Ward, *J. Phys. D*, **3**, 649 (1970).
85. R. A. Duckett, B. C. Goswami and I. M. Ward, *J. Polymer Sci.*, *Polymer Phys. Ed.*, **10**, 2167 (1972).
86. R. A. Duckett, B. C. Goswami, I. M. Ward, A. M. Zihlif, *J. Polymer Sci.*, *Polymer Phys. Ed.*, **15**, 333 (1977).
87. T. Seto and Y. Tajima, *Jap. J. Appl. Phys.*, **5**, 534 (1966).

12

Breaking Phenomena

12.1 DEFINITION OF TOUGH AND BRITTLE BEHAVIOUR IN POLYMERS

As we have seen, the mechanical properties of polymers are very greatly affected by temperature and strain rate. In general terms the load–elongation curve at a constant strain rate will change with increasing temperature as shown schematically (not necessarily to scale) in Figure 12.1. At low temperatures the load rises approximately linearly with increasing elongation up to the breaking point, when the polymer fractures in a brittle manner. At higher temperatures a yield point is observed, and the load falls before failure, sometimes with the appearance of a neck; this is ductile failure, but still at quite low strains (typically 10–20%). At still higher temperatures, providing

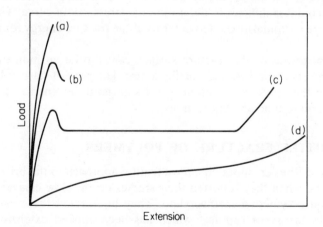

Figure 12.1. Load–extension curves for a typical polymer tested at four temperatures showing different regions of mechanical behaviour. (a) Brittle fracture, (b) ductile failure, (c) necking and cold-drawing and (d) homogeneous deformation (quasi-rubber-like behaviour).

† Part of this chapter originally appeared in 1968 as an internal report for Imperial Chemical Industries, Ltd, to whom I am grateful for permission to publish it here.

that certain conditions are fulfilled, strain hardening occurs, the neck stabilizes, and cold-drawing ensues. The extensions in this case are generally very large, i.e. up to 1000%. Finally, at even higher temperatures, homogeneous deformation is observed, with a very large extension at break. In an amorphous polymer this behaviour occurs above the glass transition temperature and the stress levels are very low.

The idea of a brittle–ductile transition is a familiar one in the discussion of the mechanical properties of metals. For polymers the situation is clearly more complicated in that there are in general four regions of behaviour and not two. However, it will still be of considerable value to discuss the factors which influence the brittle–ductile transition in polymers, and then to consider further factors which are involved in the observation of necking and cold-drawing.

The best definition of tough and brittle behaviour comes from the stress–strain curve. Brittle behaviour is designated when the specimen fails at its maximum load (as in Figure 12.1(a)). It is necessary to exclude rubbers (Figure 12.1(d)) by adding a corollary that the failure should occur at comparatively low strains (say $< 10\%$).

The distinction between brittle and ductile failure is also manifested in two other ways: (1) the energy disssipated in fracture, and (2) the nature of the fracture surface.

The energy dissipated is an important consideration for practical applications, and forms the basis of the Charpy and Izod impact tests (to be discussed further below). At the testing speeds under which the practical impact tests are conducted it is difficult to determine the stress–strain curve. Thus impact strengths are customarily quoted in terms of the fracture energy for a standard specimen.

The appearance of the fracture surface can also be an indication of the distinction between brittle and ductile failure. The present state of knowledge concerning the crack propagation is not sufficiently extensive to make this distinction more than an empirical one.

12.2 BRITTLE FRACTURE OF POLYMERS

Benbow and Roesler[1] initiated a most fruitful approach to the brittle fracture of polymers, when they reported their studies on the slow cleavage of polystyrene and polymethyl methacrylate. They interpreted these results using the Griffith theory of rupture[2] which has been applied extensively to the brittle fracture of glass and metals. The Griffith theory is the earliest statement of what has come to be known as linear elastic fracture mechanics. Although strictly it only describes the behaviour of sharp cracks in linear, perfectly elastic materials, it will be shown that it does provide an excellent framework for discussion of the brittle failure of polymers.

Griffith based his theory of rupture on two ideas. First, he considered that rupture produces a new surface area and postulated that for rupture to occur the increase in energy required to produce the new surface must be balanced by a decrease in elastically stored energy. Secondly, to explain the large discrepancy between the measured strength of materials and those based on theoretical considerations, Griffith proposed that the elastically stored energy is not distributed uniformly throughout the specimen but is concentrated in the neighbourhood of small cracks.

Fracture thus occurs due to the spreading of cracks which originate in pre-existing flaws.

In general the growth of a crack will be associated with an amount of work dW being done on the system by external forces and a change dU in the elastically stored energy U. The difference between these quantities, $dW - dU$, is the energy available for the formation of new surface. The condition for growth of a crack by an amount dc is then

$$\frac{dW}{dc} - \frac{dU}{dc} \geqslant \gamma \frac{dA}{dc}, \tag{12.1}$$

where γ is the surface free energy per unit area of surface and dA is the associated increment of surface. If the crack propagates with no displacement $\varDelta$ of the external forces $dW = 0$ and we have

$$-\left(\frac{dU}{dc}\right)_\varDelta \geqslant \gamma \frac{dA}{dc}. \tag{12.1a}$$

In this case the elastically stored energy decreases and so $-(dU/dc)_\varDelta$ is essentially a positive quantity.

Griffith calculated the change in elastically stored energy, using a solution obtained by Inglis[3] for the problem of a plate, pierced by a small elliptical crack, which is stressed at right angles to the major axis of the crack. Equation (12.1) then allows the fracture stress σ_B of the material to be defined in terms of the crack length $2c$ by the relationship

$$\sigma_B = (2\gamma E^*/\pi c)^{1/2}, \tag{12.2}$$

where E^* is the 'reduced modulus', equal to the Young's modulus E for a thin sheet in plane stress, and to $E/(1-\nu^2)$, where ν is Poisson's ratio, for a thick sheet in plane strain.

Later workers have shown that a relationship of similar form to (12.2) results when the elastic problem is generalized and extended to three dimensions[4,5].

Much recent work on fracture uses an alternative formulation of the problem due to Irwin[6]. This approach considers the stress field near an idealized crack length $2c$. In two-dimensional polar coordinates for $r \ll c$, with the x axis as the crack axis,

$$\sigma_{xx} = \frac{K_1}{(2\pi r)^{1/2}} [\cos{(\theta/2)}][1 - \sin{(\theta/2)} \sin{(3\theta/2)}],$$

$$\sigma_{yy} = \frac{K_1}{(2\pi r)^{1/2}} [\cos{(\theta/2)}][1 + \sin{(\theta/2)} \sin{(3\theta/2)}],$$

$$\sigma_{zz} = \nu(\sigma_{xx} + \sigma_{yy}) \text{ for plane strain,}$$

$$\sigma_{zz} = 0 \text{ for plane stress,}$$

$$\sigma_{xy} = \frac{K_1}{(2\pi r)^{1/2}} \cos{(\theta/2)} \sin{(\theta/2)} \cos{(3\theta/2)},$$

$$\sigma_{yz} = \sigma_{zx} = 0.$$

In these equations θ is the angle between the axis of the crack and the radius vector.

The value of this approach is that the stress field around the crack is identical in form for all types of loading situation normal to the crack, with the magnitude of the stresses (i.e. their intensity) determined by K_1 which is constant for given loads and geometry. K_1 is called the stress intensity factor, the superscript 1 indicating that we have considered loading normal to the crack. This is termed the crack opening mode I, as distinct from a sliding mode II, which we will not consider here. Although σ_{xx} and σ_{yy} clearly become infinite in magnitude as we approach the crack tip and r tends to zero, the products $\sigma_{xx}\sqrt{r}$ and $\sigma_{yy}\sqrt{r}$ and hence K_1 remain finite.

For an infinite sheet with a central crack of length $2c$ subjected to a uniform stress σ it was shown by Irwin that

$$K_1 = \sigma(\pi c)^{1/2}. \tag{12.4}$$

We then postulate that when σ reaches the fracture stress σ_B, K_1 has a critical value K_{IC} given by

$$K_{IC} = \sigma_B(\pi c)^{1/2}. \tag{12.5}$$

The fracture toughness of the material can then be defined by the value of K_{IC}, termed the critical stress intensity factor, which defines the stress field at fracture for all loadings normal to the crack and all geometries.

There is clearly a link at this stage with the earlier Griffith formulation in that equation (12.5) can be written as

$$\sigma_B = (K_{IC}^2/\pi c)^{1/2} \tag{12.6}$$

which is identical in form to equation (12.2).

In linear elastic fracture mechanics it is also useful to consider the energy available for unit increase in crack length. Following equation (12.1) above this is

$$\frac{dW}{dA} - \frac{dU}{dA} = \frac{1}{B}\left(\frac{dW}{dc} - \frac{dU}{dc}\right), \tag{12.7}$$

where B is the thickness of the specimen. This quantity is called G, the 'strain energy release rate', and it is assumed that fracture occurs when G reaches a critical value G_c.

The equivalent equation to (12.1) is then

$$G \geqslant G_c \tag{12.8}$$

and G_c is equal to 2γ of the Griffith formulation, but is generalized to include all work of fracture, not just the surface energy.

Comparison of equations (12.2) and (12.6) shows

$$G_{IC} = K_{IC}^2/E^*. \tag{12.9}$$

Although the Griffith and Irwin formulations of the fracture problems are equivalent, most recent studies of polymers have followed Irwin. Before discussing results for polymers, it is useful to show how G_c can be calculated.

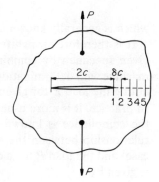

Figure 12.2. Schematic diagram of a specimen with a centre crack of length $2c$.

Consider a sheet of polymer with a crack of length $2c$ (Figure 12.2). We now define a quantity termed the compliance of the cracked sheet C which is the reciprocal of the slope of the linear load–extension curve from zero load up to the point at which crack propagation begins. At the latter point the load is P and the extension is Δ, so

$$C = \Delta/P.$$

This quantity C is not to be confused with an elastic compliance constant as defined in Sections 2.5 and 10.1. The work done in an elemental step of crack propagation is illustrated by Figure 12.3. As the crack moves from 4 to 5, the energy available for formation of new crack surface is the difference between the work done (45XY) and the increase in elastic stored energy (Triangle 05Y − Triangle 04X). It can be seen that this is the area of the shaded triangle

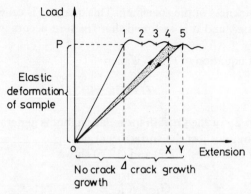

Figure 12.3. The load–extension curve for the specimen shown in Figure 12.2.

in Figure 12.3. For an increase of crack length by dc this is given by $\frac{1}{2}P^2 \, dC$ and hence

$$G_c = \frac{P^2}{2B} \frac{dC}{dc},$$ (12.10)

which is generally known as the Irwin–Kies relationship[7].

It is therefore straightforward in principle to determine G_c directly for a given specimen by combining a load–extension plot from a tensile testing machine with determination of the movement of the crack across the specimen, noting the load P for given crack lengths (points 1, 2, 3, 4, 5 in Figure 12.3). In practice, it is more usual to use test pieces of standard geometry for which the compliance is known as a function of crack length. For example, the relationship between the extension Δ (usually termed the deflection in this case) and the load P for a double cantilever beam specimen (see Figure 12.4) is given by

$$\Delta = \frac{64c^3}{EBb^3} P.$$

Hence

$$C = \frac{\Delta}{P} = \frac{64c^3}{EBb^3} \quad \text{and} \quad \frac{dC}{dc} = \frac{192c^2}{EBb^3}.$$ (12.11)

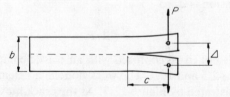

Figure 12.4. The double cantilever beam specimen.

We then have

$$G_c = \frac{P^2}{2B}\frac{dC}{dc} = \frac{P^2}{2B}\frac{192c^2}{EBb^3} \tag{12.12a}$$

or

$$G_c = \frac{3\Delta^2 b^3}{128c^4}E. \tag{12.12b}$$

The critical strain-energy release rate (or in the original Griffith terminology the fracture surface energy γ) can therefore be obtained by measurements of either the load P or the deflection Δ for given crack lengths c.

The exact equivalent formulation in terms of the critical stress intensity factor can be obtained from equation (12.9) and we have

$$K_{IC} = 4\sqrt{6}\,\frac{Pc}{Bb^{3/2}}. \tag{12.13}$$

In our discussion we have exemplified the calculations for a geometrically simple specimen only, so that the principles involved are not obscured by complex stress analysis. We now wish to follow the developments for polymers, which can conveniently take the form of a historical review.

In its simplest form, the Griffith theory and the linear elastic fracture mechanics which developed from it, ignore any contribution to the energy balance arising from the kinetic energy associated with movement of the crack. It has therefore been considered that a basic study of the brittle fracture of polymers would be likely to be most rewarding if care were taken to ensure that fracture takes place so slowly that a negligible amount of energy is dissipated in this way. With this in mind Benbow and Roesler[1] devised a method of fracture in which flat strips of polymer were cleaved lengthwise by gradually propagating a crack down the middle, i.e. crack propagation in the double cantilever beam already discussed (Figure 12.4).

Figure 12.5 shows a diagram of the Benbow and Roesler apparatus. The specimens were cut from 6.35 mm thick sheets of commercial polymethyl methacrylate; their widths varied from 25.4 to 101.6 mm and their lengths from 152.4 to 304.8 mm. A major difficulty was to ensure that the crack propagated straight along the strip. It was found empirically that the crack could be kept straight by applying a preset lengthwise compression. This experimental finding of Benbow and Roesler was subsequently explained theoretically by Cottrell[8].

The essence of Benbow and Roesler's analysis of cleavage was to equate the increase in Griffith surface energy to the decrease in elastically stored energy as the crack propagates to a larger length. In principle this is identical to the treatment outlined above for the double cantilever beam specimen where the crack length is very large. For any crack length, the equations must take a similar dimensional form, and the form of the energy balance can be

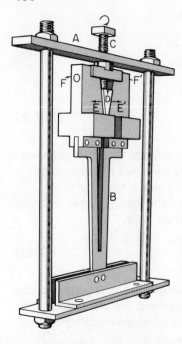

Figure 12.5. The cleavage apparatus of Benbow and Roesler. Flexure in the bar A compresses the sample B and stabilizes the crack direction. Turning the screw C moves the wedge D forward, forcing the clamps E, E' apart. Rotation of the clamps, in the plane of the specimen, is prevented by the sliding bearings F, F'.

found from consideration of similarity, any fixed value of Δ/b defining a manifold of geometrically similar systems.

It can be shown that

$$\frac{\gamma}{E} = \frac{G_c}{2E} = \phi\left(\frac{b}{c}\right) \tag{12.14}$$

(symbols as in Figure 12.4 and ϕ is a numerical function). This equation was tested by plotting experimental values of $f(\Delta^2/c)$ against $g(b/c)$, where f and g are convenient functions for a variety of different samples. For small values of b/c, i.e. for large crack length, the situation is that of the double cantilever beam and we have the explicit solution of equation (12.12b) with $\gamma/E = 3\Delta^2 b^3/128c^4$. This gives an asymptote for plots of $(\Delta^2/c)^{1/3}$ against c/b which passes through the origin. Knowing a value for the Young's modulus E, the surface energy γ can then be found.

In a subsequent paper[9], Berry adopted a slightly different approach and assumed that the force P required to bend the beam is given by an empirical formula $P = ac^{-n}\Delta$, where a is a constant. The energy balance equation was then written in terms of P, Δ, γ, c and n with n being determined by a subsidiary experiment. The surface energy can thus be obtained without a value for Young's modulus E, since this is effectively eliminated by measuring the force P. This procedure is clearly equivalent to the use of the Irwin–Kies relationship leading to equation (12.12a). It has a further advantage in that we bypass the question as to whether fracture occurs under conditions of plane stress or plane strain.

Berry's experimental procedure was also simpler from that of Benbow and Roesler in that the crack direction was determined by routing an initial groove in the sample, which has been adopted by many subsequent workers.

For a comprehensive discussion of the calculation of the fracture toughness parameters G_c and K_c for specimens with different geometries, the reader is referred to standard texts[10-12].

Both Berry and Benbow–Roesler showed that the surface energy was independent of the sample dimensions, suggesting that it is a basic material property. Berry also examined the validity of equation (12.2) for the fracture of polymethyl methacrylate and polystyrene by measuring the tensile strength of samples containing deliberately introduced cracks of known magnitude. The relationship appeared to hold within the fairly large experimental errors experienced in fracture investigations, and gave values for the surface energy comparable with those obtained by cleavage tests. Berry[9] summarized his own results and those of other workers: these are shown in Table 12.1 below.

Table 12.1. Fracture surface energies (in Kilojoules per square metre).

	Polymer	
Method	Polymethyl methacrylate	Polystyrene
Cleavage (Benbow[13])	4.9 + 0.5	25.5 ± 3
Cleavage (Svensson[14])	4.5	9.0
Cleavage (Berry[9])	1.4 ± 0.07	7.13 ± 0.36
Tensile (Berry[15])	2.1 ± 0.5	17 ± 6

At this point it is interesting to attempt a theoretical estimate of the surface energy. If it is assumed that the energy required to form a new surface originates in the simultaneous breaking of chemical bonds only, this gives an upper theoretical limit for the surface energy. Let us assume that the bond dissociation energy is 100 kcal mol^{-1} and that the concentration of molecular chains is 1 chain per 0.2 nm^2, giving 5×10^{18} molecular chains m^{-2}. It can then be shown that to form 1 m^2 of new surface requires about 1.5 J, which is two orders of magnitude less than that obtained from the cleavage and tensile measurements.

This large discrepancy between experimental and theoretical values for the surface energy is comparable to that found for metals. For metals, it was proposed by Orowan and others that the surface free energy may include a term of much greater magnitude which arises from plastic work done in deforming the metal near the fracture surface as the crack propagates. Andrews[16] has suggested that the quantity measured in the fracture of polymers should be described by $\mathcal{T}$, the 'surface work parameter', to distinguish it from a true surface energy, and has more recently proposed a generalized

408

Figure 12.6. Fracture surfaces of a cleavage sample of polymethyl methacrylate showing colour alternation (green filter). [Redrawn from Berry, in *Fracture 1959* (B. L. Averach *et al.*, eds), Wiley, New York, 1959, p. 263.]

theory of fracture which embraces not only plastic deformation but viscoelastic deformation, both of which may be important in polymers.

Andrews' generalized theory of fracture will be discussed later. At this point we will follow up Berry's observation that the largest contribution to the surface energy of a glassy polymer appears to come from a viscous flow process. Berry[17] later suggested that in polymethyl methacrylate this was related to the interference bands observed on the fracture surfaces, as seen in Figure 12.6. Berry proposed that the large surface energy term arises from work expended in the alignment of polymer chains ahead of the crack, the subsequent crack growth leaving a thin, highly oriented layer of polymeric material on the fracture surface. Following on from these ideas, Kambour[18-20] showed that there is a thin wedge of porous material at a crack tip in a glassy polymer. This is termed a craze and is shown schematically in Figure 12.7. The craze forms under plane strain conditions. The polymer is therefore not free to contract laterally and there is a consequent reduction in density. Brown and Ward[21] determined the craze profile by examining the crack tip region in polymethyl methacrylate in an optical microscope. In reflected light two sets of interference fringes were observed. These correspond to the crack and the craze respectively. It was found that the craze profile was very similar indeed to the plastic zone model proposed by Dugdale[22] for metals, as will now be described.

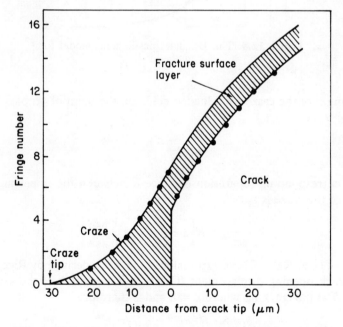

Figure 12.7. Schematic diagram of a craze. [Redrawn with permission from Brown and Ward, *Polymer*, **14**, 469 (1973). © IPC Business Press Ltd.]

We have already discussed that equations (12.3) imply that there is an infinite stress at the crack tip. In practice this clearly cannot be so, and there are two possibilities. First, there can be a yielded zone, a region where shear yielding of the polymer occurs. In principle this can occur in both thin sheets where conditions of plane stress pertain and in thick sheets where there is plane strain. Secondly, for thick specimens under conditions of plane strain, the stress singularity at the crack tip can be released by the formation of a craze. This is a line zone, in contrast to the approximately oval (plane stress) or kidney-shaped (plane strain) shear yield zones. As indicated, its shape approximates very well to the idealized Dugdale plastic zone where the stress singularity at the crack tip is cancelled by the superposition of a second stress field in which the stresses are compressive along the length of the craze (Figure 12.8). A constant compressive stress is assumed, and this is identified with the craze stress. It is not the yield stress, and crazing and shear yielding are different in nature and respond differently to changes in the structure of the polymer.

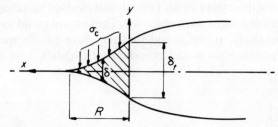

Figure 12.8. The Dugdale plastic zone model for a craze.

The length of the craze for a loaded crack on the point of propagation can be shown to be

$$R = \frac{\pi}{8} \frac{K_{IC}^2}{\sigma_c^2} \qquad (12.15)$$

and the corresponding separation distance δ between the upper and lower surfaces of the craze is

$$\delta = \frac{8}{\pi E^*} \sigma_c R \left[\xi - \frac{x}{2R} \log \left(\frac{1+\xi}{1-\xi} \right) \right], \qquad (12.16)$$

where $\xi = (1 - x/R)^{1/2}$. These expressions have been derived by Rice[23].

The crack opening displacement (COD) δ_t is the value of the separation distance δ at the crack tip where $x = 0$, and is therefore

$$\delta_t = 8\sigma_c R / \pi E^* = K_{IC}^2 / \sigma_c E^*. \qquad (12.17)$$

The fracture toughness of the polymer then relates to two parameters δ_t and σ_c, the craze stress, the product of which $\delta_t \sigma_c = G_{IC}$ the critical strain energy release rate. Direct measurements of craze shapes for several glassy polymers, including polystyrene, polyvinyl chloride and polycarbonate[24,25] have confirmed the similarity to a Dugdale plastic zone. A result of some physical significance is that the crack opening displacement is often insensitive to temperature and strain rate for a given polymer, although it has been shown to depend on molecular weight. For constant COD, the true dependence of G_{IC} on strain rate and temperature will be determined only by the sensitivity of the craze stress to these parameters. Since $G_{IC} = K_{IC}^2/E^*$ the fracture toughness K_{IC} will in addition be affected by E^*, which will also be dependent on strain rate and temperature.

The nature of crazing and the craze stress will be discussed in more detail later, but here we can note that this approach offers a deeper understanding of the brittle–ductile transition in glassy polymers in terms of competition between crazing and yielding. Because both crazing and yielding are activated processes with in general different temperature and strain rate sensitivities, it can readily be appreciated that one will be favoured over the other for some conditions and vice versa for other conditions. An additional complexity will in general rise from the nature of the stress field which may favour one process rather than the other. It is, however, important to note that the latter consideration does not enter into our consideration of the craze at the crack tip. The line of travel of the crack is a line of zero shear stress within the plane but maximum triaxial stress. From our understanding of the stress criteria for crazing to be discussed shortly, we will see that such a stress field will favour crazing and that, for long cracks where the stress field of the crack is the dominant factor, the craze length will be determined solely by the requirement that the craze grows to cancel the stress singularity at the crack tip.

In several glassy polymers[25,26] a complication occurs in that there is a thin line of material on the fracture surface where the polymer has yielded. This is called a shear lip, and is shown for polycarbonate in Figure 12.9. Bearing in mind analogous behaviour in metals, it has been proposed that the overall strain energy release rate G_c^o will be the sum of the contribution from the craze and that from the shear lips. To a first approximation we would expect

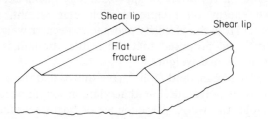

Figure 12.9. The shear lips in polycarbonate [Redrawn with permission from Fraser and Ward, *Polymer*, **19**, 220 (1978).]

412

the latter to be proportional to the volume of yielded material. If the *total* width of the shear lips on the fracture surface is w, and the shear lip is triangular in cross-section, we then have

$$G_c^\circ = G_{IC}\left(\frac{B-w}{B}\right) + \frac{\phi w^2}{2B},$$ (12.18)

where ϕ is the energy to fracture unit volume of shear lip. It has been shown that this relationship describes results for polycarbonate and polyether sulphone very well[25,26] and that ϕ corresponds quite closely to the energy to fracture in a simple tensile extension experiment.

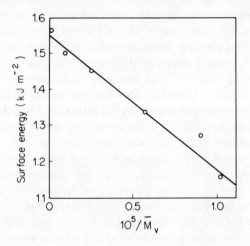

Figure 12.10. Dependence of fracture surface energy on reciprocal molecular weight ($\bar{M}_v$ is viscosity average molecular weight). [Redrawn with permission from Berry, *J. Polymer Sci. A*, **2**, 4069 (1964).]

In his studies of the fracture of polymethyl methacrylate Berry showed that the surface energy was strongly dependent on polymer molecular weight[27]. His results (Figure 12.10) fitted an approximately linear dependence of the fracture surface energy on reciprocal molecular weight, such that $\gamma = A' - B'/\bar{M}_v$ where $\bar{M}_v$ is the viscosity average molecular weight. Many years previously Flory[28] had proposed that the brittle strength is related to the number average molecular weight.

More recently Weidmann and Döll[29] have shown that the craze dimensions decrease markedly in polymethyl methacrylate at low molecular weights. In a study of the molecular weight dependence of fracture surfaces in the same polymer, Kusy and Turner[30] could observe no interference colours for a viscosity average molecular weight of less than 90 000 daltons, concluding that there was a dramatic decrease in the size of the craze. Based on craze shape studies of polycarbonate Pitman and Ward[25] reported a very high dependence of both craze stress and crack opening displacement on molecular weight and observed that both would be expected to become negligibly small for $\bar{M}_w < 10^4$.

Berry[27] speculated that the smallest molecule which could contribute to the surface energy would have its end on the boundaries of the craze region, on opposite sides of the fracture plane, and be fully extended between these points.

Kusy and Turner[31] presented a fracture model for polymethyl methacrylate in which the surface energy measured was determined by the number of chains above a critical length. Their data fitted well with their predictions, showing a limit to the surface energy at high molecular weight. The Kusy and Turner analysis did not work well for the polycarbonate data of Pitman and Ward. Moreover, the extended molecular lengths, based on the extension of a random coil would be much less than the crack opening displacement (as discussed by Haward et al.[32]) so that there is no direct correlation between the two quantities. The craze structure relates to the stretching of fibrils and the key molecular factors are the presence of random entanglements and the distance between these entanglements, not the extension of an isolated molecular chain.

12.3 THE STRUCTURE AND FORMATION OF CRAZES

We have seen how the craze at the crack tip in a glassy polymer plays a vital role in determining its fracture toughness. Crazing in polymers also manifests itself in another way. When certain polymers, notably polymethyl methacrylate and polystyrene, are subjected to a tensile test in the glassy state, above a certain tensile stress opaque striations appear in planes whose normals are the direction of tensile stress, as in Figure 12.11.

The interference bands on the fracture surfaces, which relate to the craze at the crack tip, were first observed by Berry[33] and by Higuchi[34]. Berry, and later Kambour, did much of the pioneering research to establish the nature of the surface crazes and their variation with specimen and conditions of fracture. Kambour confirmed that the PMMA fracture-surface layers were qualitatively similar to the internal crazes of this polymer, by showing that the refractive indices were the same[18]. Both surface layers and bulk crazes appear to be oriented polymer structures of low density. These are produced by orienting the polymer under conditions of abnormal constraint where it is not allowed to contract in the lateral direction, while being extended locally to strains of the order of unity, i.e. the polymer has undergone inhomogeneous cold-drawing, and a key idea is the deformation of a molecular network, as discussed in Section 11.6.3.

Detailed studies of the phenomenon of crazing in polymers have been in the following major areas: the structure of crazes, the stress or strain criteria for their formation and environmental effects. These subjects will now be discussed in turn.

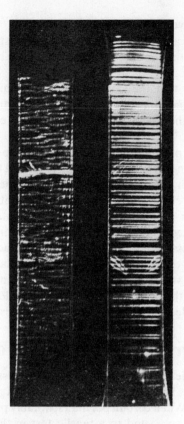

Figure 12.11. Craze formation in polystyrene.

12.3.1 The Structure of Crazes

The structure of crazes in bulk specimens was studied by Kambour[18] using the critical angle for total reflection at the craze/polymer interface to determine the refractive index of the craze. The value of the refractive index showed that the craze was roughly 50% polymer and 50% void. Another investigation involved transmission electron microscopy of polystyrene crazes impregnated with an iodine–sulphur eutectic to maintain the craze in its extended state[35,36]. The fibril structure of the craze was clearly revealed, the fibrils being separated by the voids which are responsible for the overall low density of the craze.

A rather more direct technique was developed by Beahan, Bevis and Hull[37], who examined the microstructure of microtomed thin sections of precrazed bulk polystyrene. Crazes were also examined which were formed by straining microtomed thin sections of polystyrene[38,39]. In both cases a fibril structure was observed within the craze, it being rather finer in the thin specimens. More recently Kramer, Brown and collaborators have made further detailed studies of craze structure with transmission electron microscopy[40,41], selected area electron diffraction[42] and small angle X-ray scattering[43]. It is generally agreed that the craze is constituted of cylindrical fibrils of highly oriented polymer (Figure 12.12). The fibril axes are parallel to the tensile stress direction, as expected if the fibrils consist essentially of drawn polymer.

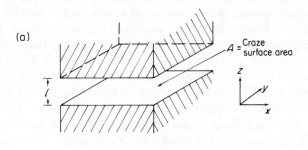

(a)

$A = $ Craze surface area

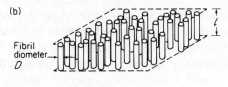

(b)

Fibril diameter D

Fibril volume fraction V_f

Figure 12.12. Schematic diagram of the forest of cylindrical fibrils oriented normal to the craze surface. [Redrawn with permission from Brown and Kramer, *J. Macromol. Sci. B*, **19**, 487 (1981) by courtesy of Marcel Dekker Inc.]

It has been concluded[44] that the extension ratio of the fibrils relates to the extensibility of a molecular network. Values for the extension ratios were estimated by Kramer and his colleagues[44,45] and also by Ward and co-workers[21,25] from analysis of optical interference patterns. These values compare reasonably well with estimates of the network extensibility from small angle neutron scattering data[44] or stress-optical measurements[46].

The studies of the craze structure by Kramer and his coworkers confirm the earlier findings of Hull and colleagues in showing that the craze structure is not uniform along its length. Although there are significant discrepancies between the displacement and the stress on the craze with those predicted by the Dugdale zone model, the latter nevertheless provides a good overall description of the mechanics of the craze and is quite adequate for most purposes.

12.3.2 Stress or Strain Criteria for Craze Formation

There is considerable interest in attempting to obtain a stress criterion for craze formation analogous to that for yield behaviour described in Chapter 11. To this end Sternstein and his collaborators[47] have used the ingenious technique of examining the formation of crazes in the vicinity of a small circular hole (1.59 mm diameter) punched in the centre of polymethyl methacrylate strips (12.7 mm × 50.8 mm × 0.79 mm) when the latter are pulled in tension. A typical pattern is shown in Figure 12.13(a). The solutions for the *elastic* stress field in the vicinity of the hole were compared with the craze pattern. It was found that the crazes grew parallel to the minor principal stress vector. As the contours of the minor principal stress vector are orthogonal to those of the major principal stress vector, this shows that the major principal stress acts along the craze plane normal and therefore parallel to the molecular orientation axis of the crazed material.

The boundary of the crazed region coincided to a good approximation with contour plots showing lines of constant major principal stress σ_1. These are shown in Figure 12.13(b) where the contour numbers are per unit of applied stress. It should be noted that at low applied stresses it is not possible to discriminate between the contours of constant σ_1 and contours showing constant values of the first stress invariant $I_1 = \sigma_1 + \sigma_2$. However, the consensus of the results is in accord with a craze-stress criterion based on the former rather than on the latter, and as we have seen the *direction* of the crazes is consistent with the former.

Sternstein, Paterno and Ongchin[48] carried this investigation one stage further by examining the formation of crazes under biaxial stress conditions. It was found that the stress conditions for crazing involved both the principal stresses σ_1 and σ_2. The results were represented in the following manner. The general biaxial stress was represented by two quantities, the first stress invariant $I_1 = \sigma_1 + \sigma_2$ and a stress bias $\sigma_b = |\sigma_1 - \sigma_2|$. The criterion for craze formation was then proposed to be

$$\sigma_b = |\sigma_1 - \sigma_2| = A + \frac{B}{I_1}, \tag{12.19}$$

where A and B are constants which depend on temperature.

Although equation (12.19) does describe the combined stress crazing data very well, it does not have any immediate interpretation in physical terms. In particular, it is difficult to impute any significance to the 'stress bias' because the largest stress bias is not $\sigma_1 - \sigma_2$, where $\sigma_2 > 0$ since $\sigma_3 = 0$. Oxborough

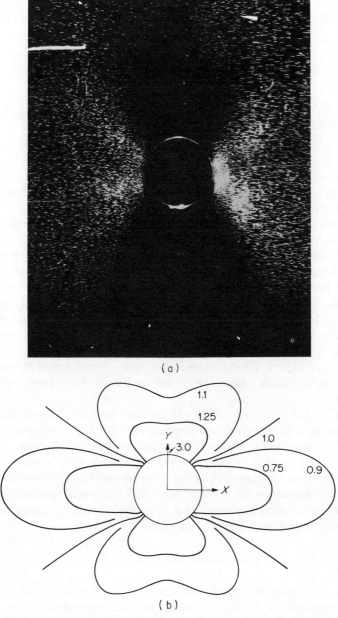

Figure 12.13. (a) Craze formation in the vicinity of a hole in a strip of polymethyl methacrylate loaded in tension. [Result obtained by L. S. A. Smith.] (b) Major principal stress contours (σ_1) for an elastic solid containing a hole. The specimen is loaded in tension in the X direction. Contour numbers are per unit of applied tensile stress. [Redrawn with permission from Sternstein, Ongehin and Silverman, Appl. Polymer Symp. **7**, 175 (1968).]

and Bowden[49] suggested instead that craze initiation relates to a critical strain e_1, which for small strains is given, for the two-dimensional stress field, by

$$e_1 = \frac{1}{E}(\sigma_1 - \nu\sigma_2),$$

where E is Young's modulus and ν Poisson's ratio.

It was proposed that the crazing criterion was

$$Ee_1 = \sigma_1 - \nu\sigma_2 = A - B/(\sigma_1 + \sigma_2). \tag{12.20}$$

In fact, the data obtained could be equally well fitted to either equation (12.19) or equation (12.20).

Equation (12.19) predicts that the stress required to initiate a craze becomes infinite when $I_1 = 0$, i.e. crazing requires a dilational stress field. There are several pieces of experimental evidence which contradict this supposition.

First, Baer and his colleagues[50] have shown that crazes initiate when the principal tensile stress σ_1 reaches a critical value irrespective of the applied hydrostatic pressure, i.e. $\sigma_1 = \sigma_c + p$, where σ_c is the craze stress at zero pressure. Moreover, it was observed that crazing could occur when the hydrostatic component of stress was negative. This result is contrary to equations (12.19) and (12.20).

Secondly, Duckett, Ward and their colleagues[51] have found that crazes can occur in torsion tests where $I_1 = 0$. Similar results have been obtained by Kitagawa[52].

Argon[53,54] has proposed a theory of crazing based on physical ideas, which introduces the influence of the deviatoric and hydrostatic components of stress as essential components of the initiation and growth mechanisms.

It is considered that crazing relates at a microscopic level to precursor micropores which are nucleated by the stress concentrations of inhomogeneous plastic deformation on the scale of 5–10 nm. The void formation associated with craze initiation is described by a porosity β, which increases with time t by a thermally activated rate process with

$$\beta = \dot{\beta}_0 t \exp\{-\Delta G^*_{\text{pore}}/kT\}, \tag{12.21}$$

where $\dot{\beta}_0$ is a frequency factor characteristic of the vibration of a region of the order of the critical radius of the activated configuration.

The formation of a round pore requires a total free energy ΔG^*_{pore}, which is the sum of two terms. The first term relates to the formation of the slip nucleus under the *local* deviatoric stress s in the vicinity of a flaw of strength K and is inversely proportional to s^{55}. The second term relates to the additional plastic deformation required to form a stable round cavity and is directly proportional to the yield stress σ_Y.

We have

$$\frac{\Delta G^*_{\text{pore}}}{kT} = \left(\frac{A}{s} + E\right)\sigma_Y, \tag{12.22}$$

where A and E are material parameters, and s is a function of both the deviatoric and hydrostatic components of stress, as well as the surface flaw concentration factor K, which is in practice a fitting constant.

Argon followed a treatment given by McClintock and Stowers to describe the dependence of the plastic expansion of such porous regions on both the hydrostatic component of stress and the local deviatoric stress s. The hydrostatic tension σ is then given by

$$\sigma = \frac{2\sigma_Y}{3} F(s, Y, \beta), \tag{12.23}$$

where the function F defines the yield locus for pore expansion.

By combining equations (12.21) and (12.23), which describe the rate process and the stress field requirements, we can then find the time t_{in} after which craze nucleus formation occurs.

In general, it will also be necessary for the pores to expand sufficiently for the craze to be of visible size. Argon proposes that this occurs by elastic unloading of the surroundings of a pore, and this growth process defines a growth time t_g. Provided, however, that the hydrostatic component of stress is not extremely small, t_g is small, and the stress criterion for craze formation is determined by the craze initiation step defined by equations (12.21)–(12.23) above.

Argon and his coworkers were also concerned with the increase of craze nuclei with time in samples subjected to different combination of macroscopic deviatoric and hydrostatic components of stress. It was shown that the theory was in reasonable agreement with experimental results for the increase in surface density of crazes with time, up to the saturation value for a particular stress field.

Finally, Argon and Salama[54] considered the growth of the crazes. As has been remarked, there is good experimental evidence that growth occurs on surfaces normal to the maximum principal tensile stress. This suggests that another mechanism must be sought for craze growth. Argon has proposed that the craze front advances by a meniscus instability mechanism in which craze tufts are produced by the repeated break-up of the concave air/polymer interface at the crack tip, as illustrated in Figure 12.14. A theoretical treatment of this model predicted that the steady state craze velocity would relate to the 5/6th power of the maximum principal tensile stress, and support for this result was obtained from experimental results on polystyrene and polymethyl methacrylate[54].

420

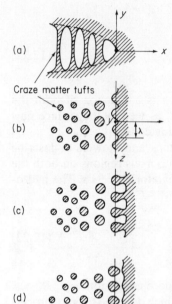

(a)

Craze matter tufts

(b)

(c)

(d)

Figure 12.14. Schematic diagram showing craze matter production by the mechanism of meniscus instability: (a) outline of a craze tip; (b) cross-section in the craze plane across craze matter tufts; (c), (d) advance of the craze front by a completed period of interface convolution. [Redrawn with permission from Argon, Hannoosh and Salama, in *Fracture* 1977, Vol. 1, Waterloo, 1977, p. 445.]

12.3.3 Crazing in the Presence of Fluids and Gases: Environmental Crazing

The crazing of polymers by environmental agents is of considerable practical importance and has been studied extensively, with notable contributions from Kambour[19,56–58], Andrews[59], Williams and Marshall[60,61] and Brown[62–64]. The subject has been reviewed by Kambour[65] and by Brown[66]. In general, environmental agents, which can be fluids or solids, reduce the stress or strain required to initiate crazing.

Kambour and coworkers[19,56–58] showed that the critical strain for crazing decreased as the solubility of the environmental agent was increased. It was also found that the critical strain decreased as the glass transition temperature of the solvated polymer decreased. Andrews and Bevan[59] adopted a slightly more formal approach, and applied the ideas of fracture mechanics. They performed fracture tests on single-edge notched tensile specimens, where a central edge crack of length c is introduced into a large sheet of polymer which is then loaded in tension. The fracture stress is related to Andrews' surface work parameter $\mathcal{T}$ (or the strain energy release rate $G_c = 2\gamma$) by an equation identical in form to equation (12.2) above. Andrews and Bevan showed that the critical stress for crack and craze propagation σ_c was indeed proportional to $c^{-1/2}$, so that $\mathcal{T}$ values could be determined. For constant experimental conditions, a range of values of $\mathcal{T}$ was obtained from which a minimum value of $\mathcal{T}$, $\mathcal{T}_0$, was estimated. From tests in a given solvent over a range of temperatures, it was found that these minimum values of $\mathcal{T}$ decreased with increasing temperature up to a characteristic temperature T_c, where $\mathcal{T}_0$ remained constant at a value $\mathcal{T}_0^*$. The values of $\mathcal{T}_0^*$ for the different solvents were shown to be a smooth function of the difference between the solubility parameters of the solvent and the polymer, reaching a minimum when this difference was zero (Figure 12.15).

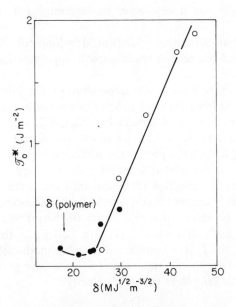

Figure 12.15. Variation of J_0^* for poly-methyl methacrylate with solubility para-meter of the solvent: ●, pure solvents; ○, water-isopropanal mixtures. [Re-drawn with permission from Andrews and Bevan, *Polymer*, **13**, 337 (1972). © IPC Business Press Ltd.]

These findings were very simply explained on the basis that the work done in producing the craze can be modelled by the expansion of a spherical cavity of radius r under a negative hydrostatic pressure p. This pressure p has two terms so that

$$p = \frac{2\gamma_\tau}{r} + \frac{2\sigma_Y}{3}\psi, \qquad (12.24)$$

where γ_τ is the surface tension between the solvent in the void and the surrounding polymer, σ_Y is the yield stress and ψ is a factor close to unity (cf. equation (12.23) above). The effect of temperature is to change the yield stress, so that with increasing temperature σ_Y falls, eventually to zero at T_c, which is the glass transition temperature of the plasticized polymer. Above T_c the fracture surface energy $\mathscr{T}_0^*$ then relates solely to the intermolecular forces represented by the surface tension γ_r.

Brown has pointed out that gases at sufficiently low temperatures make almost all linear polymers craze[62-64,66]. Parameters such as the density of the crazes and the craze velocity increase with the pressure of the gas and decrease with increasing temperature. It was concluded that the surface concentration of the absorbed gas was a key factor in determining its effectiveness as a crazing agent.

In a related, but somewhat different development, Williams and co-workers[60,61] studied the rate of craze growth in polymethyl methacrylate in methanol.

In all cases the craze growth was dependent on the initial stress-intensity factor K_0, calculated from the load and the initial notch length. Two different types of craze growth were observed. For $K_0 < K_0^*$, a specific value of K_0, the craze would decelerate and finally arrest. For $K_0 > K_0^*$ the craze would decelerate initially and finally propagate at constant speed.

It was argued that the controlling factor determining craze growth in all cases would be the diffusion of methanol into the craze. In the first case, where $K_0 < K_0^*$, the methanol is considered to diffuse along the length of the craze, and it may be shown that the length of the craze x is related to the time of growth t such that $x \propto t^{1/2}$ (Figure 12.16). In the second type of growth, where $K_0 > K_0^*$, it is considered that the methanol diffuses through the surface of the specimens, maintaining the pressure gradient in the craze and producing craze growth at constant velocity.

12.4 THE MOLECULAR APPROACH

The discussion of fracture behaviour in this chapter has been almost entirely in terms of mechanics, and even where structural ideas have been invoked, as in the case of craze formation, the development has still been at a phenomenological level.

It has long been recognized that oriented polymers (i.e. fibres) are much less strong than would be predicted on the basis of elementary considerations by assuming that fracture involves simultaneously breaking the bonds in the molecular chains across the section perpendicular to the applied stress. Calculations of this nature were originally undertaken by Mark[67] and rather more recently by Vincent[68] on polyethylene. It was found that in both cases the measured tensile strength was at least an order of magnitude less than that calculated.

We have seen one possible explanation of this discrepancy—the Griffith flaw theory of fracture. It has also been considered that there may be a general analogy between this difference between measured and calculated strengths and the difference between measured and calculated stiffnesses for oriented polymers. A general argument for both discrepancies could be that only a

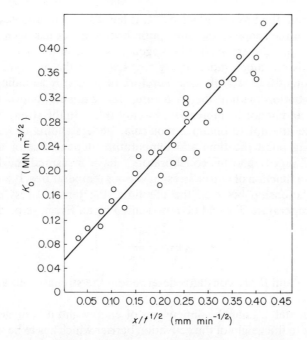

Figure 12.16. End flow craze growth behaviour for polymethyl methacrylate in methanol at 20 °C. [Redrawn with permission form Williams and Marshall, *Proc. R. Soc. A*, **342**, 55 (1975).]

small fraction of the molecular chains are supporting the applied load. In Chapter 10, we have discussed how these tie molecules (or perhaps crystalline bridges) which connect adjacent crystalline blocks play a key role in determining the axial stiffness of an oriented polymer. There has therefore been considerable interest in examining chain fracture in oriented polymers, using electron paramagnetic resonance to observe free radicals produced or infrared spectroscopy to identify such entities as aldehyde end groups which suggest chain scission. A very comprehensive survey of the results of such studies has been given by Kausch[69]. Kausch and Becht[70] have emphasized that the total number of broken chains is much too small to account by virtue of their load carrying capacity for the measured reductions in macroscopic stress. We must therefore conclude that the tie molecules which eventually break are not the main source of strength of highly oriented polymers. This conclusion is confirmed by the lack of any positive correlations between the strength of fibres and the radical concentrations at break.

Although these strong reservations have to be borne in mind, the following considerations suggest that the examination of chain fracture is not totally irrelevant to the deformation of polymers. Examination of the infrared spectra of oriented polymers under stress shows that there is a distinct shift in frequency from the unstressed state[71,72]. It has also been proposed that there is a change in the shape of the absorption line, and this has been interpreted as implying that the stress is inhomogeneous at a molecular level so that certain bonds are much more highly stressed than the average. The actual concentration of free radicals may therefore be primarily an indication of the stress distribution within the structure, for example between different microfibrils and hence tie molecules, and not relate to the strength *per se*.

A positive attempt to obtain a molecular understanding of fracture took as its starting point the time and temperature dependence of the fracture processes. Zhurkov and his collaborators[73] have measured the life time of polymers as a function of tensile stress at various temperatures. It was proposed that the relationship between the life time, the tensile stress σ_B and the absolute temperature T could be represented by an Eyring-type equation:

$$\tau = \tau_0 \exp\left\{\frac{U_0 - \beta\sigma_B}{kT}\right\},$$

where τ_0, U_0 and β are constants determining the strength characteristics of a polymer.

The parameter U_0 has the dimensions of energy and it is suggested that it corresponds to the height of the activation barrier which has to be surmounted for fracture to occur. Zhurkov and his coworkers showed that for a wide range of polymers U_0 was approximately equal to the activation energies obtained for thermal breakdown. They then showed by the technique of electron spin resonance that free radicals are produced in the fracture process in polymers and moreover that correlations can be established between the radical formation rate and the time to break.

The existence of submicrocracks in polymer has already been mentioned in connection with the Argon theory of craze initiation. Zhurkov and his collaborators[74] have used small angle X-ray scattering to establish the presence of such submicroscopic cracks. Although it has been proposed by Zakrewskii[75] that the formation of these submicrocracks is associated with a cluster of free radicals and the associated ends of molecular chains, Peterlin[76] has argued that the cracks occur at the ends of microfibrils, and Kausch[69] has concluded that the submicrocrack formation is essentially independent of chain scission.

12.5 FACTORS INFLUENCING BRITTLE–DUCTILE BEHAVIOUR: BRITTLE–DUCTILE TRANSITIONS

12.5.1 Effect of Basic Material Variables on the Brittle–Ductile Transition

Many aspects of the brittle–ductile transition in metals, including the effect of notching, which we will discuss separately, have been explained on the basis that brittle fracture occurs when the yield stress exceeds a critical value[77]. This is the Ludwik–Davidenkov–Orowan hypothesis, illustrated in Figure 12.17(a). It is assumed that brittle fracture and plastic flow are independent processes, giving separate characteristic curves for the brittle fracture stress σ_B and the yield stress σ_Y as a function of temperature at constant strain rate (as shown in Figure 12.17(b). Changing strain rate will produce a shift in these curves. It is then argued that whichever process can occur at the lower stress will be the operative one. This can be either fracture or yield. Thus the intersection of the σ_B/σ_Y curves defines the brittle–ductile transition and the material is ductile at all temperatures above this point.

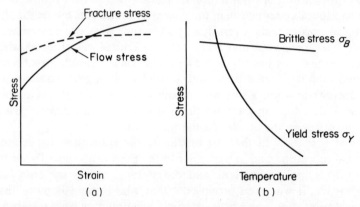

Figure 12.17. (a) and (b) Diagrams illustrating the Ludwig–Davidenkov–Orowan theories of brittle–ductile transitions.

The influence of chemical and physical structure on the brittle–ductile transition can be analysed from this simple starting point, by considering how these factors affect the brittle stress curve and the yield–stress curve respectively. As will be appreciated, this approach bypasses the relevance of fracture mechanics to brittle failure. If, however, we consider fracture initiation (as distinct from propagation of a crack) as governed by a fracture stress σ_B, this concept of regarding yield and fracture as competitive processes provides a useful starting point. We shall therefore follow this approach first and then discuss a more sophisticated viewpoint in a further section.

The brittle stress is not much affected by strain rate and temperature (e.g. by a factor of 2 in the temperature range -180 to $+20\,°C$). Vincent[78] was the first to gather evidence to support this contention, which has been borne out by more recent work[79,80]. The yield stress, on the other hand, is greatly affected by strain rate and temperature, increasing with increasing strain rate and decreasing with increasing temperature. (A typical figure would be a factor of 10 over the temperature range -180 to $+20\,°C$.) These ideas are clearly illustrated by results for polymethyl methacrylate shown in Figure

12.18(a). The brittle–ductile transition will therefore be expected to move to higher temperatures with increasing strain rate (Figure 12.18(b)). This is a well known effect with polymers and can be illustrated by varying the strain rate in a tensile test on a sample of nylon at room temperature. At low strain rates the sample is ductile and cold-draws, whereas at high strain rates it fractures in a brittle manner.

There is a further complication in varying strain rate. At low speeds we have seen that within a certain temperature range cold-drawing occurs. It is possible that at high speeds the heat is not conducted away rapidly enough. Strain hardening is therefore prevented and the specimen fails in a ductile manner. This is an isothermal–adiabatic transition; it does not affect the yield stress and therefore does not affect the brittle–ductile transition; but it does cause a considerable reduction in the energy to break and can be the situation occurring in impact tests, even if brittle fracture does not intervene. It has therefore been proposed that there are two critical velocities at which the fracture energy drops sharply as the strain rate is increased. First there is the isothermal–adiabatic transition, and secondly, at higher strain rates, the brittle–ductile transition. As we would expect, changes in ambient temperature have very little effect on the position of the isothermal–adiabatic transition, but a large effect on the brittle–ductile transition.

It was at first thought that the brittle–ductile transition was related to a mechanical relaxation and in particular to the glass transition. This is true for natural rubber, polyisobutylene and polystyrene, but is not true for most thermoplastics. It was then proposed[81] that where there is more than one mechanical relaxation, the brittle–ductile transition may be associated with a lower temperature relaxation. Although again it appeared that there might be cases where this is correct, it was soon shown that this hypothesis has no general validity. Because the brittle–ductile transition occurs at fairly high strains, whereas the dynamic mechanical behaviour is measured in the linear, low strain region, it is unreasonable to expect that the two can be directly linked. It is certain that fracture, for example, depends on several other factors such as the presence of flaws which will not affect the low-strain dynamic mechanical behaviour. This subject has been discussed extensively by Boyer[82] and by Heijboer[83].

Effect of Basic Material Variables on the Brittle–Ductile Transition[78]
Molecular Weight

Molecular weight does not appear to have a direct effect on the yield strength, but it is known to reduce the brittle strength. As long ago as 1945, Flory[84] proposed that the fracture stress of a polymer was related to the number average molecular weight $\overline{M}_n$ by the relationship

$$\text{fracture stress} = A - \frac{B}{\overline{M}_n}$$

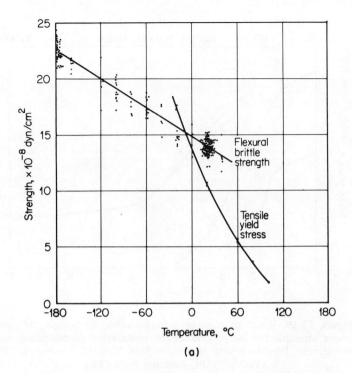

(a)

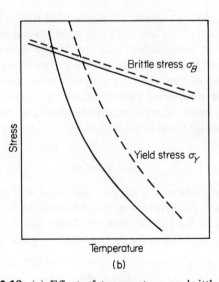

(b)

Figure 12.18. (a) Effect of temperature on brittle strength and tensile yield stress of polymethyl methacrylate. [Redrawn with permission from Vincent, *Plastics*, **26**, 141 (1961).] (b) Diagram illustrating the effect of strain rate on the brittle–ductile transition: ———, low strain rate; – – –, high strain rate.

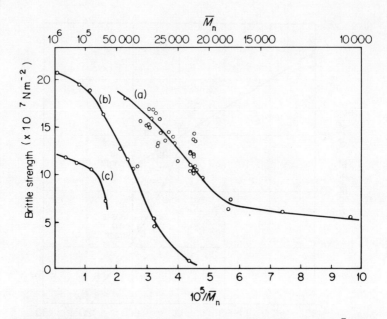

Figure 12.19. Effect of number average molecular weight, $\bar{M}_n$, on brittle strength for (a) polythene; (b) polymethyl methacrylate; (c) polystyrene. [Redrawn with permission from Vincent, *Polymer*, **1**, 425 (1960). © IPC Business Press Ltd.]

and Vincent[78] has given evidence to suggest that this relationship holds to a rough approximation for the brittle strengths of several polyethylene samples (Figure 12.19). It is remarkable that the brittle polymers and branched polyethylenes appear to be the same function of molecular weight.

The yield stresses of the different polyethylenes could differ appreciably with the degree of branching (which affects the crystallinity) so that the temperature of the brittle–ductile transition would be a complex function of at least molecular weight and branch content.

Side Groups

Vincent[78] quotes evidence to suggest that rigid side groups increase both the yield strength and the brittle strength, whereas flexible side groups reduce the yield strength and the brittle strength. There can thus be no general rule regarding the effect of side groups on the brittle–ductile transition.

Cross-linking

Cross-linking increases the yield strength but generally does not increase the brittle strength very much. The brittle–ductile transition is therefore usually raised in temperature.

Plasticizers

Plasticizers can decrease the chance of brittle failure because they usually reduce the yield stress more than they reduce the brittle strength.

Molecular Orientation

Molecular orientation is a very different basic variable from the others which we have considered. It introduces anisotropy of mechanical properties. Assuming the Orowan hypothesis of a distinction between brittle strength and yield stress, both will now depend on the direction of the applied stress. It is generally considered that on this viewpoint the brittle strength is more anisotropic than the yield stress. Thus a uniaxially oriented polymer is more likely to fracture when the stress is applied perpendicularly to the symmetry axis, than an unoriented polymer at the same temperature and strain rate. 'Fibrillation', as this is termed, is the basis of commercial processes for manufacturing synthetic fibres from polymer films.

Notch Sensitivity

It is well known that the presence of a sharp notch can change the fracture of a metal from ductile to brittle, and similar considerations apply to the behaviour of polymers. For this reason a standard impact test for a polymer is the Charpy or Izod test, where a notched bar of polymer is struck by a pendulum and the energy dissipated in fracture calculated.

A very simple explanation of the effect of notching has been given by Orowan[77]. For a deep, symmetrical tensile notch, the slip-line field is identical with that for a flat frictional punch indenting a plate under conditions of plane strain[85] (Figure 12.20).

The compressive stress on the punch required to produce plastic deformation can be shown to be $(2 + \pi)\mathbf{K}$, where $\mathbf{K}$ is the shear yield stress.

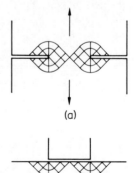

(a)

(b)

Figure 12.20. The slip-line field for a deep symmetrical notch (a) is identical with that for the frictionless punch indenting a plate under conditions of plane strain (b). [Redrawn with permission from Cottrell, *The Mechanical Properties of Matter*, Wiley, New York, 1964.]

430

This is $2.57\sigma_Y$ or $2.82\sigma_Y$ (where σ_Y is the tensile yield stress) on either the Tresca or von Mises yield criterion respectively. This shows that for an ideally deep and sharp notch in an infinite solid, the plastic constraint raises the yield stress to a value of approximately $3\sigma_Y$. This gives us the following classification for brittle–ductile behaviour, first proposed by Orowan[1]:

(1) If $\sigma_B < \sigma_Y$ the material is brittle.
(2) If $\sigma_Y < \sigma_B < 3\sigma_Y$, the material is ductile in an unnotched tensile test, but brittle when a sharp notch is introduced.
(3) If $\sigma_B > 3\sigma_Y$, the material is fully ductile, i.e. ductile in all tests, including those in notched specimens.

Vincent and others have recognized that this distinction can be applied to polymers. Their arguments have, however, been based on the more qualitative ideas of Parker[86]. Parker argued that the effect of the notch is to produce a triaxial stress system. The constraints in the contraction of a notched bar produce transverse tensions σ_2 and σ_3 in both the width and thickness directions (Figure 12.21).

For an unnotched bar, plastic flow occurs on the Tresca criterion when $\tau_{max} = \sigma_1/2$, i.e. the yield stress $\sigma_Y = 2\tau_{max}$. For a notched bar, Parker assumed that $\sigma_3 \sim (\frac{2}{3})\sigma_Y$, i.e. $(\frac{4}{3})\tau_{max}$. Yield then occurs when

$$\tau_{max} = \tfrac{1}{2}(\sigma_1 - \sigma_3) = \tfrac{1}{2}(\sigma_1 - \tfrac{4}{3}\tau_{max})$$

or

$$\sigma_1 = (\tfrac{10}{3})\tau_{max} = 1.7\sigma_Y.$$

Vincent then follows Parker in distinguishing two types of failure: triaxial tensile failure which is brittle, and shear failure which is tough or ductile. A

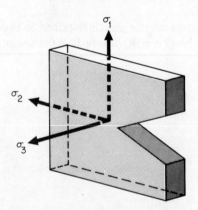

12.21. The stress system near the tip of a notch. [Redrawn with permission from Parker, *Brittle Behaviour of Engineering Structures*, Wiley, New York, 1957.]

sharp notch increases the triaxial tension relative to the shear stress, thus accentuating the possibility of brittle fracture. This explanation is similar to that given by Orowan but does not explicitly state that the brittle fracture remains unaltered by the notch and that only the yield behaviour is affected.

Vincent's σ_B–σ_Y Diagram

We may ask how relevant these ideas are to the known behaviour of polymers. Vincent[87] has constructed a σ_B–σ_Y diagram which is very instructive in this respect (Figure 12.22).

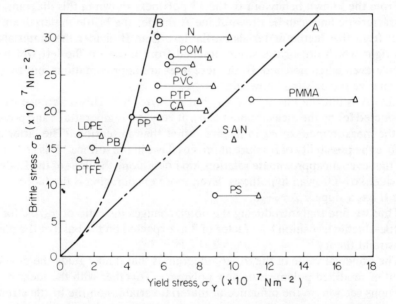

Figure 12.22. Plot of brittle stress at about $-180\,°C$ against a line joining yield-stress values at $-20\,°C$ ($\triangle$) and $20\,°C$ ($\bigcirc$) respectively for various polymers. Line A divides polymers which are brittle unnotched from those which are ductile unnotched but brittle notched, and line B divides polymers which are brittle notched, but ductile unnotched, from those which are ductile even when notched.

PMMA	Polymethyl methacrylate	PVC	Polyvinyl chloride
PS	Polystyrene	PTP	Polyethylene terephthalate
SAN	Copolymer of styrene and acrylonitrile	CA	Cellulose acetate
N	Nylon 66	PP	Polypropylene
POM	Polyoxymethylene	LDPE	Low density polyethylene
PC	Polycarbonate	PB	Polybutene-1
		PTFE	Polytetrafluoroethylene

[Redrawn with permission from Vincent, *Plastics*, **29**, 79 (1964).]

The term σ_Y was taken as the yield stress in a tensile test at a strain rate of about 50% per minute; for polymers which were brittle in tension, σ_Y was the yield stress in uniaxial compression, and σ_B was the fracture strength measured in flexure at a strain rate of 18 min^{-1} at $-180\,°C$.

The yield stresses were measured at $+20$ and $-20\,°C$, the idea being that the $-20\,°C$ values would give a rough indication of the behaviour in impact at $+20\,°C$, i.e. lowering the temperature by 40 °C is assumed to be equivalent to increasing the strain rate by a factor of about 10^5.

In the diagram the circles represent σ_B and σ_Y at $+20\,°C$; the triangles σ_B and σ_Y at $-20\,°C$. As we have already discussed, both σ_Y and σ_B are affected by subsidiary factors such as molecular weight and the degree of crystallinity so that each point can only be regarded as of first-order significance.

From the known behaviour of the 13 polymers shown in this diagram, two characteristic lines can be drawn. Line A divides the brittle materials on the right from the ductile materials on the left. Line B divides the materials on the right which are brittle when notched from those on the left which are ductile even when notched. Both these lines are approximations, but they do summarize the existing knowledge.

Line A is not the line $\sigma_B/\sigma_Y = 1$, but is $\sigma_B/\sigma_Y \sim 2$. This difference can be accounted for by the measurement of σ_B at very low temperatures and possibly by the measurement of σ_B in flexure rather than in tension. (The latter may reduce the possibility of fracture at serious flaws in the surface.) It is encouraging that even an approximate relationship holds along the lines of the Ludwig–Davidenkov–Orowan hypothesis. Even more encouraging is the fact that the line B has a slope $\sigma_B/\sigma_Y \sim 6$.

Thus we find that introducing the notch changes the ratio of σ_B/σ_Y for the brittle–ductile transition by a factor of 3 as expected on the basis of the plastic constraint theory.

The principal value of the σ_B–σ_Y diagram is that it may guide the development of modified polymers or new polymers. Together with the ideas of the previous section on the influence of material variables on the brittle strength and yield stress, it can lead to a systematic search for improvements in toughness.

12.5.2 A Theory of Brittle–Ductile Transitions Consistent with Fracture Mechanics: Fracture Transitions

The discussion of brittle–ductile transitions in the previous section assumes that brittle failure can be defined by a critical tensile stress. Although the results are very instructive, this assumption takes no account of the fact that there are what are termed size effects in the brittle behaviour of materials. In practice this means that there is a characteristic length associated with each fracture test which will determine the severity of the test, where high severity means a greater propensity for brittle failure.

The arguments which lead to an understanding of size effects in brittle fracture stem from the basic ideas of energy scaling and similarity. These ideas were first appreciated by Roesler[88] and used by Benbow and Roesler[1] in their pioneering research described in Section 12.2. Their significance with regard to brittle-ductile transitions has been recognized by Puttick[89-91], who developed a theory of fracture transitions which embraces the ideas of fracture mechanics.

To fix our ideas, consider a crack propagating in a brittle material under conditions of constant grip displacement, as in the plate with a centre crack $2c$ (Figure 12.3). According to the Griffith theory of fracture the *surface* energy of the crack is supplied by the *volume* strain energy density stored in the material. The strain energy release rate G is therefore proportional to a *length* times the strain energy density U per unit volume. For this case of a homogeneous stress field, the characteristic length is the crack length and we have

$$G = \beta c U \tag{12.25}$$

where β is a non-dimensional constant and $U = \sigma^2/2E$.

As previously discussed the fracture stress σ_B is given by

$$\sigma_B = \left(\frac{G_c E}{\pi c}\right)^{1/2}, \tag{12.26}$$

i.e. the fracture stress is determined by the material parameters G_c and E and a characteristic length which is the length of the crack.

In most real situations the stress field is inhomogeneous (i.e. finite with respect to the length of the crack) and the characteristic length is then to be identified with a characteristic length x_0 associated with the stress field, e.g. the size of a plastic zone or the length of a craze. We then have

$$G = \beta'\left(\frac{x_0}{c}\right) x_0 U, \tag{12.27}$$

where the concept of geometric similarity is invoked to enable us to conclude that the function β' depends only on (x_0/c).

In this case the fracture stress σ_B is given by

$$\sigma_B = \left(\frac{G_c E}{x_0 \beta'(x_0/c)}\right)^{1/2}, \tag{12.28}$$

i.e. the fracture stress is a function of the scale of the stress field which enters directly through x_0. The non-dimensional function β' is evaluated by the methods of fracture mechanics.

Now consider the implications of these ideas for brittle–ductile transitions. This transition is marked by the change from brittle to ductile failure because the stress reaches the yield stress in a part of the specimen.

Equation (12.28) can equally well be written as defining the critical size of the stress field in terms of the characteristic length x_0. Thus we have

$$x_0 = \frac{G_c E}{\sigma_B^2 \beta'(x_0/c)}. \qquad (12.29)$$

Now consider decreasing the characteristic length (by changing the test and hence changing the stress field) so that σ_B rises until it eventually reaches the value of the yield stress σ_Y. This causes a brittle–ductile transition, which can be defined by a critical length x_0^Y, where

$$x_0^Y = \frac{G_c E}{\sigma_Y^2 \beta'(x_0/c)}. \qquad (12.30)$$

Fracture then occurs in a plastic elastic rather than a purely elastic strain field. For example, in the double cantilever beam test (Figure 12.4) the maximum bending stress is $\sigma = (3G_c E/b)^{1/2}$ where b is the width of the beam. Hence the critical width for the transition from yielding to brittle failure is

$$b_c = 3G_c E/\sigma_Y^2. \qquad (12.31)$$

Puttick terms these transitions *lower* transitions, because they just mark the point where plastic flow commences.

A second type of transition, termed an *upper* transition, corresponds to the size of the plastic zone reaching a maximum dimension characteristic of the test. An example here is the critical size of the plastic zone at the tip of a crack in plane strain which gives

$$x_{0c} = G_c E/\sigma_Y^2. \qquad (12.32)$$

Another example is a notched bar test where, as we have seen, $\sigma_{max} \sim 2.5\sigma_Y$, and it can be shown that the critical zone size is

$$x_{0c} = G_c E/25\sigma_Y^2. \qquad (12.33)$$

We therefore see that the most acceptable approach to brittle–ductile transitions or plane strain–plane stress transitions, i.e. all types of fracture transition, is to regard each test as relating to a characteristic length in a particular test. The transition is then characterized by a critical length x_{0c}, where $x_{0c} = \alpha G_c E/\sigma_Y^2$ and α is a numerical constant whose value is determined by the stress field in the test.

In terms of material behaviour, it is the quantity $G_c E/\sigma_Y^2$ which determines brittle–ductile behaviour. In Table 12.2 the situation is summarized for some typical tests and the implications of each test are indicated.

To summarize, the choice of the particular fracture test determines α, and defines a critical length, e.g. the width of the beam in the double cantilever beam test piece or the plastic zone size at general yield of a notched bar. The fracture transition then occurs at the temperature at which the quantity $\alpha G_c E/\sigma_Y^2$ is equal to the critical length in the chosen test. For a given test

Table 12.2. Fracture transitions.

Test	α	Nature of transitions	References
Notched bar (Charpy bend)	~0.04	Upper (below to above gross yield)	Griffith and Oates[93], Puttick[91]
Plane strain fracture	~$\frac{1}{2}$	Upper (plane strain to plane stress)	Irwin[94]
Double cantilever	3	Lower (elastic to elastic plastic)	Gurney and Hunt[95]
Indentation by spherical metal ball	~25	Upper (radial fracture to no fracture)	Puttick[91]

σ_Y decreases with increasing temperature until this equality is satisfied and the transition from brittle to ductile behaviour occurs. We can now see the link between this rigorous treatment and the more simplistic approach of Vincent, described in Section 12.5.1, which is of considerable practical value. As pointed out by Puttick, it would be more accurate to replace the simplistic diagrams 12.17 and 12.18 by curves which relate the critical characteristic length as the dependent variable plotted against temperature as shown in Figure 12.23. If we fix the specimen dimension at the value given by the

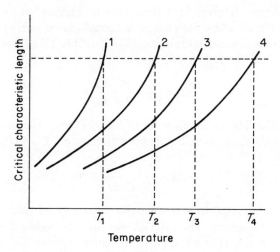

Figure 12.23. Schematic comparison of the brittle–ductile temperature transition in four different tests: 1, Hertzian indentation (lower transition); 2, plastic-elastic indentation (upper transition); 3, double cantilever beam (lower transition); 4, notched bar (upper transition). [Redrawn with permission from Puttick, *J. Phys. D*, **13**, 2249 (1980).]

436

horizontal dotted line, say 5 mm, then the transition temperatures are given by the temperatures T_1 to T_4.

12.6 IMPACT STRENGTH OF POLYMERS

A traditional measure of the fracture behaviour of a material is the Izod or Charpy impact test. The principle of both methods is to strike a small bar of polymer with a heavy pendulum swing. In the Izod test the bar is held vertically by gripping one end in a vice and the other free end is struck by the pendulum. In the Charpy test the bar is supported near its ends in a horizontal plane and struck either by a single pronged or two-pronged hammer so as to simulate a rapid three-point or four-point bend test respectively (Figure 12.24(a)).

It is customary to introduce a centre notch into the specimen so as to add to the severity of the test, as discussed in Section 12.5.1 above. The standard Charpy impact specimen has a 90° V-notch with a tip radius of 0.25 mm. For polymers a very much sharper notch is often adopted by tapping a razor blade into a machined crack tip with important consequences for the interpretation of the subsequent impact test.

We have already seen one reason for a physical difference between polymer specimens in an impact test—the presence or absence of a craze at the crack tip. In fact, the interpretation of impact tests is not straightforward and it is necessary to consider several alternatives.

(1) The polymer deforms in a linear elastic fashion up to the point of failure, which occurs when the change in stored elastic energy due to crack growth satisfies the Irwin–Kies relationship (equation (12.10) above). We then have

$$G_c = \frac{K_c^2}{E^*} = \frac{P_0^2}{2B}\frac{dC}{dc},$$

where P_0 is the load immediately prior to fracture. Since the elastically stored energy in the specimens immediately prior to failure is $U_0 = \frac{1}{2}P_0^2 C$,

$$G_c = \frac{U_0}{B}\frac{1}{C}\frac{dC}{dc} \tag{12.34}$$

U_0 is usually determined in a commercial impact tester from the potential energy lost due to impact. The total energy must be reduced by the energy associated with the kinetic energy of the sample to give U_0, and the term $[(1/C)(dC/dc)]$ is calculable from the geometry of the specimen. This approach was first proposed by Brown[96], and by Marshall, Williams and Turner[97]. It has been shown to give values for G_c which are independent of specimen geometry for impact tests on razor-notched samples of several glassy polymer, including polymethyl methacrylate, polycarbonate[98] and polyether sulphone[99]. Similar results have also been obtained for razor-notched samples of polyethylene[100].

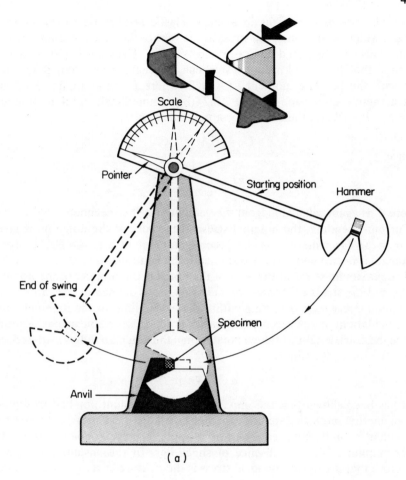

(a)

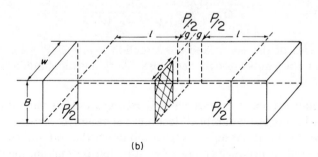

(b)

Figure 12.24. (a) Schematic drawing of a Charpy impact tester; (b) the notched Charpy impact specimen.

(2) The polymer deforms in a linear elastic fashion up to the point of failure, which occurs when the stress at the root of the notch reaches a critical value. This was suggested by Fraser and Ward[101]. The Charpy test, as undertaken in the Hounsfield impact tester, can be regarded as a four-point bend test with the bending moment $M = Pl/2$, where P is the applied load and l is a sample dimension (Figure 12.24(b)). Immediately prior to fracture, $M = M_0$, $P = P_0$ and the elastically stored energy is

$$U_0 = \tfrac{1}{2}(2M_0/l)^2 C.$$

Hence

$$M_0 = \frac{l}{2}\sqrt{\frac{2U_0}{C}},$$

where C is again calculable from the geometry of the specimen.

For pure bending, the nominal stress at the root of the notch σ_n is given by $\sigma_n = (M/I)y$, where I is the second moment of area ($= Bt^3/12$ for a rectangular beam) and y is the distance to the neutral axis.

The maximum stress at the root of the notch is the product of the nominal stress and the stress concentration factor α_k. α_k is defined as the ratio of the maximum stress σ_{max} occurring within the elastic limit to the nominal stress σ_n. Calculations of α_k for general shapes of notch are available in the literature. When the crack length c is much greater than the notch tip radius ρ, α_k reduces to the simple expression

$$\alpha_k = 2\sqrt{c/\rho}.$$

It has been shown that the impact behaviour of blunt notched specimens of polymethyl methacrylate are consistent with a critical stress at the root of the notch[101], and similar considerations apply to polycarbonate[98] and polyether sulphone[99] in the absence of shear lips. In this instance it appears therefore that the maximum local stress is the fracture criterion, independent of specimen geometry.

(3) The impact test is a measure of the energy required to propagate the crack across the specimen. In this case we have

$$G_c = \frac{U_0}{A} = \frac{U_0}{BW(1-c/W)}, \tag{12.35}$$

where the area of the uncracked cross section is $A = B(W - c)$. This is the situation for the rubber-toughened ABS (Acrylonitrile–butadiene–styrene) polymers.

An alternative impact test which is sometimes used is the falling weight impact test. In this test a dumbbell test piece is mounted vertically at the end of a large vertical rod. Annular weights fall down the rod and the minimum energy required to produce fracture is estimated by making measurements on about 20 test pieces. The reader is referred to the literature for a comprehensive account of such tests[102].

The Charpy test, and to a rather lesser extent the Izod tests, are the only impact tests which have been considered in terms of a satisfactory theoretical analysis. Even for these tests, however, there is still a gap between the engineering analysis and any accepted interpretation in physical terms. For example, although it seems likely that the brittle failure of razor-notched impact specimens is associated with the craze at the crack tip, there is no convincing numerical link between craze parameters and the fracture toughness K_{IC}, as exists for the cleavage fracture of compact tension specimens (see Section 12.2 above). Again, although the mechanics point to a critical stress criterion for some blunt notched specimens and there is an empirical correlation with the craze stress determined in other ways, the magnitude of the critical stress is very great and suggests that a more sophisticated explanation may by required. For the brittle epoxy resins, which do not show a craze at the crack tip, Kinloch and Williams[103] have suggested that the fracture of both razor-notched and blunt-notched specimens can be described by a critical stress at a critical distance ($\sim 10 \, \mu m$) below the root of the notch.

We have seen that as we change temperature and strain rate in a polymer the nature of the stress–strain curve can change remarkably. It is therefore natural to seek for correlations between the area beneath the stress–strain curve and the impact strength and between dynamic mechanical behaviour and the impact strength. Attempts to make such correlations directly have met with mixed success[104], which is not surprising in view of the complex quantitative interpretation of impact strength suggested above.

Vincent[105] has examined the statistical significance of a possible inverse correlation between impact strengths and dynamic modulus and concluded that, at best, this correlation only accounts for about two-thirds of the variance in impact strength. There are factors such as the influence of molecular weight, and details of molecular structure such as the presence of bulky side groups, which are not accounted for. Vincent[105] also reported impact tests over a wide temperature range on some polymers, notably polytetrafluorethylene and polysulphone, where peaks in brittle impact strength were observed at temperatures close to dynamic loss peaks. This suggests that in some instances it may be necessary to consider the relevance of generalized fracture mechanics (Section 12.12 below), where the viscoelastic losses occurring during loading and unloading must be taken into account.

12.6.1 High Impact Polyblends

The impact strength of many well known polymers such as polymethyl methacrylate, polystyrene and polyvinyl chloride is comparatively low. This led to the production of rubber-modified thermoplastics with high impact strength. The best known example is high impact polystyrene in which a rubber is dispersed throughout the polystyrene in the form of small aggregates or balls. Nielsen[106] lists three conditions which are required for an effective polyblend:

(1) The glass temperature of the rubber must be well below the test temperature.

(2) The rubber must form a second phase and not be soluble in the rigid polymer.

(3) The two polymers should be similar enough in solubility behaviour for good adhesion between the phases.

There are several theories which have been proposed for the action of rubbers in improving the impact strength. One is that the rubber becomes stretched during the fracture process and absorbs a great deal of energy. Another theory proposes that the rubber particles act to introduce a multiplicity of stress concentration points. Thus there are many cracks which propagate during the fracture process rather than a single crack. It is proposed that this requires greater energy due to the production of more new surfaces (relating the fracture energy to the new surface area by Griffith's theory of fracture).

Studies by Bucknell[107,108] have linked high impact strength in modified polystyrene polymer to craze formation. Bucknell and Bucknell and Smith[109] compared the force–time curves for impact specimens over a range of temperatures, with both the notched Izod impact strength and the falling weight impact strength and the nature of the fracture surface. The force–time curves, such as in Figure 12.25(a) regions similar to those observed for a homopolymer as discussed in the introduction above. Both impact strength tests also showed three regions (Figure 12.25(b) and (c)). The fracture surfaces at the lowest temperature were quite clear, whereas at high temperatures stress-whitening or craze formation occurred. These three regions were explained as follows:

(1) *Low temperatures.* The rubber is unable to relax at any stage of fracture. There is no craze-formation and brittle fracture occurs.

(2) *Intermediate temperatures.* The rubber is able to relax during the relatively slow build-up of stress at the base of the notch, but not during the fast crack propagation stage. Stress-whitening occurs only in the first (precrack) stage of fracture, and is therefore confined to the region near the notch.

(3) *High temperatures.* The rubber is able to relax even in the rapidly forming stress field ahead of the travelling crack. Stress-whitening occurs over the whole of the fracture surface. Bucknell and Smith[109] report similar results for other rubber-modified impact polymers.

12.6.2 Crazing and Stress-whitening

Bucknell and Smith[109] have remarked on the connection between crazing and stress-whitening. High impact polystyrenes are manufactured by the incorporation of rubber particles into the polystyrene. It was observed that the

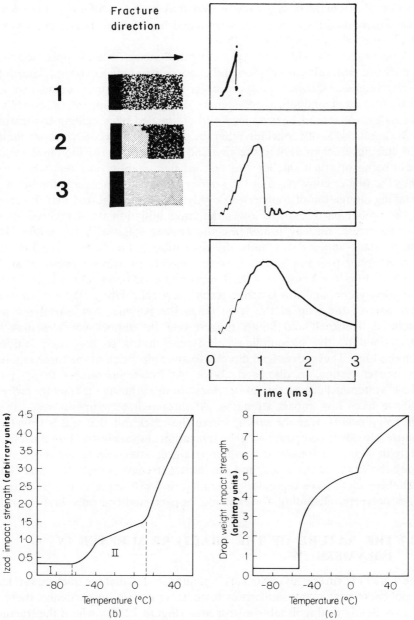

Figure 12.25. (a) Fracture surfaces of modified polystyrene notched Izod impact specimens: *top*, broken at −70 °C, type I fracture; *centre*, broken at 40 °C, type II fracture; *bottom*, broken at 150 °C, type III fracture. (b) Notched Izod impact strength of modified polystyrene as a function of temperature, showing the limits of the three types of fracture behaviour. (c) Drop weight impact strength of 2.03 mm high impact polystyrene sheet as a function of temperature. [Redrawn with permission from Bucknell, *Brit. Plast.*, **40**, 84 (1967).]

fracture of this material is usually preceded by opaque whitening of the stress area. Figure 12.11 shows a stress-whitened bar of high impact polystyrene which failed at an elongation of 35%.

A combination of different types of optical measurements (polarized light to measure molecular orientation and phase contrast microscopy to determine refractive index) showed that these stress-whitened regions are similar to the crazes formed in unmodified polystyrene. They are birefringent, of low refractive index, capable of bearing load and are healed by annealing treatments.

Bucknell and Smith concluded that the difference between stress-whitening and crazing exists merely in the size and concentration of the craze bands, these being of much smaller size and greater quantity in stress-whitening. Thus the higher conversion of the polymer into crazes accounts for the high breaking elongation of toughened polystyrene. It is suggested that the effect of the rubber particles is to lower the craze initiation stress relative to the fracture stress, thereby prolonging the crazing stage of deformation. The crazing stage appears to require the relaxation of the rubber. The function of the rubber particles is not, however, merely to provide points of stress concentration. It is known that there must be a good bond between the rubber and polystyrene, and this is achieved by chemical grafting. The rubber must bear part of the load at the stage when the polymer has crazed but not fractured. Bucknell and Smith suggest that the rubber particles may be constrained by the surrounding polystyrene matrix so that their stiffness remains high. These ideas lead directly to an explanation of the three regimes for impact testing, as discussed above. At low temperatures there is no stress-whitening because the rubber does not relax during the fracture process and we have low impact strengths. At intermediate temperatures, stress-whitening occurs near the notch, where the crack initiates and is travelling sufficiently slowly compared with the relaxation of the rubber. Here the impact strength increases. Finally at high temperatures, stress-whitening is observed along the whole of the crack, and the impact strength is high. It seems likely that these ideas have a greater generality, and will apply to the fracture of other polymers, including, for example, impact-modified polyvinyl chloride.

12.7 THE NATURE OF THE FRACTURE SURFACE IN POLYMERS

The work of Irwin and his coworkers[110] demonstrated that the fracture surfaces in polymers bear many similarities to those in metals. In particular there is often a clearly distinguishable mirror area (Figure 12.26), where the fracture commences, which is surrounded by parabolic markings due to the interference of the main crack with new cracks nucleating ahead of it (Reference 85, p. 350).

A review article[111] shows that changes in both the rate of loading and molecular weight can produce identical changes in the fracture surface of polymethyl methacrylate. These results are summarized in Figures 12.27 and 12.28 taken from this article[111]. It can be seen that the mirror-like area where

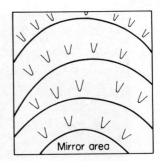

Figure 12.26. The fracture surface. [Redrawn with permission from Zandman, Publication Scientifiques et Techniques de Ministère de l'Air, Centre de Documentation de L'Armament, Paris.]

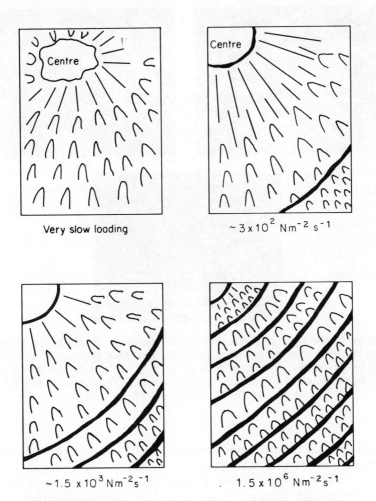

Figure 12.27. Effect of rate of loading on fracture appearance of polymethyl methacrylate. [Redrawn with permission from Wolock, Kies and Newman, in *Fracture* (B. L. Averbach *et al.*, eds), Wiley, New York, 1959, p. 250.]

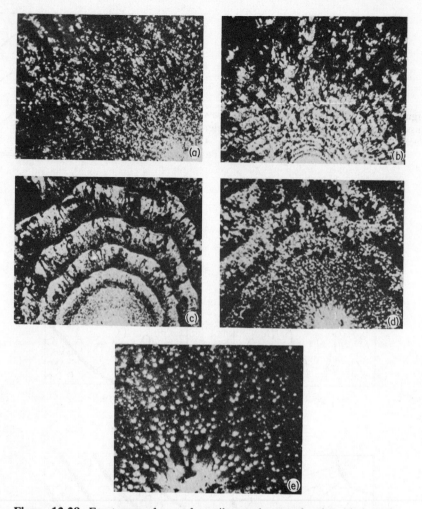

Figure 12.28. Fracture surfaces of tensile specimens of polymethyl methacrylate of various molecular weights: (a) 10^4 daltons; (b) 1.2×10^5 daltons; (c) 2×10^5 daltons; (d) 4.9×10^5 daltons; (e) 3.16×10^6 daltons. [Reproduced with permission from Wolock, Kies and Newman, in *Fracture* (B. L. Averbach *et al.*, eds), Wiley, New York, 1959, p. 250.]

the crack initates becomes more extensive either as the rate of loading is *decreased* or as the molecular weight is *increased*. This corresponds to an increase in the work done during fracture according to the idea of Bucknell and Smith for high impact polystyrene and to the experimental results of Berry for polymethyl methacrylate. It is also clear that changing the test temperature produces similar changes in the fracture surface. These changes again occur in the way which we would expect from the high impact polystyrene results, viz. that the mirror area is associated with the craze that forms prior to the occurrence of the crack.

12.8 CRACK PROPAGATION

For brittle fracture of materials, Mott[112] has shown that if the kinetic energy of the moving crack is included in the Griffith energy balance equation, it can be shown that there will be a limiting velocity for crack propagation which is proportional to the velocity of sound in the material. This has been verified for metals and polycrystals by Schardin[113] and others. It was extended to polymers by Bueche and White[114] and Schardin[112] but the results illustrated in Figure 12.29 are not so precise as for metals. Bueche and White[114]

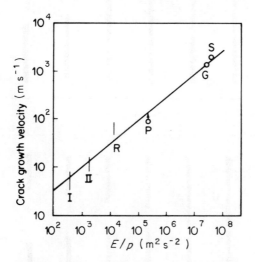

Figure 12.29. Limiting velocity of crack growth versus modulus E/density ρ. I and II are silicone rubbers; R, P, G and S are irradiated polyethylene, polymethyl methacrylate, glass and fused silica, respectively. [Redrawn with permission from Bueche and White, *J. Appl. Phys.*, **27**, 980 (1956).]

commented that although their results are consistent with the Griffith theory of pre-existing cracks, an alternative theory due to Poncelet[115], assuming that the cracks are produced by stress (the 'flaw genesis' theory), gives an almost identical numerical value for the limiting velocity.

It should be noted, however, that the mechanisms of crack propagation may be different from that of crack initation, so these experiments do not necessarily give information regarding the latter.

12.9 BRITTLE FRACTURE BY STRESS PULSES

Considerable support for the Griffith theory of fracture came from the observation that, in glass, cracks originate at the surface, and are absent in freshly prepared material, giving rise to very high observed strengths in the latter case.

The investigations of Bueche and White[114] on crack propagation in polymers were based on rapid photographic examination of the fracture. A typical set of results for a silicone rubber is shown in Figure 12.30. It is seen that the

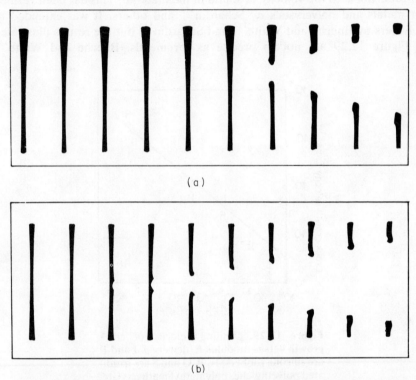

(a)

(b)

Figure 12.30. Fracture of silicone rubber at (a) 16 300 frames s^{-1} and (b) 9100 frames s^{-1}. [Reproduced with permission from Bueche and White, *J. Appl. Phys.*, **27**, 980 (1956).]

fracture can commence either at the surface or in the interior of the polymer. No obvious imperfection could be found from microscopic examination.

The question of the origin of cracks has been studied in more detail by using stress pulses to produce internal fractures. By this technique Kolsky[116] has compared the 'internal strengths' of polystyrene and polymethyl methacrylate with that of soda glass. Internal fractures were produced in cylindrical rods of polymer by detonating a small explosive charge at the centre of one of the end faces, as in Figure 12.31. Identical experiments on glass cylinders could not produce internal fracture, the glass fracturing completely across or

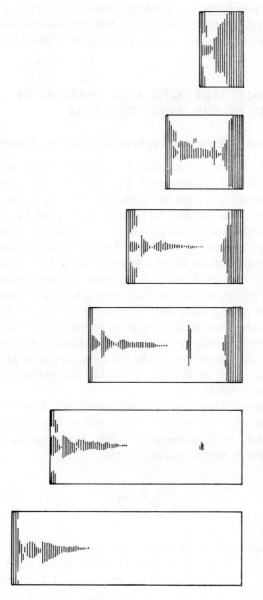

Figure 12.31. Schematic diagram of fractures in 25.4 mm diameter polymethyl methacrylate cylinders. [Reproduced with permission from Kolsky, in *Fracture* (B. L. Averbach *et al.*, eds), Wiley, New York, p. 281.]

not at all. This showed that the internal strength of glass is much greater than the surface strength.

Similar conclusions were obtained by Bowden and Field[117] in a comparison of the brittle fracture of glass and polymethyl methacrylate. In glass many changes in the fracture behaviour could be obtained by etching procedures, whereas in polymethyl methacrylate surface modifications did not affect fracture, confirming that in this case the cracks originate at flaws throughout the material.

12.10 THE TENSILE STRENGTH AND TEARING OF POLYMERS IN THE RUBBERY STATE

12.10.1 The Tearing of Rubbers: Application of Griffith Theory

The tearing of rubbers has been extensively studied by Rivlin and Thomas[118] and by Thomas[119] and his collaborators. The Griffith theory implies that the quasi-static propagation of a crack is a reversible process. Rivlin and Thomas recognized, however, that this may be unnecessarily restrictive, and that the reduction in elastically stored energy due to the crack propagation may be balanced by changes in energy other than that due to an increase in surface energy. Their approach was to define a quantity termed the 'tearing energy', which is the energy expended per unit thickness per unit increase in crack length. The tearing energy includes surface energy, energy dissipated in plastic flow processes and energy dissipated irreversibly in viscoelastic processes. Providing that all these changes in energy are proportional to the increase in crack length and are primarily determined by the state of deformation in the neighbourhood of the tip of the crack, then the total energy will still be independent of the shape of the test piece and the manner in which the deforming forces are applied.

In formal mathematical terms, if the crack increases in length by an amount dc, an amount of work $Tt\,dc$ must be done, where T is the tearing energy per unit area and t is the thickness of the sheet. Equating the work done to the change in elastically stored energy we have

$$-\left[\frac{\partial U}{\partial c}\right]_l = Tt. \tag{12.36}$$

The suffix l indicates that differentiation is carried out under conditions of constant displacement of the parts of the boundary which are not force-free. Equation (12.36) is similar in form to (12.1) above but T is defined for unit thickness of specimen and is therefore equivalent to 2γ in equation (12.1). As in the case of glassy polymers, T is not to be interpreted as a surface free energy, but involves the total deformation in the crack tip region as the crack propagates.

It is possible to choose particularly simple cases where the equation can be immediately evaluated. For example, consider the so-called 'trouser tear' experiment shown in Figure 12.32. After making a uniform cut in a rubber sheet the sample is subjected to tear under the applied forces F. The stress distribution at the tip of the tear is complex, but providing that the legs are long, is independent of the depth of the tear.

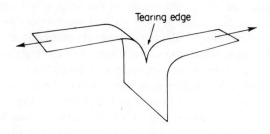

Figure 12.32. The standard 'trouser tear' experiment showing the sample before stretching (a) and when under test (b).

If the sample tears a distance Δc under the force F, the work done is given by $\Delta W = 2F\,\Delta c$. This ignores any changes in extension of the material between the tip of the tear and the legs.

Since the tearing energy $T = \Delta W/t\,\Delta c$,

$$T = 2F/t$$

and can be measured easily.

Rivlin and Thomas[118] found that two characteristic tearing energies could be defined, one for very slow rates of tearing $(T = 37\ \text{kJ m}^{-2})$ and one for catastrophic growth $(T = 130\ \text{kJ m}^{-2})$ and that both these quantities were independent of the shape of the test piece.

It is important to note that the tearing energy of a rubber does not relate directly to tensile strength. The tearing energy is the energy required to extend the rubber to its maximum elongation. It depends on the shape of the stress–strain curve together with the viscoelastic nature of the rubber. For example, we may contrast two different rubbers, the first possessing a high tensile strength but a very low elongation to fracture and very low viscoelastic losses, and the second possessing a low tensile strength but a high elongation to fracture and high viscoelastic losses. In spite of its comparatively low tensile strength the second rubber may still possess a high tearing energy.

450

12.10.2 The Tensile Strength of Rubbers

Bueche and Berry[120] have considered the relevance to rubbers of the Griffith equation

$$\sigma_B = \left(\frac{2\gamma E}{\pi c}\right)^{1/2}$$

relating tensile strength to surface energy, modulus and crack length. This equation was modified to take into account the large strain conditions occurring at the breaking extension of rubbers and used to derive values for the surface energies of a series of silicone elastomers assuming a constant critical crack size of 10^{-2} mm. As in the calculations above for glassy polymers the measured surface energies were two orders of magnitude greater than those calculated. Looked at in another way, the calculated surface energies would imply crack sizes of 10^{-4} mm which were much smaller than those indicated by direct examination of the sample. It was therefore concluded that most of the energy involved in the breaking process is dissipated in viscoelastic and flow processes.

Bueche and Berry[120] also made a direct test of the Griffith criterion by examining the relationships between tensile strength, Young's modulus and the size of the deliberately introduced cracks. Their results are summarized in Figures 12.33 and 12.34. They found a linear, rather than a square root, dependence of tensile strength on both crack length and modulus. This led

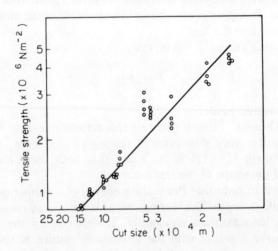

Figure 12.33. The relationship between cut size and tensile strength of a filled vulcanized silicone elastomer. [Redrawn with permission from Bueche and Berry, in *Fracture* (B. L. Averbach *et al.*, eds), Wiley, New York, 1959, p. 265.]

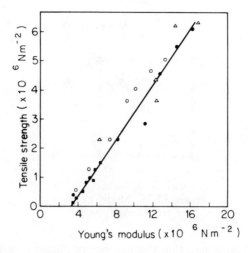

Figure 12.34. The relationship between tensile strength and Young's modulus for a series of silicone elastomers containing various fillers. $\bigcirc$, Santocel; $\bullet$, Aerosil; $\triangle$, treated Santocel; $\blacksquare$, Carrara. [Redrawn with permission from Bueche and Berry, in *Fracture* (B. L. Averbach *et al.*, eds), Wiley, New York, 1959, p. 265.]

them to reconsider their data in terms of the critical stresses developed at the breaking points for a series of irradiated silicone elastomers. It was found that the values for the critical stress were very similar, in spite of the very different ultimate extensions of these polymers. This, taken in conjunction with the surface energy calculations, led to the conclusion that, for elastomers, a critical stress criterion for rupture is preferable to the Griffith criterion.

12.10.3 Molecular Theories of the Tensile Strength of Rubbers

Most molecular theories of the strength of rubber treat rupture as a critical stress phenomenon. It is accepted that the strength of the rubber is reduced from its theoretical strength in a perfect sample by the presence of flaws. Moreover, it is assumed that the strength is reduced from that of a flawless sample by approximately the same factor for different rubbers of the same basic chemical composition. It is then possible to consider the influence on strength of such factors as the degree of cross-linking and the primary molecular weight.

Bueche[121] has considered the tensile strength of a model network consisting of a three-dimensional net of cross-linked chains (see Figure 12.35).

Consider the cube of the material with edges of length 10 mm parallel to the three chain directions in the idealized network. Assume that there are ν

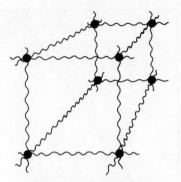

Figure 12.35. Model network
of cross-linked chains.

chains in this unit cube and that the number of chains in each strand of the network is n. There are then n^2 strands passing through each face of the cube. To relate the number n to the number of chains per unit volume of the network (which forms the link with rubber elasticity theory) we note that the product of the number of strands passing through each cube face and the number of chains in each strand will be $\frac{1}{3}\nu$ since there are three strand directions. Thus

$$n = \tfrac{1}{3}\nu, \qquad n = (\nu/3)^{1/3}. \tag{12.37}$$

Now consider a stress σ applied to the specimen parallel to one of the three strand directions. Further consider that the specimen fractures such that the strands break simultaneously at an individual fracture stress σ_c. Then

$$\sigma_B = n^2 \sigma_c$$

which from equation (12.37) can be written as

$$\sigma_B = (\nu/3)^{2/3} \sigma_c.$$

For real network ν is the number of effective chains per unit volume. It is given in terms of the actual number of chains per unit volume ν_a by the Flory relationship

$$\nu = \nu_a [1 - 2\bar{M}_c/\bar{M}_n],$$

where $\bar{M}_c$ and $\bar{M}_n$ are the average molecular weight between cross-links and the number average molecular weight of the polymer, respectively. (Note that for a network there must be at least two cross-links per chain i.e. $\bar{M}_n > 3\bar{M}_c$.)

This gives

$$\sigma_B \propto [1 - 2\bar{M}_c/\bar{M}_n]^{2/3}.$$

Bueche remarked that the variation of tensile strength with the polymer molecular weight $\bar{M}_n$, found by Flory[122] for butyl rubber, follows the predicted

$[1 - 2\bar{M}_c/\bar{M}_n]^{2/3}$ relationship. The variation of tensile strength with degree of cross-linking was also studied by Flory *et al.*[123] for natural rubber. Although there was the expected increase in tensile strength with increasing degree of cross-linking, it was also found that the strength decreased again at very high degrees of cross-linking. Flory attributed this decrease to the influence of cross-links in the crystallization of the rubber. However, a similar effect was observed for the non-crystallizing SBR rubber by Taylor and Darin[124], which led Bueche[125] to suggest an alternative explanation. He proposed that the simple model described above fails because of the assumption that each chain holds the load at fracture. Although this may be a good approximation at low degrees of cross-linking, it can be shown to be less probable at high degrees of cross-linking.

It is of considerable technological importance that the tensile strength of rubbers can be much increased by the inclusion of reinforcing fillers such as carbon black and silicone. These fillers increase the tensile strength by allowing the applied load to be shared amongst a group of chains, thus decreasing the chance of a break to propagate[126].

12.11 EFFECT OF STRAIN RATE AND TEMPERATURE

Another area in the fracture of polymers which has been studied extensively concerns the influence of strain rate and temperature on the tensile properties of elastomers and amorphous polymers. The principal experimental contribution to this area has been made by Smith and his coworkers[127–129], who measured the variation of tensile strength and ultimate strain as a function of strain rate for a number of elastomers. It was found that the results for different temperatures could be superimposed, by shifts along the strain rate axis, to give master curves for tensile strength and ultimate strain as a function of strain rate. Results of this nature are shown in Figure 12.36 which summarizes Smith's data for an unfilled SBR rubber. Remarkably, the shift factors obtained from superposition of both tensile strength and ultimate strain took the form predicted by the WLF equation for the superposition of low strain linear viscoelastic behaviour of amorphous polymers (Figure 12.37). The actual value for T_g agreed well with that obtained from dilatometric measurements.

This result suggests that, except at very low strain rates and high temperatures, where the molecular chains have complete mobility, the fracture process is dominated by viscoelastic effects. Bueche[130] has treated this problem theoretically and obtained the observed form of the dependence of tensile strength on strain rate and temperature. Later theories have attempted to obtain the time dependence of both tensile strength and ultimate strain, or the time to break at a constant strain rate[131,132].

A final point is that Smith used these and other similar data to predict what he termed the 'failure envelope' for elastomers. The failure envelope is obtained by plotting $\log \sigma_B/T$ against $\log e$ and was found to be a unique

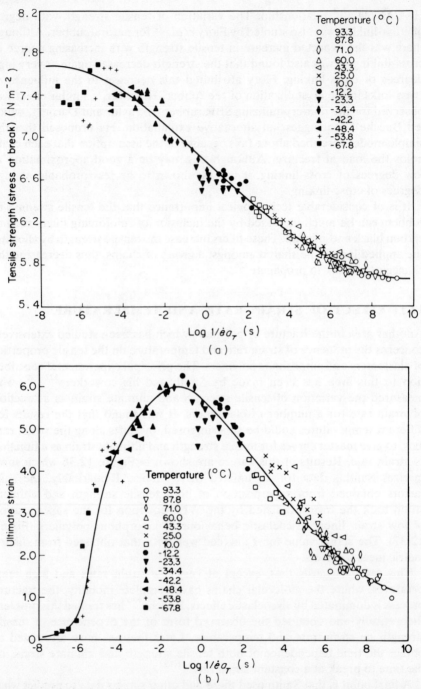

Figure 12.36. Variation of (a) tensile strength and (b) ultimate strain of a rubber with reduced strain rate $\dot{e}a_T$. Values were measured at various temperatures and rates and reduced to a temperature of 263 K. [Redrawn with permission from Smith, *J. Polymer Sci.*, **32**, 99 (1958).]

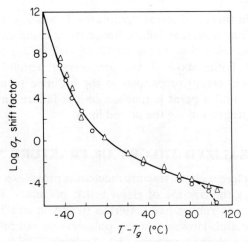

Figure 12.37. Experimental values of log a_T shift factor obtained from measurement of ultimate properties compared with those predicted using the WLF equation. △, From tensile strength; ○, from ultimate strain; —, WLF equations with $T_g = 263$ K. [Redrawn with permission from Smith, *J. Polymer Sci.*, **32**, 99 (1958).]

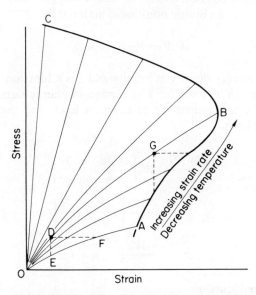

Figure 12.38. Schematic representation of the variation of stress–strain cuves with the strain rate and temperature. Envelope connects rupture points and the dotted lines illustrate stress relaxation and creep under different conditions. [Redrawn with permission from Smith and Stedry, *J. Appl. Phys.*, **31**, 1892 (1960).]

curve for all strain rates and test temperatures. It was also found[129] that the failure envelope can represent failure under more complex conditions such as creep and stress relaxation. In Figure 12.38 such failure can take place by starting from the initial stage G and progressing parallel to the abscissa (constant stress, i.e. creep) or parallel to the ordinate (constant strain, i.e. stress relaxation) until a point is reached on the failure envelope ABC, as indicated by the progress along the dotted lines.

12.12 A GENERALIZED THEORY OF FRACTURE MECHANICS

We have seen that linear elastic fracture mechanics provides a very satisfactory basis for discussing the fracture of glassy brittle polymers. The extension of fracture mechanics to rubber-like materials by Rivlin and Thomas has also proved very successful. However, many polymers do not fall into these two categories of brittle-elastic or rubber-like. Andrews[133,134] has therefore proposed a more generalized theory of fracture which is intended to form a broader base for the fracture of polymers.

Consider an infinite sheet containing a crack of length $2c$ loaded at infinity by an applied stress σ_0. This is shown in Figure 12.39(a) where X, Y are the Cartesian coordinates of the point P, referred to a fixed origin at the centre of the crack and to the undeformed state.

From dimensional considerations the local energy density W at P is given in the general case of an elastic non-linear material by

$$W(P) = W_0 f(x, y, \varepsilon_0), \tag{12.38}$$

where W_0 is the energy density at infinity and f is a function of ε_0, the strain at infinity, and $x = X/c$, $y = Y/c$. The change in energy density at P due to an increment of crack growth dc at constant load (and therefore constant ε_0, W_0) is, in the limit,

$$\begin{aligned}
\frac{dW(P)}{dc} &= W_0 \left[\frac{\partial f}{\partial x} \frac{\partial x}{\partial c} + \frac{\partial f}{\partial y} \frac{\partial y}{\partial c} \right] \\
&= \frac{-W_0}{c} \left[x \frac{\partial f}{\partial x} + y \frac{\partial f}{\partial y} \right] \\
&= \frac{-W_0}{c} g(x, y, \varepsilon_0)
\end{aligned} \tag{12.39}$$

where g is another function.

The total change in elastic stored energy in the system occurring for an increment of crack growth dc at constant load is

$$\frac{d\xi}{dc} = \sum_P \frac{dW(P)}{dc} \delta v, \tag{12.40}$$

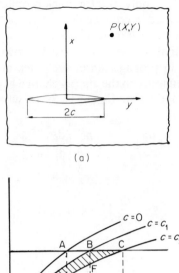

Figure 12.39. (a) The infinite sheet containing a crack of length $2c$. [Redrawn with permission from Andrews, *J. Mater. Sci.*, **9**, 887 (1974).] (b) Schematic diagram of load–elongation curves for specimens containing cracks of different lengths. [Redrawn with permission from Andrews and Billington, *J. Mater. Sci.*, **11**, 1354 (1976).]

where ξ is the total energy in the system and $\delta v = hc^2 \, \delta x \, \delta y$ is the volume element at P for an undeformed lamina of thickness h.

It follows that

$$\frac{d\xi}{dA} = \frac{-W_0 c}{4} \sum_P g(x, y, \varepsilon_0) \, \delta x \, \delta y, \qquad (12.41)$$

remembering that unit area of crack interface $A = 4ch$.

The summation is carried out in dimensionless space and is therefore independent of crack length. We have

$$-\left(\frac{d\xi}{dA}\right)_{W_0, \varepsilon_0} = k_1^*(\varepsilon_0) c W_0 \qquad (12.42)$$

where k_1^* is a function of ε_0 only. But for crack propagation we are interested in the elastic energy released in growing the crack, not the total energy change at constant load. Referring to Figure 12.39(b), this is the area of the shaded triangle, which for a small increment of crack growth is equivalent to the elastic energy released in propagation of the crack at constant extension Δ. The ratio of this latter quantity to the energy change at constant load is $p/1-p$ (ratio of OBF to (OBCD–OBC) in Figure 12.39(b), where OBF $\doteqdot$ OBC). Then

$$\left(\frac{d\xi}{dc}\right)_\Delta = -\frac{p}{1-p}\left(\frac{d\xi}{dc}\right)_{W_0,\varepsilon_0} \tag{12.43}$$

and

$$-\left(\frac{d\xi}{dA}\right)_\Delta = k_1(\varepsilon_0)cW_0 \tag{12.44}$$

where

$$k_1(\varepsilon_0) = \frac{-p}{1-p}k_1^*(\varepsilon_0).$$

For a linear elastic material $p = \frac{1}{2}$ and it can readily be shown that this equation is equivalent to the Griffith equation so that we regain equations (12.1) and (12.2).

We have $W_0 = \sigma_0^2/2E^*$, where σ_0 is the stress at infinity and

$$\frac{-d\xi}{dA} = -\frac{dU}{dA} = \gamma = \pi c\frac{\sigma_0^2}{2E^*},$$

where k_1 is replaced by π.

The fracture stress σ_B is then given by

$$\sigma_B = \sigma_0 = \left(\frac{2\gamma E^*}{\pi c}\right)^{1/2}$$

as before.

Andrews denotes the energy required to create unit area of new interface as $\mathcal{T}$, to emphasize its generality. Hence

$$\mathcal{T} = k_1(\varepsilon_0)cW_0. \tag{12.45}$$

The discussion so far parallels previous developments and equation (12.45) has been used for the fracture of rubbers where k decreases from π to about 2 as ε_0 increases from very small strains to large strains. The main point of Andrews' analysis is that the requirement of perfectly elastic deformation can be removed. We now have to consider explicitly unloading as well as loading, and to recognize that the energy density for elements unloading will be represented by a different function.

Unloading from a stress σ_p we have

$$W(P) = W_0 F(x, y, \sigma_p), \qquad (12.46)$$

where F depends on σ_p, which determines the maximum stress levels achieved. Then

$$\frac{dW(P)}{dc} = \begin{cases} \dfrac{-W_0}{c}\{g(x, y)\} & \text{for elements loading as the crack propagates,} \\[2ex] \dfrac{-W_0}{c}\{G(x, y, \sigma_p)\} & \text{for elements unloading,} \end{cases}$$

where the function G is to F as g is to f. Summing over the stress field gives

$$\frac{-d\xi}{dA} = cW_0\left[\left(\sum_{P_L} g(x, y)\,\delta x\,\delta y\right) + \left(\sum_{P_U} G(x, y, \sigma_p)\,\delta x\,\delta y\right)\right], \qquad (12.47)$$

where L, U stand for loading and unloading respectively.

Now we define a new constant of proportionality α, where $\alpha < 1$, and is the energy density recovered at P divided by that which would have been recovered from an elastic solid over the same negative stress increment during unloading. α is a function of the state of stress at P and other factors such as temperature and strain rate. We can then write

$$\left.\frac{dW(P)}{dc}\right|_U = \alpha \left.\frac{dW(P)}{dc}\right|_L$$

and equation (12.47) can be written as

$$\frac{-d\xi}{dA} = cW_0\left[\sum_{P_L} g(x, y)\,\delta x\,\delta y) + \left(\sum_{P_U} \alpha g(x, y)\,\delta x\,\delta y\right)\right]$$

$$= cW_0\left[k_1 - \sum_{P_U} \beta g(x, y)\,\delta x\,\delta y\right], \qquad (12.48)$$

where $\beta = 1 - \alpha = \beta_0(\sigma_0, T, R)g_1(x, y)$ is the hysteresis ratio, whose dependence on stress is expressed through the stress at infinity σ_0 and g_1 is another function.

Finally we can write

$$\frac{-d\xi}{dA} = cW_0\left[k_1 - \left(\beta_0\sum_{P_U} g_2(x, y)\,\delta x\,\delta y\right)\right]$$

$$= k_2(\sigma_0, T, R)cW_0. \qquad (12.49)$$

This quantity is the energy available for forming the surface crack after deduction of the energy losses throughout the bulk of the specimen. Andrews terms it a new quantity $\mathscr{T}_0$, to be compared with $\mathscr{T} = k_1 c W_0$, which is the total energy expended by the system to produce unit area of crack growth:

$$\mathscr{T} = \frac{\mathscr{T}_0 k_1}{k_2(\sigma_0, T, R)} = \mathscr{T}_0 \Phi(\sigma_0, T, R) \tag{12.50}$$

where

$$\Phi = k_1 \bigg/ \left[k_{\bar{1}} \left(\beta_0 \sum_{P_U} g_2(x, y) \, \delta x \, \delta y \right) \right] \tag{12.51}$$

is a 'loss function' which varies with the external loading σ_0 and other variables such as temperature and strain rate.

Andrews and Fukahori[135] have applied the generalized fracture mechanics theory to four polymers: styrene–butadiene rubber, ethylene–propylene diene rubber, plasticized polyvinyl chloride and low density polyethylene. They measured the propagation of the edge crack in a parallel sided strip of polymer for which equation (12.51) is modified to

$$\Phi = k_1 \bigg/ \left[k_1 - \tfrac{1}{2} \sum_{P_U} \beta g \, \delta x \, \delta y \right].$$

Combining equations (12.50) and (12.51) we have

$$\frac{\mathscr{T}}{\mathscr{T}_0} = k_1 \bigg/ \left[k_1 - \tfrac{1}{2} \sum_{P_U} \beta g \, \delta x \, \delta y \right], \tag{12.52}$$

and thus

$$k_1(1 - \mathscr{T}_0/\mathscr{T}) = \tfrac{1}{2} \sum_{P_U} \beta g \, \delta x \, \delta y. \tag{12.53}$$

since $\mathscr{T}_0/\mathscr{T} \ll 1$ we take natural logarithms to get

$$\ln k_1 - \mathscr{T}_0/\mathscr{T} = \ln \left[\tfrac{1}{2} \sum_{P_U} \beta g \, \delta x \, \delta y \right]. \tag{12.54}$$

Hence a plot of

$$\ln \left[\tfrac{1}{2} \sum_{P_U} \beta g \, \delta x \, \delta y \right] \quad \text{against} \quad \mathscr{T}^{-1}$$

should give a straight line of negative slope $-\mathscr{T}_0$ with an intercept of $\ln k_1$ as $\mathscr{T}^{-1} \to 0$. The fracture energy $\mathscr{T}$ was determined for each material at three different crack velocities. The corresponding k_1 term was determined by measuring the increase in compliance with crack length. This gives the apparent stored energy change $-d\xi/dc$ from which k_1 can be determined using equation (12.45) above. The term

$$\sum_{P_U} \beta g\, \delta x\, \delta y$$

requires two types of experiment. First the hysteresis ratio β is determined as a function of the stress at infinity σ_0 by stress–strain cycling of dumbbell samples to various load levels σ_0. Then

$$\sum_{P_U} \beta g\, \delta x\, \delta y$$

is evaluated by printing grids on the parallel strip edge crack samples, and evaluating the change in energy density as the crack propagates and some regions are loaded and others unloaded.

Results for the four materials are shown in Figure 12.40 and the corresponding values of $\mathscr{T}_0$ in Table 12.3. These values are now approaching theoretical

Table 12.3. Values of $\mathscr{T}_0$ obtained by Andrews and Fukahori[135]

Polymer	$\mathscr{T}_0(Jm^{-2})$
Styrene-butadiene rubber	65
Ethylene-propylene diene rubber	65
Plasticized polyvinylchloride	100
Low denisty polyethylene	200

values calculated by Lake and Thomas[136] on the basis that the minimum energy of fracture would be that required to break unit area of interatomic bonds across the fracture plane. In detail this involves assumptions regarding the molecular network in each polymer. The experimental data support the the general form of the theory and these preliminary results suggest that this may be a useful direction in which to seek a fundamental understanding of polymer fracture at a molecular and structural level. It does, however, remain to be established that the quantities k_1 and

$$\sum_{P_U} \beta g\, \delta x\, \delta y$$

can in general be evaluated with sufficient accuracy for this approach to be of wide application, so that accurate values for $\mathscr{T}_0$ can be determined.

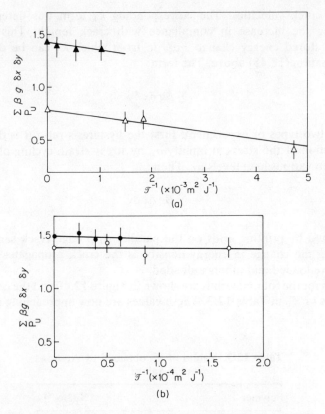

Figure 12.40. Plots show agreement of experimental data points with equation (12.65): (a) for styrene–butadiene (△) and ethylene propylene diene (▲) rubbers; (b) for polyethylene (●) and polyvinyl chloride (○). [Redrawn with permission from Andrews and Fukohari, *J. Mater. Sci.*, **12**, 1307 (1977).]

12.13 FATIGUE IN POLYMERS

A common form of failure of materials in practical use is by fatigue, where the failure occurs due to the cyclic application of stresses which are below that required to cause yield or fracture when a continuously rising stress is applied. The effect of such cyclic stresses is to initiate microscopic cracks at centres of stress concentration within the material or on the surface, and subsequently to enable these cracks to propagate, leading to eventual failure.

Early studies of fatigue in polymers concentrated on stress cycling of unnotched samples, to produce S versus N plots similar to those which have proved so useful for characterizing fatigue in metals (S being the maximum loading stress, N the number of cycles to failure). An example of this type of plot for a polymer[137] is shown in Figure 12.41. A major aspect of such a

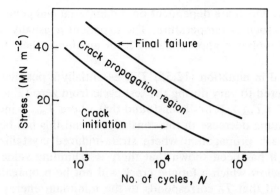

Figure 12.41. Fatigue response of polyvinyl chloride: relationship between applied stress σ and number of cycles to failure N, for both initiation of fatigue cracks and final failure. [Redrawn with permission from Manson and Hertzberg, *CRC Crit. Rev. Macromol. Sci.*, **1**, 433 (1973).]

test is the question of adiabatic heating, which can lead to failure by thermal melting. Clearly the test frequency will be important in this respect, and there will be a critical frequency above which thermal effects become important.

Stress cycling tests in unnotched samples do not readily distinguish between crack initiation and crack propagation. Further progress requires a similar approach to that adopted in fracture studies, viz. the introduction of very sharp initial cracks so as to examine crack propagation utilizing fracture mechanics concepts.

The first studies of fatigue in polymers of this quantitative nature concentrated on rubbers, where Thomas[138], and later Lake and Thomas[136] and Lake and Lindley[139], applied the tearing energy concept of fracture proposed by Rivlin and Thomas to fatigue crack propagation. Thomas showed that the fatigue crack growth rate could be expressed in the form of an empirical relationship,

$$\frac{dc}{dN} = AT^n,$$ (12.55)

where c is the crack length, N is the number of cycles, and T is the surface work parameter, which is analogous to the strain energy release rate G in linear elastic fracture mechanics. For a single edge notch specimen

$$T = 2k_1 cU,$$ (12.56)

where c is the crack length and $U = \sigma^2/2E$ is the stored energy density for a linear elastic material. k_1 is a constant which varies from π at small extensions (the linear elastic value) to approximately unity at large extensions[140]. A and

n are constants which are dependent on the material and generally vary with test conditions such as temperature. The exponent n usually lies between 1 and 6 and for rubber is approximately 2 for anything other than very small dc/dN.

As expressed in equation (12.55) T is essentially a positive quantity and can be considered to vary during the test cycle from zero ($T = T_{min} = 0$) to a finite value ($T = T_{max}$). It has been found that where T_{min} is increased, there is a corresponding decrease in the constant A, and this has been attributed to reduced crack propagation where strain-induced crystallization occurs. Furthermore, it has been shown that there is a limiting value of $T = T_0$, a fatigue limit below which a fatigue crack will not be propagated. Lake and Thomas showed that T_0 corresponds to the minimum energy required per unit area to extend the rubber at the crack tip to its breaking point. As pointed out by Andrews[141] it can be considered that initiation requires either that the material contains intrinsic flaws of magnitude c_0 or that flaws of this size are produced during the test itself, with c_0 defined by equation (12.56), where $T_0 = k_1 c_0 W$. Andrews and Walker[142] carried this approach one stage further, incorporating a generalized form of fracture mechanics along similar lines to the discussion of Section 12.12 above, to analyse the fatigue behaviour of low density polyethylene. This was viscoelastic in the range of interest so that the more generalized fracture mechanics was required to deal with unloading as well as loading during crack propagation. The fatigue characteristics were predicted from the crack growth data using a single fitting constant, the intrinsic flaw size c_0. It was further suggested that c_0 had a size corresponding to the spherulite dimensions, and that interspherulite boundary cracks constituted the intrinsic flaws.

For glassy polymers, fracture mechanics has been the usual starting point[143-146]. The fatigue crack growth rate is usually expressed in the form of an empirical relationship

$$\frac{dc}{dN} = A'(\Delta K)^m, \tag{12.57}$$

where c is the crack length, N the number of cycles, ΔK the range of the stress intensity factor (i.e. $K_{max} - K_{min}$, where K_{min} is generally zero), and A' and m are constants depending on the material and test conditions. For $K_{min} = 0$, equation (12.57) is clearly identical in form to equation (12.55), which is generally adopted for rubbers. Recall from Section 12.2.1 that the strain energy release rate $G = K^2/E$ for plane stress. Then

$$G = 2T = K_{max}^2/2E = (\Delta K)^2/2E,$$

and equations (12.55) and (12.57) are formally equivalent if $m = 2n$. Equation (12.57) is also the most general form of the law proposed by Paris[147,148] for predicting fatigue crack growth rates in metals. The general situation for glassy polymers is illustrated schematically in Figure 12.42(a) with some typical

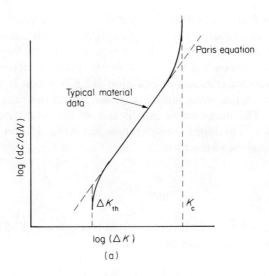

(a)

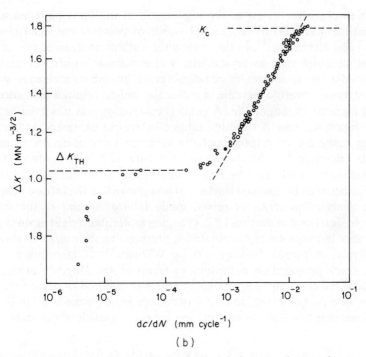

(b)

Figure 12.42. (a) Schematic diagram of fatigue crack growth rate dc/dN as a function of the range of stress intensity factor ΔK. (b) Fatigue crack growth characteristics for a vinyl urethane polymer. [Redrawn with permission from Harris and Ward, *J. Mater. Sci.*, **8**, 1655 (1973).]

results shown in Figure 12.42(b). These differ in two respects from the Paris equation. First, analogous to the case of rubbers, there is a distinct threshold value of ΔK, denoted by ΔK_{th}, below which no crack growth is observed. Secondly, as ΔK approaches the critical stress intensity factor K_c, the crack accelerates. A further criticism of equation (12.57) is that it does not allow for the influence of the mean stress, as distinct from the range of the stress intensity factor. The mean stress usually has an important influence on the crack growth rate. The latter consideration led Arad, Radon and Culver[149] to suggest an equation of the form

$$\frac{dc}{dN} = \beta \lambda^n \tag{12.58}$$

where $\lambda = (K_{max}^2 - K_{min}^2)$. This is equivalent to equation (12.57) because the cycle strain energy release rate ΔG is given by

$$\Delta G = \frac{1}{E}(K_{max}^2 - K_{min}^2).$$

A comprehensive review of the application of the Paris equation and its modified form (12.58) to fatigue behaviour of polymer has been given by Manson and Hertzberg[150]. In this review the authors considered the effect of physical variables such as crystallinity and molecular weight. In particular they noted a strong sensitivity of fatigue crack growth to molecular weight. In polystyrene a fivefold increase in molecular weight resulted in a more than 10-fold increase in fatigue life. A general correlation was observed between the fracture toughness K_c and the fatigue behaviour expressed as the stress intensity range ΔK corresponding to an arbitrary value of dc/dN (chosen as 7.6×10^{-7} m cycle^{-1}). This is shown in Figure 12.43. A study of fatigue behaviour in polycarbonate by Pitman and Ward[151] has also brought out the similarity between fatigue and fracture. It was shown that the fatigue behaviour can be analysed in terms of mixed mode failure. Similar to the fracture behaviour described in Section 12.2, changing molecular weight again changed the balance between energy dissipated in propagating the craze and shear lips respectively. A recent development by Williams[152,153] attempts to model fatigue crack propagation behaviour in terms of the Dugdale plastic zone analysis of the crack tip. Each fatigue cycle is considered to reduce the craze stress in one part of craze, so that a two-stage plastic zone is established. It was shown that this leads to an equation for crack growth of the form

$$\frac{dc}{dN} = \beta'[K^2 - \alpha K_c^2], \tag{12.59}$$

which gives a good fit to experimental data for polystyrene over a substantial range of temperatures.

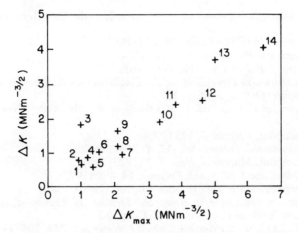

Figure 12.43. Relationship between the stress intensity range ΔK, corresponding to an arbitrary value of dc/dN, 7.6×10^{-7} m cycle^{-1} and the maximum stress intensity factor range ΔK_{max}, observed at failure for a group of polymers. The polymers are (1) cross-linked polystyrene, (2) polymethyl methacrylate, (3) polyvinyl chloride, (4) low density polyethylene, (5) polystyrene, (6) polysulphone, (7) high impact polystyrene, (8) acrylonitrile–butadiene–styrene resin, (9) chlorinated polyether, (10) polyphenylene oxide, (11) nylon 6, (12) polycarbonate, (13) nylon 66, (14) poyvinylidene fluoride. [Redrawn with permission from Manson and Hertzberg, *CRC Crit. Rev. Macromol. Sci.*, **1**, 433 (1973).]

Both Williams and Pitman and Ward conclude that it is difficult to assign physical significance to the parameters in the Paris equation. Further developments in this area will require a more distinctly physical approach.

REFERENCES

1. J. J. Benbow and F. C. Roesler, *Proc. Phys. Soc. B*, **70**, 201 (1957).
2. A. A. Griffith, *Phil. Trans. Roy. Soc.*, **221**, 163 (1921).
3. G. E. Inglis, *Trans. Inst. Naval Architect.* **55**, 219 (1913).
4. R. A. Sack, *Proc. Phys. Soc.*, **58**, 729 (1946).
5. H. A. Elliott, *Proc. Phys. Soc.*, **59**, 208 (1947).
6. G. R. Irwin, *J. Appl. Mech.*, **24**, 361 (1957).
7. G. R. Irwin and J. A. Kies, *Welding J. Res. Suppl.*, **33**, 1935 (1954).
8. B. Cottrell, *Int. J. Fract. Mech.*, **2**, 526 (1966).
9. J. P. Berry, *J. Appl. Phys.*, **34**, 62 (1963).
10. W. F. Brown and J. F. Srawley, ASTM STP 410, 1966.
11. J. F. Srawley and B. Gross, NASA Report E-3701, 1967.

468

12. J. G. Williams, *Stress Analysis of Polymers*, 2nd edn, Ellis Horwood, Chichester, 1980.
13. J. J. Benbow, *Proc. Phys. Soc.*, **78**, 970 (1961).
14. N. L. Svensson, *Proc. Phys. Soc.*, **77**, 876 (1961).
15. J. P. Berry, *J. Polymer Sci.*, **50**, 313 (1961).
16. E. H. Andrews, in *Proceedings of the Conference on the Physical Basis of Yield and Fracture, Oxford, 1966*, p. 127.
17. J. P. Berry, in *Fracture* (B. L. Averbach *et al.*, eds), Wiley, New York, 1959, p. 263.
18. R. P. Kambour, *Polymer* **5**, 143 (1964).
19. R. P. Kambour, *J. Polymer Sci. A2*, **4**, 349 (1966).
20. R. P. Kambour, *Macromol, Rev.*, **7**, 1 (1973).
21. H. R. Brown and I. M. Ward, *Polymer*, **14**, 469 (1973).
22. D. S. Dugdale, *J. Mech. Phys. Solids*, **8**, 100 (1960).
23. J. R. Rice, in *Fracture—An Advanced Treatise* (H. Liebowitz, ed.), Academic Press, New York and London, 1968, Chap. 3.
24. W. Döll and G. W. Weidmann, *Colloid Polymer Sci.*, **254**, 205 (1976).
25. G. L. Pitman and I. M. Ward, *Polymer* **20**, 895 (1979).
26. P. J. Hine, R. A. Duckett and I. M. Ward, *Polymer*, **22**, 1745 (1981).
27. J. P. Berry, *J. Polymer Sci. A*, **2**, 4069 (1964).
28. P. J. Flory, *J. Amer. Chem. Soc.*, **67**, 2048 (1945).
29. W. Döll and G. W. Weidmann, *Progr. Colloid Polymer Sci.*, **66**, 291 (1979).
30. R. P. Kusy and D. T. Turner, *Polymer*, **18**, 391 (1977).
31. R. P. Kusy and D. T. Turner, *Polymer*, **17**, 161 (1976).
32. R. N. Haward, H. E. Daniels and L. R. G. Treloar, *J. Polymer Sci., Polymer Phys. Ed.*, **16**, 1169 (1978).
33. J. P. Berry, in *Fracture* (B. L. Averbach *et al.*, eds), Wiley, New York, 1959, p. 263.
34. M. Higuchi, *Rept. Res. Inst. Appl. Mech.* (*Japan*), **6**, 173 (1959).
35. R. P. Kambour and A. S. Holik, *J. Polymer Sci. A2*, **7**, 1393 (1969).
36. R. P. Kambour and R. R. Russell, *Polymer*, **12**, 237 (1971).
37. P. Beahan, M. Bevis and D. Hull, *Phil. Mag.*, **24**, 1267 (1971).
38. P. Beahan, M. Bevis and D. Hull, *J. Mater. Sci.*, **8**, 169 (1972).
39. P. Beahan, M. Bevis and D. Hull, *Polymer*, **14**, 96 (1973).
40. B. D. Lauterwasser and E. J. Kramer, *Phil. Mag.*, **39**, 469 (1979).
41. A. M. Donald and E. J. Kramer, *Phil. Mag.*, **43**, 857 (1981).
42. H. R. Brown, *J. Polymer Sci., Polymer Phys. Ed.*, **17**, 143 (1979).
43. H. R. Brown and E. J. Kramer, *J. Macromol. Sci. B*, **19**, 487 (1981).
44. A. M. Donald, E. J. Kramer and R. A. Bubeck, *J. Polymer Sci., Polymer Phys. Ed.*, in press.
45. A. M. Donald and E. J. Kramer, *Polymer*, in press.
46. F. Rietsch, R. A. Duckett and I. M. Ward, *Polymer*, **20**, 1135 (1979).
47. S. S. Sternstein, L. Ongchin and A. Silverman, *Appl. Polymer Symp.*, **7**, 175 (1968).
48. S. S. Sternstein and L. Ongchin, *Amer. Chem. Soc. Polymer Preprints*, **10**, 1117 (1969).
49. P. B. Bowden and R. J. Oxborough, *Phil. Mag.*, **28**, 547 (1973).
50. K. Matsushige, S. V. Radcliffe and E. Baer, *J. Mater. Sci.*, **10**, 833 (1974).
51. R. A. Duckett, B. C. Goswami, L. S. A. Smith, I. M. Ward and A. M. Zihlif, *Brit. Polymer J.* **10**, 11 (1978).
52. M. Kitagawa, *J. Polymer Sci., Polymer Phys. Ed.*, **14**, 2095 (1976).
53. A. S. Argon and J. G. Hannoosh, *Phil. Mag.*, **36**, 1195 (1977).
54. A. S. Argon, J. C. Hannoosh and M. M. Salama, in *Fracture 1977*, Vol. 1, Waterloo, Canada, 1977, p. 445.

55. A. S. Argon, *Pure Appl. Chem.*, **43**, 247 (1975).
56. G. A. Bernier and R. P. Kambour, *Macromolecules*, **1**, 393 (1968).
57. R. P. Kambour, C. L. Gruner and E. E. Romagosa, *J. Polymer Sci.*, **11**, 1879 (1973).
58. R. P. Kambour, C. L. Gruner and E. E. Romagosa, *Macromolecules*, **7**, 248 (1974).
59. E. H. Andrews and L. Bevan, *Polymer*, **13**, 337 (1972).
60. G. P. Marshall, L. E. Culver and J. G. Williams, *Proc. Roy. Soc. A*, **319**, 165 (1970).
61. J. G. Williams and G. P. Marshall, *Proc. Roy. Soc. A*, **342**, 55 (1975).
62. N. Brown and Y. Imai, *J. Appl. Phys.*, **46**, 4130 (1975).
63. Y. Imai and N. Brown, *J. Mater. Sci.*, **11**, 417 (1976).
64. N. Brown, B. D. Metzger and Y. Imai, *J. Polymer Sci., Polymer Phys. Ed.*, **16**, 1085 (1978).
65. R. P. Kambour, in *Proceedings of the International Conference on the Mechanics of Environment Sensitive Cracking Materials, 1977*, p. 213.
66. N. Brown, in *Methods of Experimental Physics*, Vol. 16, Part C (R. A. Fava, ed.), Academic Press, New York, 1980, p. 233.
67. H. Mark, *Cellulose and its Derivatives*, Interscience Publishers, New York, 1943.
68. P. I. Vincent, *Proc. Roy. Soc. A*, **282**, 113 (1964).
69. H. H. Kausch, *Polymer Fracture*, Springer-Verlag, Berlin, 1978.
70. H. H. Kausch and J. Becht, *Rheol. Acta.* **9**, 137 (1970).
71. S. N. Zhurkov, I. I. Novak, A. I. Slutsker, V. I. Vettegren, V. S. Kuksenko, S. I. Veliev, M. A. Gezalov and M. P. Vershina, in *Proceedings of the Conference on the Yield, Deformation and Fracture of Polymers, Cambridge, 1970*.
72. R. P. Wool, *J. Polymer Sci.*, **13**, 1795 (1975).
73. S. N. Zhurkov and E. E. Tomashevsky, in *Proceedings of the Conference on the Physical Basis of Yield and Fracture*, Oxford, 1966 p. 200.
74. S. N. Zhurkov, V. S. Kuksenko and A. I. Slutsker, in *Proceedings of the Second International Conference on Fracture, Brighton, 1969*, p. 531.
75. V. A. Zakrevskii and V. Ye. Korsukov, *Polymer Sci. USSR*, **14**, 1064 (1972).
76. A. Peterlin, *Int. J. Fracture*, **11**, 761 (1975).
77. E. Orowan, *Rept. Prog. Phys.*, **12**, 185 (1949).
78. P. I. Vincent, *Polymer*, **1**, 425 (1960).
79. J. M. Stearne and I. M. Ward, *J. Mater. Sci.*, **4**, 1088 (1969).
80. P. L. Clarke, PhD thesis, Leeds University, 1981.
81. E. A. W. Hoff and S. Turner, *Bull. Amer. Soc. Test. Mater.*, **225**, TP208 (1957).
82. R. F. Boyer, *Polymer Eng. Sci.*, **8**, 161 (1968).
83. J. Heijboer, *J. Polymer Sci. C*, **16**, 3755 (1968).
84. P. J. Flory, *J. Amer. Chem. Soc.*, **67**, 2048 (1945).
85. A. H. Cottrell, *The Mechanical Properties of Matter*, Wiley, New York, 1964, p. 327.
86. E. R. Parker, *Brittle Behaviour of Engineering Structures*, Wiley, New York, 1957.
87. P. I. Vincent, *Plastics*, **29**, 79 (1964).
88. F. C. Roesler, *Proc. Phys. Soc. B*, **69**, 981 (1981).
89. K. E. Puttick, *J. Phys. D*, **11**, 595 (1978).
90. K. E. Puttick, in *Proceedings of the 3rd International Conference on Mechanical Behaviour of Materials*, Vol. 3, Pergamon Press, Oxford, 1979, p. 11.
91. K. E. Puttick, *J. Phys. D*, **13**, 2249 (1980).
92. J. A. Kies and A. B. J. Clark, in *Proceedings of the 2nd International Conference on Fracture, Brighton, 1969*, Paper 42.
93. J. R. Griffiths and G. Oates, in *Proceedings of the 2nd International Conference on Fracture, Brighton, 1969*, Paper 19.
94. G. R. Irwin, *Handbuch Phys.*, **6**, 551 (1958).

470

95. C. Gurney and J. Hunt, *Proc. Roy. Soc. A*, **229**, 508 (1967).
96. H. R. Brown, *J. Mater. Sci.*, **8**, 941 (1973).
97. G. P. Marshall, J. G. Williams and C. E. Turner, *J. Mater. Sci.*, **8**, 949 (1973).
98. R. A. W. Fraser and I. M. Ward, *J. Mater. Sci.*, **12**, 459 (1977).
99. P. J. Hine, PhD thesis, Leeds, 1981.
100. R. W. Truss, R. A. Duckett and I. M. Ward, *Polym. Eng. Sci.* (in press).
101. R. A. W. Fraser and I. M. Ward, *J. Mater. Sci.*, **9**, 1624 (1974).
102. P. I. Vincent, in *Physics of Plastics* (P. D. Ritchie, ed.) Iliffe Books, London, 1965, p. 174.
103. A. J. Kinloch and J. G. Williams, *J. Mater. Sci.*, **15**, 987 (1980).
104. R. M. Evans, H. R. Nara and R. G. Bobalek, *Soc. Plast. Engrs J.*, **16**, 76 (1960).
105. P. I. Vincent, *Polymer*, **15**, 111 (1974).
106. L. E. Nielsen, *Mechanical Properties of Polymers*, Reinhold, New York, 1962.
107. G. B. Bucknell, *Brit. Plast.*, **40**, 84, (1967).
108. G. B. Bucknell, *Brit. Plast.*, **40**, 118 (1967).
109. G. B. Bucknell and R. R. Smith, *Polymer*, **6**, 437 (1965).
110. J. A. Kies, A. M. Sullivan and G. R. Irwin, *J. Appl. Phys.*, **21**, 716 (1950).
111. I. Wolock, J. A. Kies and E. B. New
     al., eds), Wiley, New York, 1959, p. 250.
112. N. F. Mott, *Engineering*, **165**, 16 (1948).
113. H. Schardin, in *Fracture* (B. L. Averbach *et al.*, eds), Wiley, New York, 1959, p. 297.
114. A. M. Bueche and A. V. White, *J. Appl. Phys.*, **27**, 980 (1956).
115. E. F. Poncelet, *Metals Technology*, **11**, 1684 (1944).
116. H. Kolsky, in *Fracture* (B. L. Averbach *et al.*, eds) Wiley, New York, 1959, p. 281.
117. F. P. Bowden and J. E. Field, *Proc. Roy. Soc. A*, **282**, 331 (1964).
118. R. S. Rivlin and A. G. Thomas, *J. Polymer Sci.*, **10**, 291 (1953).
119. A. G. Thomas, *J. Polymer Sci.*, **18**, 177 (1955).
120. A. M. Bueche and J. P. Berry, in *Fracture* (B. L. Averbach *et al.*, eds), Wiley, New York, 1959, p. 265.
121. F. Bueche, *Physical Properties of Polymers*, Interscience Publishers, New York, 1962, p. 237.
122. P. J. Flory, *Ind. Eng. Chem.*, **38**, 417 (1946).
123. P. J. Flory, N. Rabjohn and M. C. Shaffer, *J. Polymer Sci.*, **4**, 435 (1949).
124. G. R. Taylor and S. Darin, *J. Polymer Sci.*, **17**, 511 (1955).
125. F. Bueche, *J. Polymer Sci.*, **24**, 189 (1957).
126. F. Bueche, *J. Polymer Sci.*, **33**, 259 (1958).
127. T. L. Smith, *J. Polymer Sci.*, **32**, 99 (1958).
128. T. L. Smith, *Soc. Plast. Engrs J.*, **16**, 1211 (1960).
129. T. L. Smith and P. J. Stedry, *J. Appl. Phys.*, **31**, 1892 (1960).
130. F. Bueche, *J. Appl. Phys.*, **26**, 1133 (1955).
131. F. Bueche and J. C. Halpin, *J. Appl. Phys.*, **35**, 36 (1964).
132. J. C. Halpin, *J. Appl. Phys.*, **35**, 3133 (1964).
133. E. H. Andrews, *J. Mater. Sci.*, **9**, 887 (1974).
134. E. H. Andrews and E. W. Billington, *J. Mater. Sci.*, **11**, 1354 (1976).
135. E. H. Andrews and Y. Fukahori, *J. Mater. Sci.*, **12**, 1307 (1977).
136. G. J. Lake and A. G. Thomas, *Proc. Roy. Soc. A*, **300**, 108 (1967).
137. S. J. Hutchinson and P. P. Benham; *Plast. Polymer*, April 1970, 102.
138. A. G. Thomas, *J. Polymer Sci.*, **31**, 467 (1958).
139. G. J. Lake and P. B. Lindley, in *Proceedings of the Conference on the Physical Basis of Yield and Fracture*, Oxford, 1966, p. 176.
140. H. W. Greensmith, *J. Appl. Polymer Sci.*, **7**, 993 (1963).
141. E. H. Andrews, in *Testing of Polymers*, Vol. 4 (W. E. Brown, ed.), Wiley, New York, 1968, p. 237.

142. E. H. Andrews and B. J. Walker, *Proc. Roy. Soc. A*, **325**, 57 (1971).
143. H. F. Borduas, L. E. Culver and D. J. Burns, *J. Strain Analysis*, **3**, 193 (1968).
144. R. W. Hertzberg, H. Nordberg and J. A. Manson, *J. Mater. Sci.*, **5**, 521 (1970).
145. S. Arad, J. C. Radon and L. E. Culver, *J. Mech. Eng. Sci.*, **13**, 75 (1971).
146. J. S. Harris and I. M. Ward, *J. Mater. Sci.*, **8**, 1655 (1973).
147. P. C. Paris, in *Fatigue, an Interdisciplinary Approach*, Syracuse University Press, Syracuse, N.Y., 1964, p. 107.
148. P. C. Paris and F. Erdogan, *J. Basic Eng., Trans. ASME*, **85**, 528 (1963).
149. S. Arad, J. C. Radon and L. E. Culver, *Polymer Eng. Sci.*, **12**, 193 (1972).
150. J. A. Manson and R. W. Hertzberg, *CRC Crit. Rev. Macromol. Sci.*, **1**, 433 (1973).
151. G. L. Pitman and I. M. Ward, *J. Mater. Sci.*, **15**, 635 (1980).
152. J. G. Williams, *J. Mater. Sci.*, **12**, 2525 (1977).
153. Y. W. Mai and J. G. Williams, *J. Mater. Sci.*, **14**, 1933 (1979).

Index